Inaugural Section Special Issue

Inaugural Section Special Issue: Key Topics and Future Perspectives in Natural Hazards Research

Editor

Deodato Tapete

MDPI • Basel • Beijing • Wuhan • Barcelona • Belgrade • Manchester • Tokyo • Cluj • Tianjin

Editor
Deodato Tapete
Italian Space Agency (ASI)
Italy

Editorial Office
MDPI
St. Alban-Anlage 66
4052 Basel, Switzerland

This is a reprint of articles from the Special Issue published online in the open access journal *Geosciences* (ISSN 2076-3263) (available at: https://www.mdpi.com/journal/geosciences/special_issues/natural_hazards_research).

For citation purposes, cite each article independently as indicated on the article page online and as indicated below:

LastName, A.A.; LastName, B.B.; LastName, C.C. Article Title. *Journal Name* **Year**, *Volume Number*, Page Range.

ISBN 978-3-03943-833-4 (Hbk)
ISBN 978-3-03943-834-1 (PDF)

Contents

About the Editor

Deodato Tapete, Ph.D. in Earth Sciences, is a Researcher in Earth Observation and Data Analytics at the Italian Space Agency (ASI). With more than 10 years of research experience in Earth sciences, natural hazards and remote sensing, Dr. Tapete specializes in synthetic aperture radar imaging and interferometry for deformation monitoring, hazard assessment and archaeological prospection, as well as in the assessment of anthropogenic impacts on the environment, natural resources and cultural heritage. His publications in top-ranked journals (e.g., *Remote Sensing of Environment, Remote Sensing*) are highly cited across the community. He is a Fellow of the Higher Education Academy (FHEA). He serves as an Associate Editor of 'Environmental Remote Sensing' of *Remote Sensing* (ISSN 2072-4292) journal, and as Editor-in-Chief of the 'Natural Hazards' Section of *Geosciences* (ISSN 2076-3263; CODEN: GBSEDA) journal.

Preface to "Inaugural Section Special Issue: Key Topics and Future Perspectives in Natural Hazards Research"

Since early 2018, the "*Natural Hazards*" Section of *Geosciences* has aimed to publish pure, experimental, or applied research that is focused on advancing methodologies, technologies, expertise, and capabilities to detect, characterize, monitor, and model natural hazards and assess their associated risks. This stream of geoscientific research has nowadays reached a high degree of specialization and represents a multi-disciplinary research realm.

To inaugurate this section, the Special Issue "*Key Topics and Future Perspectives in Natural Hazards Research*" was launched.

The call for papers was initially opened a year and a half ago, and since then, the Special Issue has collected 10 selected papers, covering the following hot topics of natural hazards research:

(i) trends in publications and research directions at international level;
(ii) the role of Big Data in natural disaster management;
(iii) assessment of seismic risk through the understanding and quantification of its three components (i.e., hazard, vulnerability and exposure/impact);
(iv) climatic/hydro-meteorological hazards (i.e., drought, hurricanes);
(v) scientific analysis of past incidents and disaster forensics (i.e., the Oroville Dam 2017 spillway incident).

In the hope that readers will use these contributions to learn new methodologies and take inspiration for new research and applications, I would like to express my sincere gratitude to all the authors, editors and reviewers for their commitment during this editorial project.

My special thanks go to Mr Richard Li, *Geosciences* Managing Editor, for his dedication to this project and his valuable collaboration in the setup, promotion, and management of the Special Issue.

Deodato Tapete
Editor

Editorial

Key Topics and Future Perspectives in Natural Hazards Research

Deodato Tapete

Italian Space Agency (ASI), Via del Politecnico snc, 00133 Rome, Italy; deodato.tapete@asi.it

Received: 3 January 2020; Accepted: 7 January 2020; Published: 9 January 2020

Abstract: Since early 2018 the "*Natural Hazards*" Section of *Geosciences* journal has aimed to publish pure, experimental, or applied research that is focused on advancing methodologies, technologies, expertise, and capabilities to detect, characterize, monitor, and model natural hazards and assess their associated risks. This stream of geoscientific research has reached a high degree of specialization and represents a multi-disciplinary research realm. To inaugurate this section, the Special Issue "*Key Topics and Future Perspectives in Natural Hazards Research*" was launched. After a year and half since the call for papers was initially opened, the special issue is now completed with the editorial introducing the collection of 10 selected papers covering the following hot topics of natural hazards research: (i) trends in publications and research directions at international level; (ii) the role of Big Data in natural disaster management; (iii) assessment of seismic risk through the understanding and quantification of its three components (i.e., hazard, vulnerability and exposure/impact); (iv) climatic/hydro meteorological hazards (i.e., drought, hurricanes); and (v) scientific analysis of past incidents and disaster forensics (i.e., the Oroville Dam 2017 spillway incident). The present editorial provides a summary of each paper of the collection within the current context of scientific research on natural hazards, pointing out the salient results and key messages.

Keywords: natural hazard; earthquake; drought; hurricane; dam spillway; shear-wave velocity; psychology; disaster management; big data; cyber-infrastructure

---

1. Introduction

In early 2018 the journal *Geosciences* was re-organized into six sections, including one focused on "*Natural Hazards*". I accepted the journal invitation to serve as the Editor-in-Chief of this section and defined its profile.

The section is dedicated to the publication of pure, experimental, or applied research that aims to advance methodologies, technologies, expertise, and capabilities to detect, characterize, monitor, and model natural hazards and assess their associated risks. This stream of geoscientific research has reached a high degree of specialization and represents a multi-disciplinary research realm. In this context, the Natural Hazards section is open to geoscientific studies of natural hazards that are carried out using ground investigations, in situ instrumentations and remote sensing. However, the section also accepts methodological papers proposing workflows and routines for modelling and forecasting, as well as cross-cutting articles dealing with the different aspects of hazard assessment and management. The latter encompass hazard mitigation, emergency management, post-disaster recovery, the scientific communication of hazards, and capacity building.

The emphasis of the section is on hazards that are predominantly associated with natural processes and phenomena, including environmental, geological or geophysical, hydro-meteorological, atmospheric, climatological, oceanographic, and biological hazards.

However, areas of interest include: research investigating the role played by human action in (co-)triggering natural hazards and/or exacerbating their impact on the environment; natural hazards that display slow kinematics and increasingly manifest in time (for example land subsidence), or require long-term observations and measurements to be detected and characterized; and, of course, hazards with an abrupt onset or that quickly spread and affect large areas.

So far, the Natural Hazards section has been very successful, with more than 260 published papers, 13 currently open special issues and other 18 already closed, which definitely make the section the biggest of the journal.

To inaugurate this new section, in May 2018 I launched the *"Inaugural Section Special Issue: Key Topics and Future Perspectives in Natural Hazards Research"*, in the hope to capture the state-of-the-art of natural hazards research through a collection of review and research papers that could outline where we were as a research community and future perspectives. I envisaged that the following topics would have been covered:

- new findings in understanding triggering and propagation mechanisms;
- revision of previous hypotheses or hazard scenarios;
- development of new conceptualizations and methodologies;
- testing of new data, sensors, and techniques for investigation, monitoring, change detection, and multi-temporal or back-analysis;
- development of new or refined models for forecasting;
- integration of hazard assessment in risk analysis;
- design of new materials and solutions for risk mitigation;
- exploration of new ways to communicate hazards and increase awareness;
- study of societal impacts of natural hazards.

2. Facts and Figures of the Special Issue

A total of 21 submissions were received for consideration of publication in the special issue from May 2018 to September 2019. After editorial checks and the peer-review process involving external and independent experts in the field, the acceptance rate was 48%. The published special issue, therefore, contains a collection of 10 research articles. The acceptance rate is in line with the current trend of the journal, and proves that the whole editorial process was very rigorous and selective.

Figure 1a shows the countries where the study areas of the papers published in the special issue are located, compared with the countries of the author's affiliation. Almost all the papers are about study areas that are located in the same country as the affiliation of at least one of the authors. In the majority of the cases, papers are the outcome of an international collaboration between different academia and research institutions. In one case only, the study area country does not coincide with the author's affiliation country. This reflects the typology of the paper, i.e., the scientific analysis of a recent disaster that the paper authors chose to investigate because it was an interesting case study of a given incident type occurred under certain contextual conditions [1].

With regard to study areas, four papers do not have a specific test site, while other four focus on selected cities, towns and villages. Two papers, instead, propose analyses of natural hazards at national and transnational levels (Figure 1b).

The article metrics after one year and half from the publication of the first paper are encouraging. Some of the papers have already attracted attention across the community. Yu et al. [2] has gained 16 citations, Emmer [3] 10 and Chieffo and Formisano [4] 6. Not surprisingly, the first two papers with the highest number of citations are a review and a bibliometrics paper. They provide insights into the realm of big data in natural disaster management and the trends in research on natural hazards worldwide, respectively. The third most cited paper, instead, has been referenced in the recent literature presumably owing to the methodological approach to seismic risk assessment presented therein, potentially replicable in other contexts.

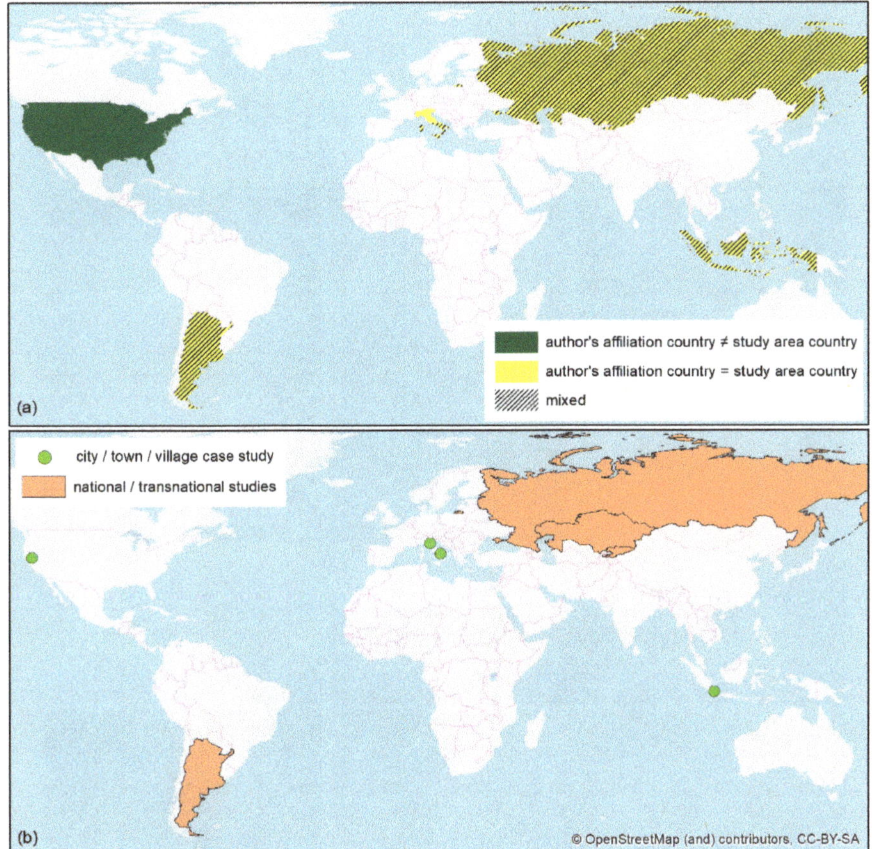

Figure 1. (**a**) Spatial distribution of the countries where the study areas of the papers published in the special issue are located versus the country where the affiliation institutions of the authors are based; (**b**) geographic distribution of the study areas distinguished by typology ("city/town/village case study" and "national/transnational studies").

3. Overview of the Published Papers

Figure 2 provides a pictorial composition summarising the main research topics that are covered by the papers published in the special issue:

- Trends in natural hazards research;
- Big Data in natural disaster management;
- Seismic risk: hazard, vulnerability and exposure/impact;
- Climatic/hydro-meteorological hazards;
- Scientific analysis of past incidents and disaster forensics.

Figure 2. Natural hazards research topics covered by the papers published in the special issue: (**a**) trends in natural hazards research; (**b**) Big Data in natural disaster management; (**c**) earthquakes and seismic risk; (**d**) drought; (**e**) hurricanes; (**f**) analysis of disaster incidents (Oroville Dam spillway incident, California, USA, February 2017); (**g**) electricity utilities and infrastructure. Photo source: Wikimedia Commons.

An overview of each paper is provided in the following sections.

3.1. Trends in Natural Hazards Research

The scientometrics and bibliometrics paper by Emmer [3] is an excellent well-structured overview of the research on different types of natural hazards worldwide, and is based on the analysis of a set of more than 580,000 items published between 1900 and 2017 and recorded in the Clarivate Analytics Web of Science database. The author describes precisely how he searched and then classified the research items according to a classification of natural hazards that he adapted from [5,6]. In this study two general categories of natural hazards are covered: (i) climatic/hydro-meteorological; and (ii) geological/geomorphic. Floods, storms, drought, hurricanes are examples falling in the first category, while earthquakes, slope movements, erosion, volcanic activity of the second. The analysis is focused around spatio-temporal patterns (geographies) of the research and selected scientometrical characteristics. The author also compares the geographic focus of the studies with the events/fatalities/damages recorded in the two freely available global databases of natural disasters MunichRE NatCatSERVICE and SwissRE Sigma Explorer.

Among the key outcomes, it is worth mentioning the following:

- The overall number of research items has been increasing dramatically over time, with floods, earthquakes, storms and droughts being the most frequently researched natural hazards;
- At the same time, the ratio between research items focusing on (i) and those on (ii) has remained relatively stable over time, with climatic/hydro-meteorological dominating in all studied periods;
- Nevertheless, significant differences exist between individual countries, thus suggesting a regionalism reflecting the most common and pressing natural hazards at regional to local scales. Research on (i) dominates in the majority of countries, while research on (ii) in Italy, Japan and France. An increased share of research on droughts is, instead, observed in India and Australia;
- From the global perspective, the research on individual types of natural hazards is in all cases dominated by researchers from the USA, followed by researchers from China, with the exception of research on volcanic activity, where Italy is ranked second;
- The majority of research items are written by author(s) from one country. Only one fourth are international research items. The two clusters of bi-lateral cooperation are: (1) European (leading countries England, Germany, France, Italy) and (2) circum-Pacific (leading countries USA, China, Canada, Australia, Japan);
- Both NatCatSERVICE and Sigma Explorer report increasing extents of damage caused by natural disasters yearly, apparently reflecting changing conditions, increasing population pressure and vulnerability globally, as well as possible inconsistencies in disaster databases on a global level;
- Major disasters (e.g., 2008 Wenchuan earthquake) are capable of causing an increase in research publications focused on that particular type of natural hazard.

3.2. *Big Data in Natural Disaster Management*

In natural hazards research and development (R&D), there is no doubt that "Big Data" have nowadays become a key component. They are contributing to change the way natural hazards are studied and natural disasters are managed. This is what Yu et al. [2] are sure about in their systematic review of the scientific literature aiming to analyze the role of Big Data in natural disaster management.

The authors created a sample of peer-reviewed journal articles published from 2011 to 2018, starting from an initial collection in the research catalogue Google Scholar, and then manually skim-reading each publication to reach the final body of relevant literature. Three aspects are specifically reviewed in the paper: (1) the major sources of Big Data; (2) the associated achievements in different disaster management phases; and (3) emerging technological topics associated with leveraging this new ecosystem of Big Data to monitor and detect natural hazards, mitigate their effects, assist in relief efforts, and contribute to the recovery and reconstruction processes.

In the current context that no standard definition of "Big Data" in disaster management is available, the authors acknowledge that some documents exist at national and international levels that outline the emerging data collections and a catalogue of sources of Big Data in disaster resilience. However, it is the authors' opinion that the concept of Big Data goes beyond the datasets themselves, regardless of their size. Big Data should be considered as the integration of diverse data sources and the capability to analyze and use the data (usually in real time) to the benefit of the population and society faced with a given disaster.

The authors find that satellite imagery, crowd-sourcing, and social media data serve as the most popular data for disaster management. Not surprisingly, satellite remote-sensing technology is primarily used for post-disaster damage assessment through change detection, and to respond through operational assistance. At the same time, the use of aerial imagery captured via unmanned aerial vehicles

(UAVs) is becoming more common owing to the efficiency of such instrumentations in situational awareness to provide much faster higher spatial resolution data compared to satellite imagery.

In this regard, this finding aligns with the current trend in operational services for disaster risk management. For example, the Copernicus Emergency Management Service (Copernicus EMS) has been assessing the feasibility and the associated benefits of the use of airborne sensors on-board drones or planes in its operational workflow, through specific agreements with plane and drone operators for on-demand rapid data acquisition and processing services in Europe. Significant advantages can be achieved when these data sources are used in combination with satellite Earth Observation data, for both mapping and validation tasks [7]. However, Yu et al. rightly point out that some challenges are still open (e.g., short battery life; unforeseen behaviour in different atmospheric conditions; limited scope of pilot training for users; and legislation that severely limits the use of UAVs in most countries).

Similar critical comments are made by the authors with regard to crowd-sourcing and social media, by highlighting opportunities and current limitations that not only scientists but also first responders and policy/decision-makers need to account for. Although this aspect is only briefly investigated in the paper, it is evident that more research and technological transfer are required to develop effective analytics (even of "real-time" type, in the case of social media), to extract valuable and reliable information that can be used as validated inputs in the disaster management cycle and be integrated with authoritative data, such as terrain and census.

Yu et al. have also mapped the major data sources and the application fields within the four distinct phases of the disaster management cycle, i.e., "mitigation", "preparedness", "responses", and "recovery".

"Machine learning" and "Cyber-infrastructure" emerge as the two evolutionary technologies that may facilitate disaster management in various ways. The review of the existing technologies and recent achievements proves that this is currently a vibrant research field, but with the major weakness of a technological gap between research and society. Big Data and cyber-infrastructure still pose significant challenges in terms of data volume, fast data transfer, intuitive data visualization and, more generally, efficient data management, that require financial and information technology (IT) resources, as well as expert users and translators of these data into information valuable for end users.

3.3. *Seismic Risk: Hazard, Vulnerability and Exposure/Impact*

The seismic risk is the most investigated natural risk in this special issue, with 5 out of the total 10 papers exploring either hazard or vulnerability or exposure/impact, i.e., the three components of the seismic risk equation.

3.3.1. Hazard

To characterize the seismic hazard and risk at a given location, the knowledge of local geology is mandatory. This is even more important in those sites, such as large cities and megacities, that are heavily urbanized and where rapid urbanisation may have happened with poor knowledge of the subsurface and little adherence to earthquake-proof building codes, so as an epicentral hit has the potential to cause 1 million fatalities [8]. The majority of cities are built on sedimentary basins that, on the one hand, offer flat topography for building development and fertile soils and groundwater resource for subsistence but, on the other hand, have potential to cause amplification and resonance of seismic wave motion in the case of earthquakes.

This is the geological context of the Jakarta Basin, in Indonesia, that is investigated by Ry et al. [9]. The authors tested a relatively new and simple technique to map shallow seismic structure using body-wave polarization. This technique proved to be a cost-effective alternative to the use of borehole and active source surveys, wherever three-component seismometers are operated.

To this end, Ry et al. exploited two dense, temporary broadband seismic networks covering Jakarta city and its surroundings during two distinct periods: 96 stations from October 2013 to February 2014; 143 stations between April and October 2018. This second deployment provided coverage just outside Jakarta in order to reveal the extent of the basin edge. Signals from 56 earthquakes with a good signal-to-noise ratio (SNR), varying from local to regional and teleseismic earthquakes, were recorded and evaluated. By applying the polarization technique to these earthquake signals, the apparent half-space shear-wave velocity (V_s^{ahs}) beneath each station was obtained, providing spatially dense coverage of the sedimentary deposits and the edge of the basin.

The results showed that spatial variations in V_s^{ahs} are compatible with previous studies, and appear to reflect the average shear-wave velocity (V_s) of the top 150 m. The authors were also able to extend this information beyond the city limits of Jakarta to what was thought to be the basin edge. The understanding of the complete geometry of the Jakarta sedimentary basin is crucial to develop a more accurate ground-motion simulation for hypothetical earthquake scenarios that can characterize the seismic risk in Jakarta.

3.3.2. Vulnerability

Giuffrida et al. [10] focus on the seismic vulnerability in small inland urban centres. This is a rather important topic that, sometimes, has received less attention than the seismic risk in cities and densely populated agglomerations. Areas that are marginal to large cities, or are located in mountainous regions or locations outside of main infrastructure and transportation networks, may be exposed to the risk of depopulation. If hit by earthquakes, these minor centres could be definitely abandoned soon after the seismic event. Therefore, in the process of restoration and redevelopment of minor historic centres, land and urban planning needs to encompass seismic vulnerability reduction policies, while preserving the integrity and cultural identity of the main buildings.

In this regard, the study by Giuffrida et al. is contextualized in the current Italian legislation framework. The Emergency Limit Condition (ELC) was introduced in 2012 as a municipal-scale analysis set up based on the Civil Protection Plans, and aims to guarantee that the emergency management system works during the post-earthquake phase. In particular, the ELC represents the limit condition for which, after the seismic event, all the functions (including residence) of the urban settlement are lost, except for the strategic functions necessary for the emergency management, their accessibility and connection with the territorial context.

The authors propose an approach to seismic vulnerability reduction made of three main stages: (1) knowledge of the typological, constructive, and technological features of the buildings; (2) analysis of the possible damage in case of an earthquake; and (3) planning of actions to reduce the vulnerability of buildings, with a cost modelling tool to define the trade-off between the extension and intensity of the vulnerability reduction works, given the available budget. The case study is the old town of Brisighella, in the province of Ravenna, Emilia-Romagna region, central Italy, for which the authors provide an evaluation of the vulnerable assets, in terms of both the human and urban capitals. This evaluation allows the authors to compare the costs of the seismic retrofit with the advantages of safety and to provide further evidence to formalize the equalization model.

A small rural municipality is also the case study of the paper by Chieffo and Formisano [4]. The authors assessed the seismic vulnerability and damage of the old masonry building compounds in a sector of the historic centre of Senerchia, in the province of Avellino, Campania region, southern Italy. The authors describe how they classified the inspected building aggregates by construction typology using the CARTIS form, i.e., the method developed by the PLINIVS research centre of the University of Naples "Federico II" in collaboration with the Italian Civil Protection Department.

The work aimed to evaluate the effects of local amplification varying the topographic class and the type of soils foreseen in the Italian "Updating of Technical Standards for Construction" NTC18 issued in 2018. The influence of soil conditions was considered by implementing a macroelement model of a typical masonry aggregate of the investigated study area, with the goal to plot damage scenarios expected under different earthquake moment magnitudes and site-source distances. In practice, the authors assessed the global seismic vulnerability of the building sample using the macroseismic method according to the European macroseismic EMS-98 scale, and identified the buildings most susceptible to seismic damage. Then, they developed 12 damage scenarios by means of an appropriate seismic attenuation law and analysed them with regard to local induced hazard effects. According to these damage scenarios, the site effects lead to a damage increment variable from 2% to 50%, which was much more marked at the smallest considered distance. In addition, local seismic effects were considerable for larger magnitudes. Seismic amplification factors due to the soil condition increase the occurrence probability of attaining the largest damage thresholds.

3.3.3. Economic and Social Impacts

Population and the possible fatalities caused by an earthquake of a given magnitude are the first element at risk and consequence that are calculated in a quantitative seismic risk assessment. However, earthquakes are natural disasters that can also cause enormous economic damage if they affect the integrity and functioning of buildings, infrastructure and utilities.

The paper by Iakubovskii et al. [11] focuses on electricity utilities and it is original in that the proposed mathematical model is applied at transnational level to evaluate the reliability of the interconnected electricity supply system of three countries of the Eurasian Economic Union (EAEU)—Russia, Kazakhstan, and Kyrgyzstan—under the threat of earthquakes. The EAEU currently consists of Armenia, Belarus, Kazakhstan, Kyrgyzstan, and Russia, and is currently facing a process of integration of its energy markets due to the planned creation of a common electricity market by 2019. The question around the reliability of electricity supply in the wake of natural hazards gained the attention of the EAEU states in the framework of energy strategies. Even though such interruptions are not frequent, each power outage or blackout affects several millions of people and causes vast economic losses and damages. Therefore, the authors call for the development of coordinated policies and risk management strategies to deal with electricity outage risks in the EAEU.

In their work, Iakubovskii et al. implement a modified version of the simplified reliability assessment approach, based on the so-called "N—i criterion". The authors determine which elements of the system are susceptible to failure due to an earthquake of a given magnitude. The results of the scenario analysis of earthquakes and their impacts on the reliability of the power supply system highlight that the energy security of the EAEU region is affected by the existence of interconnections that are vulnerable. The interconnections where disruptions of the electricity supply will have high impacts are situated between Kazakhstan and Kyrgyzstan, as well as between the isolated "West" node of Kazakhstan and Russia. Power supply interruptions at these lines can seriously influence the stability of the electricity transmission system, and lead to huge economic losses in the affected regions.

Going back to the human component of exposure and impact, Raccanello et al. [12] rightly recall that natural disasters have a potentially highly traumatic impact on psychological functioning of the affected population, specifically on children. This paper is extremely interesting since it investigates the psychological representation of earthquakes in children's minds, a topic that has been given scarce attention in the literature so far. The study of how people, and in particular children, perceive and represent natural disasters not only provides new elements of knowledge from the theoretical point of view of this discipline but also, from an applied perspective, provides the foundations for improving the risk awareness of youngest generations in order to promote adequate preparedness, in particular through emotional prevention.

The authors involved a convenience sample of 128 primary school children, from the second to the fourth grades, coming from a variety of socio-economic status levels, in north-eastern Italy. Most of the children had never experienced earthquakes directly and only a small percentage of children had experienced them. None of the children who had experienced earthquakes at least once reported any damage. The participants were asked to complete a written definition task and an online recognition task. The answers were analysed through the Rasch model.

In the children's representation of earthquakes, natural elements such as geological ones, were the most salient, followed by man-made and then by person-related elements. Older children revealed a more complex representation of earthquakes, and this was detected through the online recognition task.

The authors conclude that Italian primary school children possess a basic knowledge of earthquakes, are able to differentiate elements pertaining to different domains, and give particular relevance to the core geological issues characterizing earthquakes. At the same time, children showed an initial awareness of the different elements to which damage can be associated (e.g., the macroscopic and external consequences for the structures which are built by the individuals). Human behaviors such as escaping, biological consequences such as being hurt, or affective reactions assume similar relevance. By contrast, age differences were not so well detected in the written task compared to the recognition one. The better differentiation in the latter may be due to the more intuitive nature of the recognition task requiring a lower amount of cognitive resources, as well as the fact that the use of information and communication technology (ICT) instruments may stimulate children to engage with the task.

3.4. Climatic/Hydro-Meteorological Hazards

The "*Natural Hazards*" Section encourages submissions about hydro-meteorological, atmospheric, climatological and oceanographic hazards. Differently from the statistics of the whole research publication realm presented in [3], in *Geosciences* journal these types of natural hazards have been so far less researched than the geological/geomorphic ones. Therefore, it is interesting that 2 of the 10 papers published in this inaugural special issue relate to drought and weather prediction in the context of the simulation of hurricanes.

With regard to drought, it is well known that the deficit or inadequate timing of precipitation over an extended period of time leads to water scarcity that, insidiously, propagates in time through the hydrological cycle, and causes serious damage to socioeconomic and environmental systems. Naumann et al. [13] recall that, in the last 20 years, several drought periods were reported across Argentina, with at least three devastating events in 2006, 2009 and 2011. The authors focus their analysis on a systematic quantification of the drought risk at national scale: (i) as a function of long-term hazard, exposure and vulnerability; and (ii) dynamically as a combination of changes in drought conditions and the exposed assets. Media news, official reports at national and provincial levels, and the DesInventar disaster loss database were used as data sources to identify drought impacts. By combining drought hazard indicators and exposure layers, the authors analysed spatial and temporal patterns of exposure towards a better understanding of the interaction between changes in population structure and regional climate variability. Assets exposed to droughts were identified with several records of drought impacts and declarations of farming emergencies. Indeed, dry periods had detrimental impacts on very diverse sectors including agriculture and livestock production, but also inland river transportation and hydropower production.

If suitable medium-range forecasts became available, the method by Naumann et al. could be helpful for monitoring the probability of impact occurrence from the onset of a drought, or even before, and therefore could provide scientific input to trigger pro-active measures in order to cope with and mitigate the potential impacts of droughts.

Predictability is also the keyword of the study published by Shen [14]. The author builds upon previous studies wherein he achieved realistic 30-day simulations of multiple African easterly waves (AEWs) and an averaged African easterly jet (AEJ), as well as of hurricane Helene (2006) from Day 22 to Day 30. In the present paper, Shen further analyzes such extended predictability based on recent understandings of chaos and instability within Lorenz models and the generalized Lorenz model (GLM), and concludes that a statement of the theoretical predictability of two weeks is not universal. He also shows new insight into chaotic and non-chaotic processes revealed by the GLM and that there is the potential for extending prediction lead times at extended range scales. Examples of simulation of hurricanes with larger errors are also discussed, leading to the conclusion that the predictability of hurricane formation may be better near the Cape Verde Islands that are closer to the continent (e.g., for hurricane Helene) than over the Atlantic Ocean (e.g., for hurricane Florence). Shen projects his future work on refining the model to better examine the validity of the mechanism in explaining the recurrence of multiple AEWs.

3.5. Scientific Analysis of Past Incidents and Disaster Forensics

The paper by Koskinas et al. [1] reminds us that, when disasters happen (either natural or human-induced or as a combination of the two) and gain the attention of the general public, it is quite frequent that questions arise about the causes (and responsibilities). Scientific investigations aim to provide objective and evidence-based description and interpretation of the event, in the hope that this knowledge would strengthen the society capabilities to prevent similar events in future or, if not possible otherwise, to improve the disaster management and contribute to building resilience to hazards.

On the other hand, forensic investigation applied to the study of natural hazard disasters is a relatively new discipline. Despite what the term "forensic" could suggest, these types of investigations are not intended to seek or assign legal responsibility, but rather to understand which factors and how they contributed to the gestation and occurrence of a disaster in order to prevent and mitigate disaster risk [15].

Koskinas et al. present the outcomes of their hydroclimatic analysis of the Oroville Dam's catchment in California where, in February 2017, a huge spillway incident occurred at the local namesake dam (Figure 2f). Heavy rainfall during the 2017 California floods damaged the main spillway of the dam on 7 February, prompting the evacuation of more than 180,000 people living downstream along the Feather River and the relocation of a fish hatchery.

Figure 3 compares cloud-free satellite multispectral images collected by Copernicus Sentinel-2 constellation on 1 December 2016 and 11 March 2017. The GIF animations accessible in the Supplementary Materials of this editorial highlight the impact of rainfalls causing the rise of the reservoir surface level and the extent of the damage to the main spillway, still visible one month after the event.

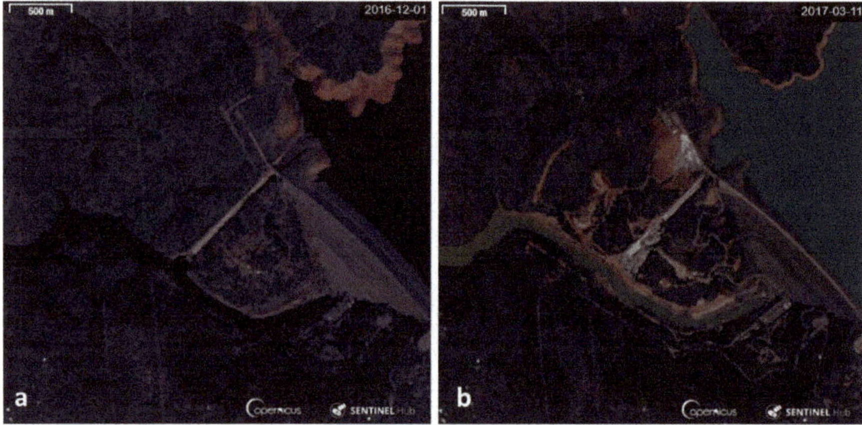

Figure 3. Oroville dam, California, USA, (**a**) before and (**b**) after the spillway incident occurred in February 2017, in two satellite multispectral images collected on 1 December 2016 and 11 March 2017, respectively (contains Copernicus Sentinel-2 data 2016–2017). The full time-lapse is provided in the Supplementary Materials.

The incident was massively covered by the media and an independent forensic team was tasked with determining the causes. Koskinas et al. chose to back-analyse this incident because it represents an interesting case study of dam failure that occurred under standard operating conditions, yet at an unfortunate time. The hydroclimatic analysis of the catchment is conducted, along with a review of related design and operational manuals. Based on summary characteristics of the 2017 floods, the authors outline possible causes in order to understand which factors contributed more significantly.

They conclude that the event was most likely the result of a structural problem in the dam's main spillway and detrimental geological conditions. However, the analysis of surface level data also reveals operational issues that were not present during previous larger floods. The authors suggest that a discussion should be promoted about flood control design methods, specifications, and dam inspection procedures, and how these can be improved to prevent the future occurrence of similar events.

Supplementary Materials: The following GIF animations are available online at http://www.mdpi.com/2076-3263/10/1/22/s1: time-lapse of cloud-free satellite multispectral images collected by Copernicus Sentinel-2 constellation from 1 December 2016 to 11 March 2017, showing the evolution of the Oroville Dam 2017 spillway incident (created in Sentinel Hub EO Browser and exported under CC BY 4.0 license).

Funding: This research received no external funding.

Acknowledgments: The Guest Editor thanks all the authors, *Geosciences*' editors, and reviewers for their great contributions and commitment to this Special Issue. A special thank goes to Richard Li, *Geosciences*' Assistant Editor, for his dedication to this project and his valuable collaboration in the setup, promotion and management of the Special Issue.

Conflicts of Interest: The author declares no conflict of interest.

References

1. Koskinas, A.; Tegos, A.; Tsira, P.; Dimitriadis, P.; Iliopoulou, T.; Papanicolaou, P.; Koutsoyiannis, D.; Williamson, T. Insights into the Oroville dam 2017 Spillway incident. *Geosciences* **2019**, *9*, 37. [CrossRef]
2. Yu, M.; Yang, C.; Li, Y. Big data in natural disaster management: A review. *Geosciences* **2018**, *8*, 165. [CrossRef]
3. Emmer, A. Geographies and scientometrics of research on natural hazards. *Geosciences* **2018**, *8*, 382. [CrossRef]
4. Chieffo, N.; Formisano, A. Geo-hazard-based approach for the estimation of seismic vulnerability and damage scenarios of the old city of senerchia (Avellino, Italy). *Geosciences* **2019**, *9*, 59. [CrossRef]
5. Burton, I.; Kates, R.W. The Perception of Natural Hazards in Resource Management. *Nat. Resour. J.* **1964**, *3*, 412–441.
6. Wisner, B. *Handbook of Hazards and Disaster Risk Reduction*; Routledge: London, UK, 2012.
7. Drones and Planes in Support of Copernicus: Examples from the Emergency Management Service—Copernicus In Situ Component. Available online: https://insitu.copernicus.eu/news/drones-and-planes-in-support-of-copernicus-examples-from-the-emergency-management-service (accessed on 31 December 2019).
8. Bilham, R. The seismic future of cities. *Bull. Earthq. Eng.* **2009**, *7*, 839–887. [CrossRef]
9. Ry, R.V.; Cummins, P.; Widiyantoro, S. Shallow shear-wave velocity beneath jakarta, indonesia revealed by body-wave polarization analysis. *Geosciences* **2019**, *9*, 386. [CrossRef]
10. Giuffrida, S.; Trovato, M.R.; Circo, C.; Ventura, V.; Giuffrè, M.; Macca, V. Seismic vulnerability and old towns. A cost-based programming model. *Geosciences* **2019**, *9*, 427. [CrossRef]
11. Iakubovskii, D.; Komendantova, N.; Rovenskaya, E.; Krupenev, D.; Boyarkin, D. Impacts of earthquakes on energy security in the Eurasian economic union: Resilience of the electricity transmission networks in Russia, Kazakhstan, and Kyrgyzstan. *Geosciences* **2019**, *9*, 54. [CrossRef]
12. Raccanello, D.; Vicentini, G.; Burro, R. Children's Psychological Representation of Earthquakes: Analysis of Written Definitions and Rasch Scaling. *Geosciences* **2019**, *9*, 208. [CrossRef]
13. Naumann, G.; Vargas, W.M.; Barbosa, P.; Blauhut, V.; Spinoni, J.; Vogt, J.V. Dynamics of socioeconomic exposure, vulnerability and impacts of recent droughts in Argentina. *Geosciences* **2019**, *9*, 39. [CrossRef]
14. Shen, B.W. On the predictability of 30-day global mesoscale simulations of african easterly waves during summer 2006: A view with the generalized lorenz model. *Geosciences* **2019**, *9*, 281. [CrossRef]
15. Mendoza, M.T.; Schwarze, R. Sequential Disaster Forensics: A Case Study on Direct and Socio-Economic Impacts. *Sustainability* **2019**, *11*, 5898. [CrossRef]

Article

Geographies and Scientometrics of Research on Natural Hazards

Adam Emmer

The Czech Academy of Sciences, Global Change Research Institute, 603 00 Brno, Czech Republic; aemmer@seznam.cz or emmer.a@czechglobe.cz

Received: 14 September 2018; Accepted: 16 October 2018; Published: 18 October 2018

Abstract: This contribution aims to reveal patterns of research on natural hazards worldwide, based on the analysis of the Clarivate Analytics Web of Science database. A set of 588,424 research items published between 1900 and 2017 is analyzed, covering different types of natural hazards. Two categories of natural hazards are distinguished in this study: (i) geological/geomorphic (earthquakes, slope movements, erosion, volcanic activity, and others); and (ii) climatic/hydro-meteorological (floods, storms, drought, hurricane, and others). General trends, the geographical focus, and the involvement and cooperation between individual countries are revealed, pointing out certain patterns (e.g., hotspots of research) and trends (e.g., changing publishing paradigm). Further, a global overview of research on natural hazards is confronted with disastrous events, fatalities, and losses of MunichRE and SwissRE global databases of natural disasters.

Keywords: natural hazards; disaster; scientometrics; bibliometrics; citation analysis; NatCatSERVICE; Sigma Explorer

1. Introduction

Natural disasters claim lives and result in damages of billions of USD yearly [1,2]. Research on various types of natural hazards is thus well-justified, especially in the context of changing conditions (e.g., climatological [3]) and increasing population pressure and vulnerability globally (e.g., [4]). Furthermore, the number of events and extent of damage caused by natural disasters are reported to have gradually increased in the past decades [5]. While a worldwide overview of the occurrence of different types of natural hazards and disasters is covered by numerous global [1,5], as well as regional [6], databases, and some types of natural hazards have been the subject of previous scientometrics studies (e.g., tsunamis [7,8], earthquakes [9,10], landslides [11,12], or lake outburst floods [13]), the research on natural hazards and their geographies has not yet been mapped from the global perspective.

Hence, the main objective of this work is to provide detailed insights into the research on different types of natural hazards worldwide, specifically focusing on analyzing spatiotemporal patterns (geographies) of research and selected scientometrical characteristics. The results of this study should provide a comprehensive overview, focusing on both the scientific community and the practitioners. Observed trends and patterns of research on natural hazards are put into the context of the number of events and fatalities, as well as the extent of damage, caused by natural disasters worldwide.

2. Data and Methods

2.1. Web of Science Database, Classification of Natural Hazards, and Dataset Building

The Clarivate Analytics Web of Science (WOS) Core Collection database (www.webofknowledge.com) covers more than 13,000 highly credible journals and numerous books, and includes over 100 mil.

research items (articles, reviews, book chapters, etc.) for the period 1900–present [14]. The dataset analyzed in this study was built in June 2018, using the classification scheme of natural hazards and carefully designed search chains (see Table 1). The number of research items in the WOS database is gradually increasing over time; however, since the analyzed period is 1900–2017, it is assumed that the majority of journals and publishers have already released their 2017 and older publications to the WOS.

Table 1. Classification of natural hazards used in this study (adapted from [15,16]), WOS search query chains, and the number of research items found (June 2018). Note that numbers of research items found for individual categories (climatic/hydro-meteorological; geological/geomorphic), as well as for the total, do not correspond with the sum of research items found for individual types of natural hazards, because some items are focusing on two or more types of natural hazards simultaneously (multi-hazard items, see Section 3.1.2).

Categories (Bold) and Types of Natural Hazards	WOS Advanced Search Query Chain #	Research Items Found (1900–2017)
Climatic/hydro-meteorological	TS = ((flood *) OR (drought * OR (water scarcity)) OR (hurricane * OR typhoon * OR cyclone *) OR ("sea level rise" OR "rising sea level") OR (wildfire *) OR (desertification) OR (tornado *) OR ("heat wave *" OR heatwave *) OR (dust AND storm *) OR (TS = ((storm * OR (rainfall AND (extreme OR heavy)))) NOT TS = dust)	342,250
- Flood (FL)	TS = (flood *)	123,227
- Storm (ST)	TS = (storm * OR (rainfall AND (extreme OR heavy))) NOT TS = (dust)	91,298
- Drought (DR)	TS = (drought * OR (water scarcity))	83,214
- Hurricane (HU)	TS = (hurricane * OR typhoon * OR cyclone *)	42,097
- Sea level rise (SR)	TS = ("sea level rise" OR "rising sea level")	12,725
- Wildfire (WF)	TS = (wildfire *)	11,086
- Desertification (DE)	TS = (desertification)	5763
- Heat wave (HW)	TS = ("heat wave *" OR heatwave *)	4903
- Tornado (TO)	TS = (tornado *)	4384
- Dust storm (DS)	TS = (dust AND storm *)	4264
Geological/geomorphic	TS = ((earthquake *) OR ((slope AND movement *) OR landslide * OR avalanche * OR rockfall * OR rockslide * OR "debris flow *") OR (erosion * AND (soil * OR rock * OR water OR wind)) OR (volcan * AND (activ * OR eruption * OR explos *)) OR (subsidence *) OR (tsunami *))	273,459
- Earthquakes (EQ)	TS = (earthquake *)	109,123
- Slope movements (SM)	TS = ((slope AND movement *) OR landslide * OR avalanche * OR rockfall * OR rockslide * OR "debris flow *")	57,846
- Erosion (ER)	TS = (erosion * AND (soil * OR rock * OR water OR wind))	57,269
- Volcanic activity (VO)	TS = (volcan * AND (activ * OR eruption * OR explos *))	40,469
- Subsidence (SU)	TS = (subsidence *)	21,777
- Tsunami (TS)	TS = (tsunami *)	13,478
TOTAL	TS = ((earthquake *) OR ((slope AND movement *) OR landslide * OR avalanche * OR rockfall * OR rockslide * OR "debris flow *") OR (erosion * AND (soil * OR rock * OR water OR wind)) OR (volcan * AND (activ * OR eruption * OR explos *)) OR (subsidence *) OR (tsunami *) OR (flood *) OR (drought * OR (water scarcity)) OR (hurricane * OR typhoon * OR cyclone *) OR ("sea level rise" OR "rising sea level") OR (wildfire *) OR (desertification) OR (tornado *) OR ("heat wave *" OR heatwave *) OR (dust AND storm *)) OR (TS = ((storm * OR (rainfall AND (extreme OR heavy)))) NOT TS = dust)	588,424

TS = Topic; OR, AND, NOT are Boolean operators; * covers all words with given root (e.g., "flood*" covers "flood", "floods", "flooding", "flooded", . . .).

Many classification schemes of natural hazards and disasters exist, reflecting different needs and purposes. In line with the classification scheme of natural hazards presented by [15,16], the term 'natural hazard' in this study covers various types of events of two general categories: (i) climatic/hydro-meteorological natural hazards; and (ii) geological/geomorphic natural hazards. Biological hazards such as fungal, bacterial, and viral diseases and infestations or astronomical (extraterrestrial) hazards are not the subject of this study.

Various types of natural hazards are firstly classified according to the categories (see Table 1) and secondly according to the number of research items found:

- Major types of natural hazards (over 40,000 research items found; four major types of natural hazards for each of category–climatic/hydro-meteorological (FL, ST, DR, HU) and geological/geomorphic (EQ, SM, ER, VO)).
- Other types of natural hazards (between 4000 and 40,000 research items found; two other types of geological and geomorphic natural hazards (SU, TS), six other types of climatic hydro-meteorological hazards (SR, WF, DE, HW, TO, DS)). Other types of natural hazards are grouped into one category for some of the analysis (CO, GO; see Section 2.2).
- Not considered types of natural hazards (natural hazards not reaching the threshold of 4000 research items at WOS (1900–2017) are not considered in this study). These are, for example: shifting sand (TS = (shift * AND sand); 3127 research items), blizzard and snowstorms (TS = (blizzard * OR snowstorm * OR (snow storm *)); 2704 research items), hailstorms (TS = (hailstorm * OR (hail AND storm *)); 1289 research items), lahars (921 research items), . . .).

It is obvious that the classification of some specific types of natural hazards (e.g., slope movements covering many different types of processes) may be ambiguous, especially when considering links and interactions (hazard chains) between individual types of natural hazards [17]. For example, debris flows or avalanches might be classified both as geological/geomorphic hazards and hydrological hazards, depending on the classification scheme used. Since it is not possible to further distinguish these in detail, considering the scope and extent of this study, all diverse types of slope movements are classified as geological/geomorphic hazards, referring to the traditional way of assigning all gravitational processes into this category [15].

The final classification scheme of natural hazards used in this study is presented in Table 1. While the WOS search query chains are easy to define for some types of natural hazards (e.g., subsidence), others are more complicated, especially considering terminological richness (e.g., different types of slope movements). Clearly, different search chains could lead to different results. To avoid (minimize) potential distortion of obtained results, each of the search query chains was carefully designed to cover a given type of natural hazard and checked with the most frequently represented WOS categories for the fit with expected ones, such as geoscience multidisciplinary, water resources, meteorology atmospheric sciences, geochemistry geophysics, geology, engineering geological, etc. The search query chains were defined after several iterations in more complicated cases (e.g., volcanic activity).

2.2. Dataset Analysis

Using pre-defined search chains (see Section 2.1), a dataset consisting of 588,424 research items is built and analyzed from various perspectives. Firstly, research on different types of natural hazards in time is analyzed, focusing on the share of individual types of natural hazards and the comparison between two categories—climatic/hydro-meteorological natural hazards and geological/geomorphic natural hazards (see Section 3.1.1). Special attention is paid to the multi-hazard research items focusing on more than one type of natural hazard. The overall share and the number of multi-hazard studies between each of the two individual types of natural hazards are analyzed using the WOS Advanced Search tool, based on a combined search among two sub-datasets (two types of natural hazards), focusing on links between them and within/between general categories of natural hazards (see Section 3.1.2).

Secondly, the geographical focus of research (defined by the name of the states and countries in some specific cases such as United Kingdom) is analyzed using content analysis [18] of titles and abstracts (see Section 3.2), revealing geographical hotspots of research on natural hazards. The numbers of research items geographically focusing on individual countries are assigned manually, using the WOS Advanced Search tool (see also [13]). Thirdly, the dataset is analyzed from the perspective of the country of researchers—affiliations (see Section 3.3), again using the WOS Results Analysis tool (Countries/Regions bookmark). This tool is also used to analyze research on individual types of natural hazards in the top 10 research countries, and to analyze the top 10 research countries for individual types of natural hazards (see Section 3.3.1). Bilateral cooperation between the top 25 research countries is analyzed in Section 3.3.2. The numbers of joint research items between two individual countries (cooperation matrix) are assigned manually, using the WOS Advanced Search tool. The results are visualized using the Network Visualization Tool of VOSviewer software version 1.6.9. [19] with the following parameters: Scale 2.0 (weighted by total link strength), Label size variation 1.0, Line size variation 0.7. Fourthly, the citations (see Section 3.4) are analyzed using the WOS Citation Report tool. Since only up to 10,000 research items can be analyzed using this tool, natural hazard types with more than 10,000 research items (see Table 1) had to be analyzed manually by parts. Selected impact indicators [20] such as the H index [21] and i100 index (number of research items within the dataset with 100 or more citations) are obtained from the citation analysis.

2.3. Databases of Disasters

Two freely available global databases of natural disasters are used to compare the focus of natural hazard studies with recorded events (see Section 4.1)/fatalities (see Section 4.2)/damages (see Section 4.3). These are MunichRE NatCatSERVICE [1] and SwissRE Sigma Explorer [5]. Both of these databases include information about damages and fatalities caused by individual types of natural hazards. The NatCatSERVICE analyses tool is comprised of 935 catastrophic events and 17,320 relevant events (935 catastrophic events) for the period 1980–2017, while Sigma Explorer is comprised of 5505 events for the period 1970–2018 (see Table 2). It is important to note that these databases consist of only a specific subset of real events and geographical coverage, event-specific coverage may differ in time, and total numbers of fatalities and damages caused are likely underestimated. It is thus necessary to carefully interpret observed trends [22,23].

Table 2. Basic characteristics of two databases of natural disasters used in this study.

	Database of Catastrophes	
	MunichRE NatCatSERVICE	**SwissRE Sigma Explorer**
Geographical coverage	global	global
Time span	1980–2017	1970–2018
No. of events	17,320 relevant (935 catastrophes)	5505 (5421 for period 1970–2017)
Data used	Number of events, losses and fatalities by natural hazard types and by years	Number of events (cumulative), losses and fatalities by natural hazard types and by years
Link	https: //natcatservice.munichre.com/	http: //www.sigma-explorer.com/
Reference	[1]	[5]

3. Results

3.1. Research on Different Types of Natural Hazards

3.1.1. General Trends

The overall number of research items has been increasing dramatically over time (see Figure 1), reaching a total of 588,424 research items published (until 2017), of which floods, earthquakes, storms, and droughts are the most frequently researched natural hazard types. While a total of 2307 research items was published between 1900 and 1949 (0.4% of all), 30,720 research items (5.2% of all) were published between 1950 and 1989, and 94,068 between 1990 and 1999 (16.3% of all). Over half of all research items have been published since 2010 (2010–2017). A comparable trend is observed in the WOS category Physical Geography (half of all research items in this category published since 2009), a not so strong increase is observed in WOS categories Environmental Sciences, Water Resources, Multidisciplinary Geosciences, Meteorology & Atmospheric Sciences, and Geology (half of all research items published since 2007, 2006, 2004, 2003, and 2001, respectively), and an even less strong increase in the category Multidisciplinary Sciences (half of all research items published since 1994). These figures indicate the prominent role and increasing interest of scientists in researching natural hazards in recent years.

While the overall number of research items published has increased rapidly in past decades, the ratio between research items focusing on the geological/geomorphic natural hazards and research items focusing on the climatic/hydro-meteorological natural hazards has remained relatively stable over time (see Figure 1). The research items focusing on climatic/hydro-meteorological natural hazards dominated in all studied periods, with 51.4% in 1900–1949 to 57.9% in 2010–2017. Significant differences, however, exist between individual countries (see Section 3.2).

The share of research items focusing on individual types of natural hazards is also relatively stable, with marginal changes being detected since the 1990s (see also Figure 1). The most significant changes in the whole studied period 1900–2017 are the decreasing share of research on earthquakes (from 30.4% in 1900–1949 to 15.7% in 2010–2017) and of research on storms (from 23.1% in 1950–1969 to 12.5% in 2010–2017). The most significant increasing share is observed for droughts (from 6.7% in 1900–1949 to 14.1% in 2010–2017), slope movements (from 2.9% in 1900–1949 to 7.9% in 2010–2017), and other climatic/hydro-meteorological natural hazards (from 3.7% in 1900–1949 to 7.2% in 2010–2017). A very stable share is observed for floods (minimum 17.3% in 1900–1949, maximum 18.3% in 2000–2009).

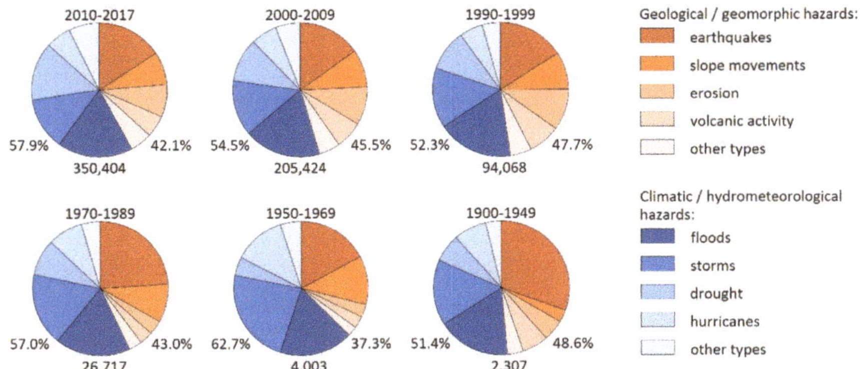

Figure 1. The share of major types of natural hazards in different time periods. The numbers show the total number of research items and the percentages show the share between research items dealing with the geological/geomorphic natural hazards and research items dealing with the climatic/hydro-meteorological natural hazards.

3.1.2. Multi-Hazard Research Items and Links Between Types of Natural Hazards

Obvious disproportion is observed between the sum of research items found for individual types of natural hazards (n = 682,923) and the overall number of research items found in the WOS database when using all individual search chains (see Table 1) simultaneously (n = 588,424), meaning that 94,499 research items found (16.1% of all) deal with two or more types of natural hazards (so called multi-hazard research items). The share of multi-hazard research items on the number of research items found for individual types of natural hazards varies from 14.5% (droughts), 17.0% (volcanic activity), and 17.4% (earthquakes) to 37.5% (other climatic/hydro-meteorological hazards), 38.1%

(erosion), and 42.5% (other geological/geomorphic hazards). Both climatic/hydro-meteorological natural hazards and geological/geomorphic natural hazards are addressed in 26,013 research items (i.e., 27.5% of all multi-hazard items).

It is further shown that the most significant links (>5% share of multi-hazard items on the total of research items for the two given types of natural hazards) exist between storms and hurricanes (research items, 7.9%), storms and other climatic/hydro-meteorological hazards (6.1%), earthquakes and other geological/geomorphic hazards (5.8%), and storms and floods (5.2%; see Table 3). Table 3 also reveals stronger links between natural hazards of one category—climatic/hydro-meteorological or geological/geomorphic (darker colors in the upper left and the lower right sector of the table).

Table 3. Number of multi-hazard research items (see the text for explanation) between individual types of natural hazards (classes: $\leq$ 999; 1000–2499; 2500–4999; $\geq$ 5000) and the share of multi-hazard research items on the total number of research items found for the two types of natural hazards involved (classes: $\leq$ 0.9%; 1.0–2.4%; 2.5–4.9%; $\geq$ 5.0%). Increased number (share) of research items between two types of natural hazards is indicated by a darker color. The last row shows the share of multi-hazard research items within individual types of natural hazard. See Table 1 for the explanation of abbreviations of natural hazard types.

		Number of Multi-Hazard Research Items between Two Types of Natural Hazards									
	X	**EQ**	**SM**	**ER**	**VO**	**GO**	**FL**	**ST**	**DR**	**HU**	**CO**
Share of multi-hazard research items between two types of natural hazards [%]	EQ	X	5299	724	4189	7935	2657	1403	231	1742	581
	SM	3.3	X	3035	1959	1898	1982	410	170	220	329
	ER	0.4	2.7	X	1307	2276	3451	2501	302	873	549
	VO	2.9	2.0	1.4	X	1314	4829	4337	1245	576	3104
	GO	5.8	2.5	1.4	2.6	X	2631	1593	301	1122	1490
	FL	1.2	1.9	2.7	1.2	1.7	X	10,827	6492	3865	3725
	ST	0.7	1.6	2.9	0.3	1.2	5.2	X	3307	10,104	8023
	DR	0.1	0.2	0.9	0.1	0.3	3.2	1.9	X	792	2675
	HU	1.2	0.9	0.6	0.3	1.5	2.4	7.9	0.6	X	1736
	CO	0.4	0.5	3.2	0.4	2.0	2.3	6.1	2.2	2.1	X
Share of multi-hazard research items [%]		17.4	17.0	25.2	38.1	42.5	24.3	35.6	14.5	35.4	37.5

3.2. Geographical Focus of Research on Natural Hazards

Based on the search in abstracts and titles, geographical focus (defined by the name of the country) can be assigned to 378,851 research items (64.4% of all). The research on natural hazards has been performed in all 195 countries of the world, ranging from 82,525 research items focusing on the USA to three research items focusing on Sao Tome and Principe, and Antigua and Barbuda (see Figure 2). The USA are followed by China (31,403 research items), Japan (18,006 research items), Australia (17,193 research items), and Canada (14,423 research items). In Europe, 11,980 research items are focusing on Italy (6th worldwide), 9646 on the U.K. (8th worldwide), and 9329 on Spain (10th worldwide). In Asia, China and Japan are followed by India (11,856 research items, 7th worldwide), and in Latin America, Mexico (9331 research items, 9th worldwide) is followed by Brazil (7101 research items, 12th worldwide). South Africa, with 3802 research items, is the most researched African state (19th worldwide).

The apparent hotspot where only a limited amount of research on natural hazards is performed is identified in Africa—a total of 24,800 research items (6.5% of all with assigned geographical focus) are geographically focusing on Africa. Considering the number of inhabitants of individual countries [24] with more than 1000 research items, the highest share of research items per 1000 inhabitants is observed in Iceland (8.3 research items per 1000 inhabitants), New Zealand (1.8 research items per 1000 inhabitants), Mongolia and Australia (both 0.7 research items per 1000 inhabitants), and Norway (0.5 research items per 1000 inhabitants), while 0.005 research items per 1000 inhabitants is observed in Nigeria; 0.006 research items per 1000 inhabitants is observed in Nigeria; 0.009 research items per

1000 inhabitants is observed in India; and 0.01 research items per 1000 inhabitants is observed in Pakistan, Sudan, Bangladesh, Iraq, Indonesia, and Viet Nam.

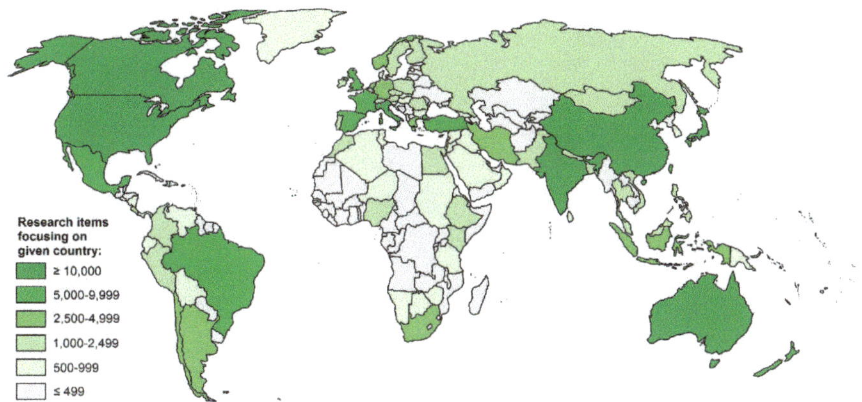

Figure 2. Geographical focus of research on natural hazards.

Considering the area of individual countries [25] with more than 1000 research items, it is observed that the highest number of research items per km^2 is reached in Taiwan (0.180 research items/km^2), followed by Israel (0.066 research items/km^2), The Netherlands (0.063 research items/km^2), and Switzerland (0.057 research items/km^2). Naturally, the lowest number is observed in large countries such as Russia (0.00018 research items/km^2), Brazil (0.00083 research items/km^2), and Colombia (0.00087 research items/km^2).

3.3. Research by the Countries of the Authors (Affiliations)

3.3.1. General Overview and Research in Top 10 Research Countries

The research on natural hazards is dominated by authors affiliated with institutions located in the USA (177,671 research items (co-)authored, i.e., 30.2%), followed by authors affiliated with institutions located in China (67,350 research items (co-)authored, i.e., 11.4%). Authors located in five other countries (co-)authored over 30,000 research items for each country (England, Japan, Germany, Italy, and France) and authors from three other countries (co-)authored over 20,000 research items (Australia, Canada, and India). Altogether, the authors affiliated with the institutions located in these top 10 natural hazard research countries (co-)authored 71.3% of all research items in the analyzed dataset. Researchers from the top 25 research countries (12 from Europe, 7 from Asia, 4 from Americas, and 2 from Australia) altogether (co-)authored 85.8% of all research items.

Significant differences exist between research on different types of natural hazards in the top 10 research countries. Figure 3 shows that research on climatic/hydro-meteorological natural hazards dominates in the majority of countries (up to 69.2% share in case of Australia), while research on geological/geomorphic natural hazards dominates in Italy (65.2%), Japan (65.1%), and France (51.8%). The explanation for these differences might be related to the natural hazards affecting individual countries, e.g., increased share of research on earthquakes in Japan (40.7% compared to world's average 18.5%) or increased share of research on slope movements in Italy (18.2% compared to world's average 9.8%; see also Section 4.1). An increased share of research on droughts is observed in India and Australia (21.6% and 20.5%, respectively, compared to world's average of 14.1%).

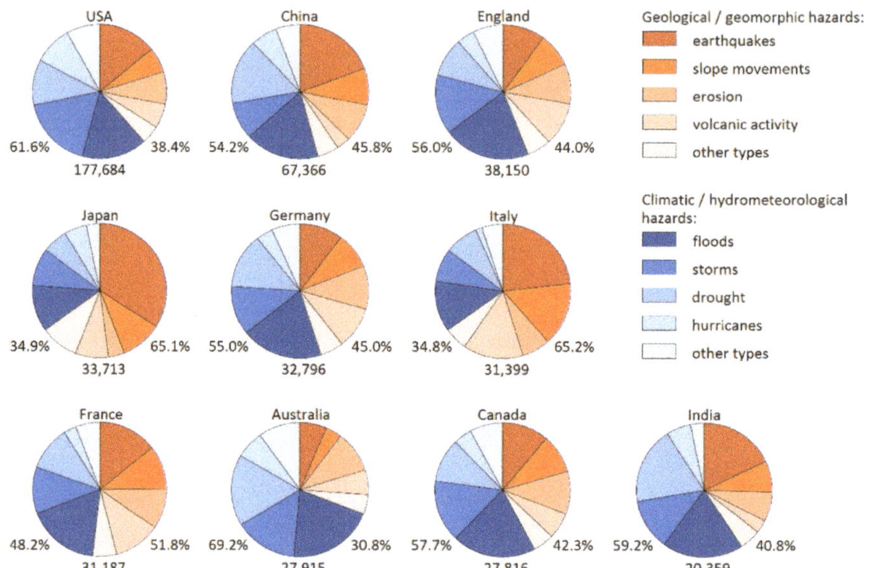

Figure 3. The share of major types of natural hazards among the top 10 research countries. The numbers show the total number of research items and the percentages show the share between research items dealing with the geological/geomorphic natural hazards and research items dealing with the climatic/hydro-meteorological natural hazards.

Considering the GDP of the top 10 research countries [26], it is observed that nine of them are at the same time ranked among the countries with the highest GDP (the only exception is Australia, which is ranked 13th, while Brazil is ranked 8th). The ratio between the number of inhabitants [24] of individual countries with reference to the world's population and the share of research items (co-)authored shows that researchers from the USA (co-)authored 30.2% of research items while the population of the country is only 4.3% of the world's population. An even more extreme ratio is observed in Australia with 4.7% research items (co-)authored and 0.3% of population, England (6.5% of research items (co-)authored and 0.7% of population), and Canada (4.7%; 0.5%). On the other hand, researchers from China (co-)authored 11.4% of research items while the population of the country is 18.1% of world's population. Researchers from India (co)-authored 3.5% of research items while the population of the country is 17.3% of the world's population. The research performance of individual countries is thus highly unbalanced.

From the global perspective, the research on individual types of natural hazards is in all cases dominated by researchers from the USA, followed by researchers from China, with the exception of research on volcanic activity, where Italy is ranked second (see Table 4). Spain is ranked 5th for research on droughts, 7th for research on other climatic/hydro-meteorological hazards, 8th for erosion, and 10th for research on slope movements; Taiwan is ranked as 6th for research on hurricanes; Russia is ranked 9th for research on volcanic activity and 10th for research on earthquakes; Switzerland is ranked 9th for research on slope movements; and The Netherlands is ranked 10th for research on erosion, as well as climatic/hydro-meteorological hazards.

Table 4. Top 10 research countries for individual types of natural hazard (see Table 1 for abbreviations).

3.3.2. Cooperation between Countries

The majority of research items are written by the author(s) from one country and only slightly above one fourth of all research items (27.3%) are characterized as international research items. Significant differences, however, exist between individual countries. Among the top 25 research countries, the higher share of international research items is observed in Scandinavian countries (Sweden, Norway), Switzerland, Scotland, and The Netherlands, while the lower share is observed in India, China, the USA, Iran, and Japan.

Two clusters of bi-lateral cooperation are identified (see Figure 4)—European (leading countries England, Germany, France, Italy) and circum-Pacific (leading countries USA, China, Canada, Australia, Japan). Considering the amount of joint research items between two countries, the most intense scientific cooperation exists between the USA and China (8642 joint research items), the USA and England (6343 joint research items), and the USA and Canada (5750 joint research items). Additionally, 3000+ joint research items exist between the USA and Germany, the USA and France, the USA and Australia, the USA and Japan, and the USA and Italy.

Considering the share of joint research items between two countries on the total number of research items of these two countries, the highest share of joint research items exists between Switzerland and Germany (4.5%; 2010 joint research items), England and Scotland (4.5%; 1999 joint research items), and England and Germany (3.9%; 2760 joint research items). More than 3.0% of joint research items further exist between several often neighboring countries, such as Germany and France, Italy and France, England and France, Switzerland and France, the USA and China, Germany and The Netherlands, Spain and Italy, Norway and Sweden, and Spain and France (see Figure 4).

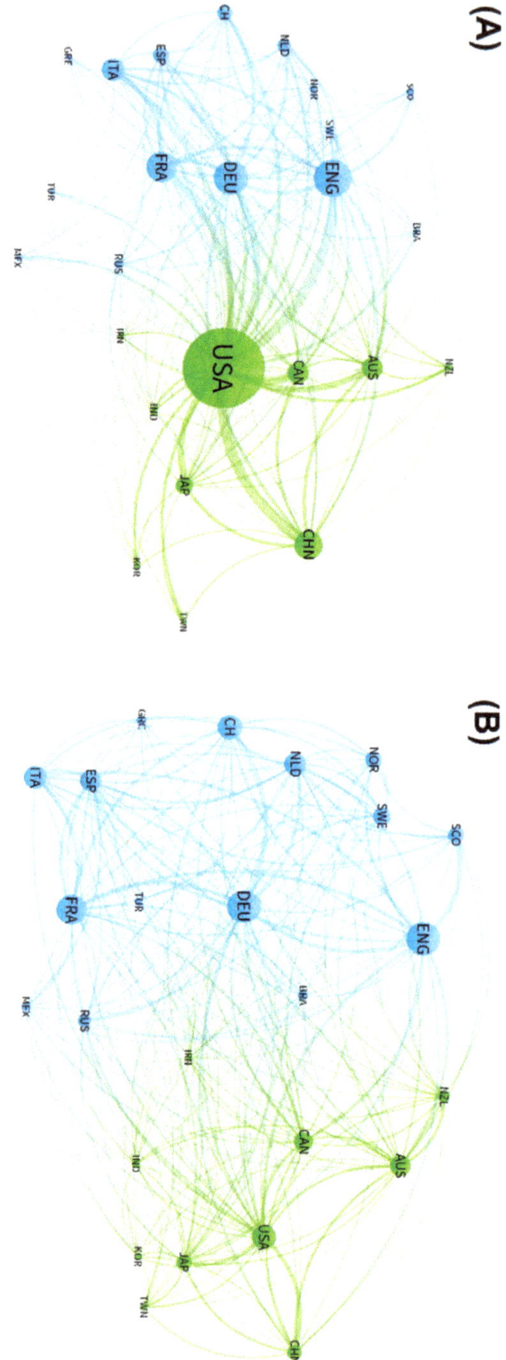

Figure 4. Cooperation between top 25 research countries. Part (**A**) shows absolute number of joint research items between two countries (varying from 9 to 8642); part (**B**) shows relative share of joint research items between two countries on the total number of research items of these two countries (varying from 0.06% to 4.5%). Two apparent clusters are distinguished (see the text).

3.4. Citations

A total of 588,424 research items analyzed have obtained 9996,579 citations (1900–2017), resulting in an average of 17.0 citations per item (in 8/2018). The H index of the whole dataset is 561 and the i100 index is 17,414 (see Table 5). A clear relationship exists between the number of research items and citations obtained:

$$Ct = 15.6 \cdot It + 59{,}147 \quad (R^2 = 0.95) \tag{1}$$

where *Ct* is the total of citations obtained by an individual type of natural hazard and *It* is the total number of research items for a given natural hazard. A comparison of average citations per item between climatic/hydro-meteorological natural hazards and geological/geomorphic natural hazards shows a slightly higher average of climatic/hydro-meteorological natural hazards (17.2 vs. 16.7 citations per item). Average citations per item, however, significantly vary among individual types of natural hazards from 9.9 citations per item (tornado) and 11.6 citations per item (tsunami) to 23.9 citations per item (volcanic activity) and 24.8 citations per item (heat wave).

Table 5. Total number of citations obtained by research items of individual types of natural hazards, average citations per item, H index, and i100 index (in August 2018).

Categories and Types of Natural Hazards	Research Items Found (1900–2017)	Citations (Total) (1900–2017)	Citations/Item	H Index (8/2018)	i100 Index (8/2018)
Climatic/hydro-meteorological	342,250	5,884,967	17.2	494	10,603
- Flood (FL)	123,227	1,868,551	15.1	313	3161
- Storm (ST)	91,298	1,501,880	16.5	301	2658
- Drought (DR)	83,214	1,719,385	20.7	354	3478
- Hurricane (HU)	42,097	621,611	14.8	224	1058
- Sea level rise (SR)	12,725	290,771	22.9	194	595
- Wildfire (WF)	11,086	222,717	20.1	160	417
- Desertification (DE)	5763	102,177	17.7	131	209
- Heatwave (HW)	4903	108,654	24.8	141	247
- Tornado (TO)	4384	48,623	9.9	87	67
- Dust storm (DS)	4264	100,621	23.6	128	194
Geological/geomorphic	273,459	4,573,732	16.7	414	7684
- Earthquakes (EQ)	109,123	1,530,718	14.0	294	2587
- Slope movements (SM)	57,846	866,888	15.0	243	1420
- Erosion (ER)	57,269	1,081,608	18.9	264	1824
- Volcanic activity (VO)	40,469	968,054	23.9	255	1794
- Subsidence (SU)	21,777	413,471	18.9	184	659
- Tsunami (TS)	13,478	156,268	11.6	128	228
TOTAL	588,424	9,996,579	17.0	561	17,414

The highest H index (354) is observed for drought, which is ranked 4th in terms of the total number of research items, but 2nd in the total number of citations obtained (see Table 5). A logarithmic relationship is observed between H indexes and the total number of research items of a given type of natural hazard (*It*), and total number of citations obtained (*Ct*):

$$H_{index} = 61.9 \cdot \ln It\text{-}412.6 \quad (R^2 = 0.89) \tag{2}$$

$$H_{index} = 64.4 \cdot \ln Ct\text{-}621.9 \quad (R^2 = 0.95) \tag{3}$$

where *Ct* is the total number of citations obtained by an individual type of natural hazard and *It* is the total number of research items of a given natural hazard. The i100 indexes are proportionally related (linear relationship) to the amount of research items focusing on individual types of natural hazards (Pearson coefficient 0.996) and to the total of citations obtained by them (Pearson coefficient

0.999). The relationships between the number of items published and citations obtained (H indexes; Equations (1)–(3)) are in line with [27] and recent observations of [28].

4. Discussion: Research on Natural Hazards and Reported Disasters—Global Overview

4.1. Research on Natural Hazards and Their Occurrence in Time

The occurrence of natural hazards based on MunichRE NatCatSERVICE data [1] in the period 1980–2017 indicates an increasing number of reported events (a total of 17,320 events, of which 1532 (8.8%) are classified as geological/geomorphic natural hazards). These figures provide an average of 194 research items published in the WOS database per one geological/geomorphic event in the NatCatSERVICE database and an average of 24 research items published in the WOS database per one climatic/hydro-meteorological event in the NatCatSERVICE database. Specifics of the database building (especially classification scheme and the definition of disasters; see also Section 2.3), however, need to be considered carefully when interpreting these observations (see [22,23]). While the cumulative number of reported events exhibits an increasing linear trend for both climatic/hydro-meteorological and geological/geomorphic natural hazards, a rather exponential trend is observed for the increasing number of research items published (see Figure 5). This exponential trend could be explained as a result of the increasing interest of the research community in natural hazard science (see also the comparison in Section 3.1.1) or as a result of a generally changing publishing paradigm (see also [13]), as well as possible other motivation for research on natural hazards, such as changing strategic natural hazards-/disasters-related policies [29,30] and/or research funding priorities [31,32].

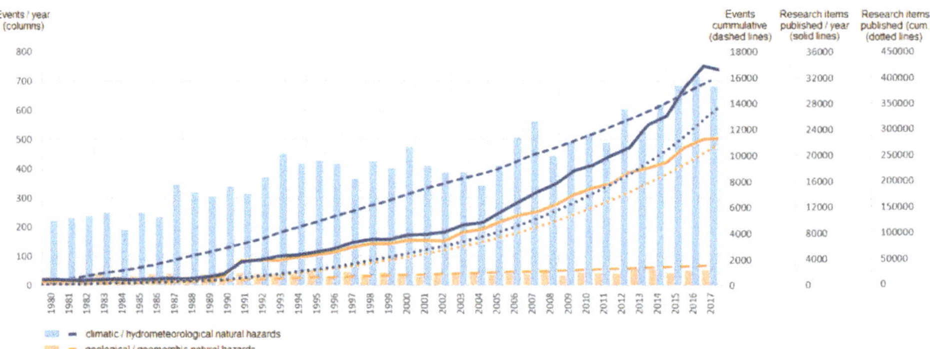

Figure 5. Research on natural hazards and the occurrence of natural hazards in 1980–2017 (number of events based on [1]).

4.2. Research on Natural Hazards and Fatalities Caused

Fatalities caused by natural hazards in both global databases [1,5] significantly vary yearly, reflecting the occurrence of major disasters (e.g., major earthquakes, tsunamis, and floods; see Figure 6). Furthermore, the cumulative number of fatalities caused by natural hazards is increasing with a rather linear trend interrupted by episodic imprints of major disasters, such as the 2004 Indian Ocean earthquake and tsunami, or the 2010 Haiti earthquake. A total of 1,723,648 fatalities are attributed to disasters in the period 1980–2017 (of which 848,437, i.e., 49.2% to geological/geomorphic ones [1]). A total of 2,530,566 fatalities are observed for the period 1970–2017 (of which 1,356,146; i.e., 53.6% are attributed to geological/geomorphic hazards [5]). Some of the major disasters are also recognizable in research on natural hazards, clearly not only motivated by the number of fatalities caused. An example is again the 2004 disaster, which led to the increase of research items on tsunamis, from 154 research items published in 2004 to 624 research items published in 2005 (+305%).

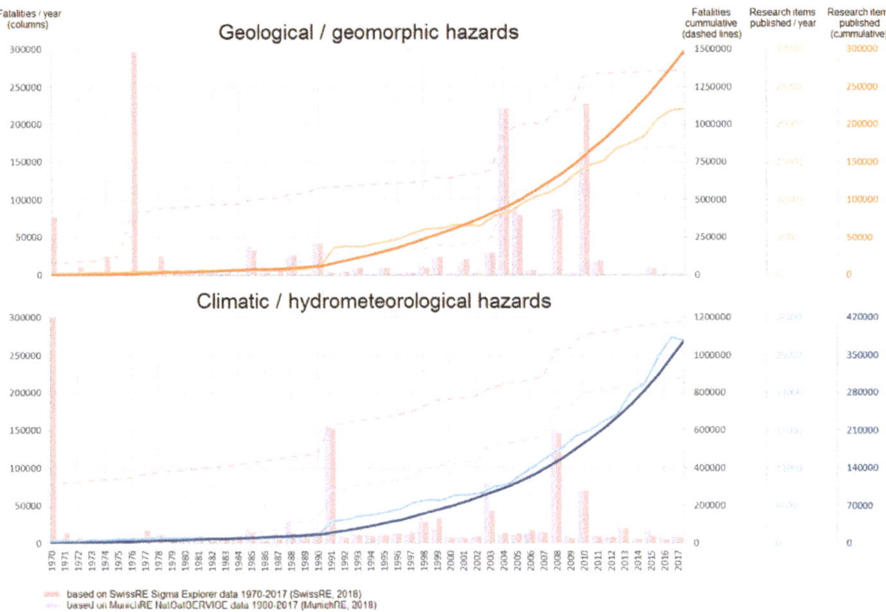

Figure 6. Research on geological/geomorphic and climatic/hydro-meteorological natural hazards and fatalities caused (fatalities estimations based on [1,5]).

4.3. Research on Natural Hazards and Losses Caused by Natural Hazards

Both NatCatSERVICE [1] and Sigma Explorer [5] report increasing extents of damage caused by natural disasters yearly, apparently reflecting changing conditions (e.g., [3]), increasing population pressure and vulnerability globally [4], as well as possible inconsistencies in disaster databases on a global level [22,23] (see also Section 2.3). Overall losses caused by natural disasters were estimated to be 4614 billion USD in the period 1980–2017 [1] and 4697 billion USD in the period 1970–2017 [5]. The share of geological/geomorphic disasters varies from 20.3% to 25.9% among the two different databases used. While major disasters are capable of significantly influencing the overall statistics of losses (especially in case of geological/geomorphic disasters), the amount of research items published does not clearly reflect them on a scale of natural hazard categories (see Figure 7). Imprints of large disasters are, however, recognizable on a scale of individual types of natural hazards (see [10]). The 2008 Wenchuan earthquake (the second costliest geological/geomorphic disaster in the database) caused an increase in research on earthquakes (from 3952 research items published in 2008 to 4743 research items published in 2009), as well as slope movements (from 2470 research items published in 2008 to 2868 research items published in 2009).

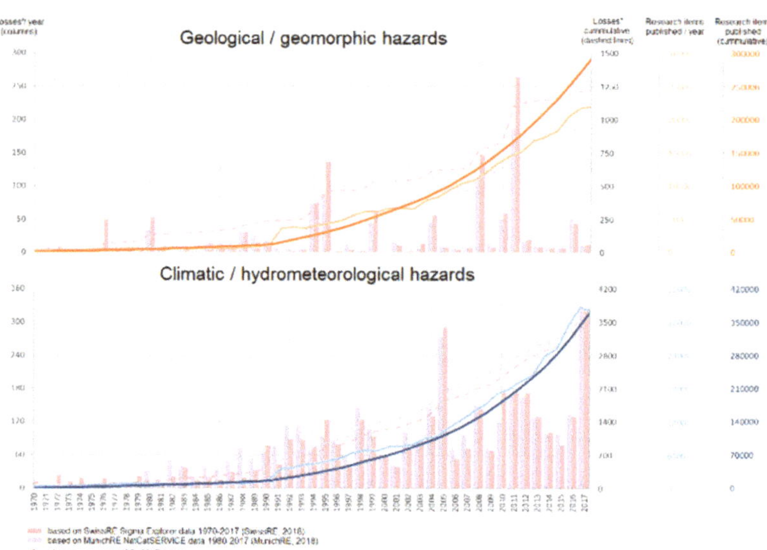

Figure 7. Research on geological/geomorphic and climatic/hydro-meteorological natural hazards and losses caused (losses estimations based on [1,5]).

5. Conclusions

This study provides insight into the global characteristics of research on natural hazards from the perspective of scientometry and the view of geography. Key findings of this study can be summarized as follows:

- Research on climatic/hydro-meteorological hazards is prevailing over the research on geological/geomorphic hazards globally (55.8%/44.2%), while significant differences exist between individual countries, reflecting the main types of natural hazards affecting these countries.
- The total amount of research items published is increasing exponentially for all types of natural hazards, with over the half of all research items published in 2010–2017 (analyzed period 1900–2017).
- In terms of the total number of research items published, leading hazards are floods, earthquakes, storms, and droughts; the share of individual types of natural hazards in research on natural hazards is rather stable over time.
- Research on all types of natural hazards is dominated by researchers from the USA, followed by researchers from China; the list of top 10 research countries for individual types of natural hazards further includes England, Japan, Germany, Italy, France, Australia, Canada, India, Spain, Taiwan, Russia, Switzerland, and The Netherlands.
- The majority of research items are written by the author(s) from one country; strong cooperation in natural hazard research exists between the top 10 research countries and geographically neighboring countries; two clusters of cooperation in natural hazard research are revealed—European and circum-Pacific.
- The analysis of geographical focus reveals hotspots of natural hazards research (USA, China, Japan, Australia, Canada, Italy, India; > 10,000 research items focusing on each country), as well as hotspots of a limited amount of research on natural hazards in Africa and Central Asia.

- Research on heat waves and volcanic activity obtained the highest average number of citations per research item (24.8 and 23.9 citations/item), while research on tornadoes and tsunamis yielded the lowest (9.9 and 11.6 citations/item).
- The exponentially increasing amount of published research items does not correspond with the number of events reported, or fatalities or losses claimed (linear increasing trend; based on two global databases of natural disasters), rather indicating a changing paradigm in publishing in natural hazard science and/or possible other motivation for research on natural hazards (changing policies, funding priorities).
- The share of research on climatic/hydro-meteorological hazards and geological/geomorphic hazards roughly corresponds to the share on the total number of fatalities claimed, but not damages caused globally (dominance of climatic/hydro-meteorological disasters).
- Major disasters such as the 2004 Indian Ocean tsunami or 2008 Wenchuan earthquake are capable of imprinting into the global statistics of research on individual types of natural hazards.

Author Contributions: This study was designed and executed by Adam Emmer.

Funding: This work was supported by the Ministry of Education, Youth and Sports of the Czech Republic within the National Sustainability Programme I (NPU I), grant number LO1415.

Acknowledgments: The author thanks two anonymous reviewers for their constructive comments and valuable suggestions which helped to improve this work.

Conflicts of Interest: The author declares no conflict of interest.

References

1. MunichRE NatCatSERVICE. Available online: https://natcatservice.munichre.com/ (accessed on 18 October 2018).
2. Hewitt, K. Environmental disasters in social context: Toward a preventive and precautionary approach. *Nat. Hazards* **2013**, *66*, 3–14. [CrossRef]
3. Intergovernmental Panel on Climate Change (IPCC). *Contribution of Working Group I to the Fifth Assessment Report of the Intergovernmental Panel on Climate Change*; Cambridge University Press: Cambridge, UK; New York, NY, USA, 2013; 1522p.
4. Birkmann, J. *Measuring Vulnerability to Natural Hazards*; United Nations University Press: New Delhi, India, 2006; 524p.
5. SwissRE Sigma Explorer. Available online: http://www.sigma-explorer.com/ (accessed on 18 October 2018).
6. DesInventar—Inventory System of the Effects of Disasters. Available online: https://www.desinventar.org/ (accessed on 18 October 2018).
7. Chiu, W.T.; Ho, Y.S. Bibliometric analysis of tsunami research. *Scientometrics* **2007**, *73*, 3–17. [CrossRef]
8. Sagar, A.; Kademani, B.S.; Garg, R.G.; Kumar, V. Scientometric mapping of Tsunami publications: A citation based study. *Malays. J. Libr. Inf. Sci.* **2010**, *15*, 23–40.
9. Liu, X.J.; Zhan, F.B.; Hong, S.; Niu, B.B.; Liu, Y.L. A bibliometric study of earthquake research: 1900–2010. *Scientometrics* **2012**, *92*, 747–765. [CrossRef]
10. Qian, G. Scientometrics Analysis on the Research Field of Wenchuan Earthquake. *Disaster Adv.* **2012**, *5*, 704–707.
11. Wu, X.L.; Chen, X.Y.; Zhan, F.B.; Hong, S. Global research trends in landslides during 1991–2014: A bibliometric analysis. *Landslides* **2014**, *12*, 1215–1226. [CrossRef]
12. Mikoš, M. The bibliometric impact of books published by the International Consortium on Landslides. *Landslides* **2018**, *15*, 1459–1482. [CrossRef]
13. Emmer, A. GLOFs in the WOS: Bibliometrics, geographies and global trends of research on glacial lake outburst floods (Web of Science, 1979–2016). *Nat. Hazards Earth Syst. Sci.* **2018**, *18*, 813–827. [CrossRef]
14. Clarivate Analytics Web of Science. Available online: www.webofknowledge.com (accessed on 18 October 2018).
15. Burton, I.; Kates, W.R. The Perception of Natural Hazards in Resource Management. *Nat. Resour. J.* **1964**, *3*, 412–441.

16. Wisner, B.; Gaillard, J.C.; Kelman, I. *The Routledge Handbook of Hazards and Disaster Risk Reduction*; Routledge, Taylor and Francis Group: London, UK; New York, NY, USA, 2012; 875p.
17. Gill, J.C.; Malamud, B.D. Reviewing and visualizing the interactions of natural hazards. *Rev. Geophys.* **2014**, *5*, 680–722. [CrossRef]
18. Hsieh, H.F.; Shannon, S.E. Three Approaches to Qualitative Content Analysis. *Qual. Health Res.* **2005**, *15*, 1277–1288. [CrossRef] [PubMed]
19. Van Eck, N.J.; Waltman, J. Software survey: VOSviewer, a computer program for bibliometric mapping. *Scientometrics* **2010**, *84*, 523–538. [CrossRef] [PubMed]
20. Waltman, L. A review of the literature on citation impact indicators. *J. Infometrics* **2016**, *10*, 365–391. [CrossRef]
21. Hirsch, J.E. An index to quantify an individual's scientific research output. *Proc. Natl. Acad. Sci. USA* **2005**, *102*, 16569–16572. [CrossRef] [PubMed]
22. Kron, W.; Steuer, M.; Low, P.; Wirtz, A. How to deal properly with a natural catastrophe database—Analysis of flood losses. *Nat. Hazards Earth Syst. Sci.* **2012**, *12*, 535–550. [CrossRef]
23. Wirtz, A.; Kron, W.; Low, P.; Steuer, M. The need for data: Natural disasters and the challenges of database management. *Nat. Hazards* **2014**, *70*, 135–157. [CrossRef]
24. World Population Review. Available online: http://worldpopulationreview.com/countries/ (accessed on 18 October 2018).
25. United Nations Statistics Division, Environmental Indicators—Total Area. Available online: https://unstats.un.org/unsd/environment/totalarea.htm (accessed on 18 October 2018).
26. The World Bank Data—GDP Stats. Available online: https://data.worldbank.org/indicator/NY.GDP.MKTP.CD?year_high_desc=true (accessed on 18 October 2018).
27. Wuchty, S.; Jones, B.F.; Uzzi, B. The increasing dominance of teams in production of knowledge. *Science* **2007**, *316*, 1036–1039. [CrossRef] [PubMed]
28. Sandstrom, U.; Van den Besselaar, P. Quantity and/or Quality? The Importance of Publishing Many Papers. *PLoS ONE* **2016**, *11*, e0166149. [CrossRef] [PubMed]
29. Birkmann, J.; Buckle, P.; Jaeger, J.; Pelling, M.; Setiadi, N.; Garschagen, M.; Fernando, N.; Kropp, J. Extreme events and disasters: A window of opportunity for change? Analysis of organizational, institutional and political changes, formal and informal responses after mega-disasters. *Nat. Hazards* **2010**, *55*, 637–655. [CrossRef]
30. Gall, M.; Nguyen, K.H.; Cutter, S.L. Integrated research on disaster risk: Is it really integrated? *Int. J. Disaster Risk Reduct.* **2015**, *12*, 255–267. [CrossRef]
31. Roy, N.; Thakkar, P.; Shah, H. Developing-World Disaster Research: Present Evidence and Future Priorities. *Disaster Med. Public Health Prep.* **2011**, *5*, 112–116. [CrossRef] [PubMed]
32. Ebi, K.L.; Semenza, J.C.; Rocklov, J. Current medical research funding and frameworks are insufficient to address the health risks of global environmental change. *Environ. Health* **2016**, *15*, 108. [CrossRef] [PubMed]

Review

Big Data in Natural Disaster Management: A Review

Manzhu Yu *, Chaowei Yang and Yun Li

NSF Spatiotemporal Innovation Center, George Mason University, 4400 University Drive, Fairfax, VA 22030, USA; cyang3@gmu.edu (C.Y.); yli38@gmu.edu (Y.L.)

* Correspondence: myu7@gmu.edu

Received: 12 March 2018; Accepted: 3 May 2018; Published: 5 May 2018

Abstract: Undoubtedly, the age of big data has opened new options for natural disaster management, primarily because of the varied possibilities it provides in visualizing, analyzing, and predicting natural disasters. From this perspective, big data has radically changed the ways through which human societies adopt natural disaster management strategies to reduce human suffering and economic losses. In a world that is now heavily dependent on information technology, the prime objective of computer experts and policy makers is to make the best of big data by sourcing information from varied formats and storing it in ways that it can be effectively used during different stages of natural disaster management. This paper aimed at making a systematic review of the literature in analyzing the role of big data in natural disaster management and highlighting the present status of the technology in providing meaningful and effective solutions in natural disaster management. The paper has presented the findings of several researchers on varied scientific and technological perspectives that have a bearing on the efficacy of big data in facilitating natural disaster management. In this context, this paper reviews the major big data sources, the associated achievements in different disaster management phases, and emerging technological topics associated with leveraging this new ecosystem of Big Data to monitor and detect natural hazards, mitigate their effects, assist in relief efforts, and contribute to the recovery and reconstruction processes.

Keywords: big data; disaster management; review

---

1. Introduction

Natural disasters can be defined as a combination of natural hazards and vulnerabilities that endanger vulnerable communities that are incapable of withstanding the adversities arising from them [1]. Human beings invariably face threats of natural as well as human-made disasters, which often lead to massive damages, human suffering, and negative economic impacts. The main characteristics of natural disasters are unpredictability, availability of limited resources in impacted areas, and dynamic changes in the environment [2]. Unpredictability implies that severe impacts on people and property during natural disasters cannot be predicted with acceptable accuracy [3]. The issue of limited resources emerges because unpredictability makes it difficult to allocate adequate resources in advance. Dynamic changes in the environment result because it is difficult to make predictions about the movement of people and the damages that may occur because of the natural disaster. It is difficult to predict such changes based on data that pertains to normal periods [4]. Introducing disaster management policies and applying appropriate levels of information technology and equipment offer immense potential in enhancing the capabilities of disaster management policies. In addition, the evolving trends have opened massive technological resources for reducing disaster risks [5].

Big data is defined as the technological paradigm that allows researchers to conduct an efficient analysis of vast quantities of data that is made available through the current practices [6,7]. It is the collection of scientific and engineering methods and tools that help in making the best of massive amounts of available data. Big data addresses not only storage issues, but also issues related to accessibility, distribution, analysis, and effective visual presentation of data and analysis. Big data has now become a crucial element of communication, which complements the conventional exchange of intentional and explicit messages; such as first responders talking over a voice connection; or an announcement of a text message through which warning is given to citizens faced with the threat of an approaching natural disaster [8,9]. More precisely, communication also entails understanding and monitoring the entire body of public and openly available communication such as messages and content that is publicly exchanged on social media. In such situations, people may be exchanging messages in reporting their condition to their loved ones or making appeals for help. However, big data allows researchers to conduct a detailed analysis of all communications which provides valuable information that has a general validity for the population at large; such as information about a disease outbreak.

In general, the disaster management cycle comprises four distinct phases, which are "mitigation", "preparedness", "responses", and "recovery". The goal of the mitigation phase is to minimize the effects of a disaster (building warning codes and risk zones, risk analysis, public education). The main focus of the preparedness phase is on planning how to respond to a disaster. It includes preparedness plans, emergency exercises, and training, but also the Early Warning System development and implementation. Response activities pertain to providing the required disaster management services to save lives and safeguard property and protect the environment during disaster management situations. "Recovery" is the process of returning systems to normal levels after a disaster.

In recent years, the literature on disaster management mostly focused on the potential that lies in using specific kinds of data for natural disaster management [10–12]. It is in this context that this paper makes a review of major big data sources, the associated achievements in different disaster management phases, and emerging technological topics associated with leveraging this new ecosystem of Big Data to monitor and detect natural hazards, mitigate their effects, assist in relief efforts, and contribute to the recovery and reconstruction processes.

2. Review Methodology

In the literature review process, we followed a systematic approach on selecting papers related to "big data" and "disaster management". Firstly, we searched in Google Scholar with the two key words and obtained 4432 results. Secondly, we manually selected papers that were most relevant to our topic. This step led to a total number of 223 articles. Thirdly, we filtered the list of articles to only include journal articles, leading to 149 articles from 101 journals. It is very likely, though, that we missed several articles that are on the same topic. These articles were reviewed individually for analysis. Starting from two articles in 2011, the number of articles in the review field started to grow gradually, with nine articles in 2012, 15 articles in 2013, 18 articles in 2014, 46 articles in 2015, 25 articles in 2016, 21 articles in 2017, and 12 articles in 2018 (Figure 1). The peak of the topic "big data" in combination with "disaster management" occurred in the year 2015, and researchers might be directed to other related topics or emerging technologies benefiting disaster management. The rest of this paper will mainly discuss three perspectives based on the 149 articles: (1) major data sources; (2) big data contributions in different disaster phases; and (3) emerging technologies benefiting from big data and disaster management.

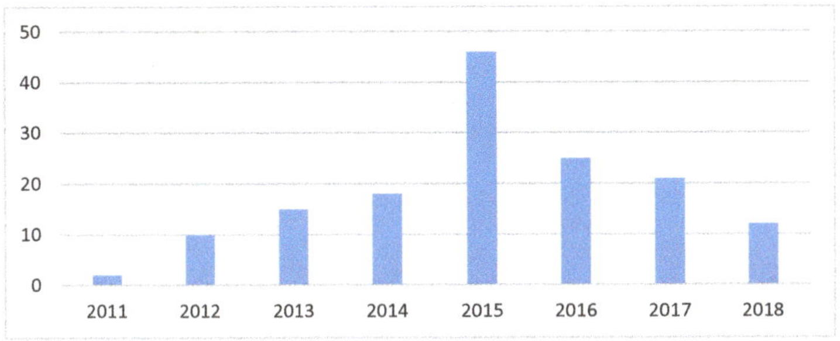

Figure 1. Distribution of reviewed articles by year of publication.

3. Major Data Sources

To the best of this authors' knowledge, there is no standard definition of "big data" in disaster management. A Federal Geographic Data Committee (FGDC) report [13] reviewed the emerging data collections, including real-time spatiotemporal data (e.g., GPS data), broadening of participation (e.g., Volunteer Geographic Information and social media), small satellites, and unmanned aircraft vehicles (UAVs). A United Nation (UN) report [14] illustrated examples of big data sources in disaster resilience, including exhaust data (mobile-based, financial transactions, transportation, online traces), digital content (social media and crowd-sourcing), and sensing data (physical sensing devices and remote sensing). In this paper, we define big data as the integration of diverse data sources and the capability to analyze and use the data (usually in real time) to benefit the population that participates in the disaster situation. The concept of big data is beyond the datasets themselves, regardless of their size.

In this section, we endeavor to review the major big data sources—especially the emerging ones —for disaster management, including satellite imagery, aerial imagery and videos from unmanned aerial vehicles (UAVs), sensor web and Internet of Things (IoT), airborne and terrestrial Light Detection and Ranging (LiDAR), simulation, spatial data, crowdsourcing, social media, and mobile GPS and Call Data Records (CDR). Figure 2 illustrates the distribution of reviewed articles by different data sources and their year of publication. It can be observed that an increase in article numbers is shown in almost all types of major data sources during 2014–2016, when the topic "big data" is popular in "disaster management", and a decrease in 2017. Satellite imagery, crowdsourcing, and social media data serve as the most popular data for disaster management.

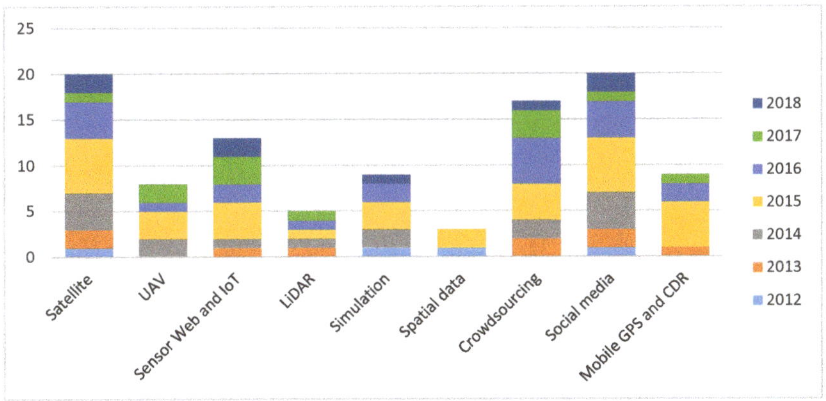

Figure 2. Distribution of reviewed articles by major data sources and year of publication.

3.1. Satellite Imagery

Satellite remote sensing technology provides qualitative and quantitative opportunities in the context of varied functions such as assessing post-disaster damage, responding through operational assistance [15–17], and risk reduction. The most remarkable contribution of remote sensing imagery is post-disaster damage assessment through change detection [18]. The application of remote sensing in disaster management is conducted through methods such as higher resolution, multidimensional, and multi-technique.

The recently evolved high resolution satellite imagery facilitates the collection of detailed texture information before and after a natural disaster for change detection. Such information is crucial to attain after the occurrence of disasters. Information collected through high resolution satellite imagery may pertain to the structural deformation of land areas, directional changes and creation of water bodies, and details about damaged building stock in the disaster affected area [19,20]. Consequently, rescue methods can be suggested and initiated for the immediate location and recovery of bereaved persons and for locating the corresponding area accurately by using high-resolution remote sensing techniques [19]. In addition to the two-dimensional information, it is also possible to produce three-dimensional images along with height information in the form of stereo images from satellites, which helps in identifying and measuring the intensity of damage [21,22]. Three-dimensional analysis allows detecting collapsed buildings and affected regions by comparing the differences in the height of buildings and estimating the heights and volumes of piled up debris with the use of pre- and post-disaster stereo image pairs [23–25]. To deal with the limitations of optical satellite imagery in cloud cover, rain conditions, and at nighttime, active sensors such as the synthetic aperture radar (SAR) can be effectively used to enlarge its observational capability during a natural disaster [18,26].

Despite identifying impacts or change brought about by disasters, satellite imagery has been intensively used for disaster risk reduction, including identifying human settlements [27], assessing flood risk [28], and landslide risk reduction [29].

3.2. UAV-Based Aerial Imagery and Videos.

Aerial imagery captured via unmanned aerial vehicles (UAVs) is playing an increasingly important role in disaster response, due to its efficiency in situational awareness. Aerial imagery can be captured with a high spatial resolution and processed much faster in comparison to satellite imagery [30]. Besides the improvements offered by oblique imagery acquired from piloted platforms, UAVs provide additional advantages [31]. These include fully controlled flight, VHR imagery of up to a 2 cm resolution that allows the detection of fine cracks, and the large degree of image overlap, which allows the generation of very detailed point clouds. With the use of UAVs, first responders can better understand which structures were affected by the given event and can determine the extent of the damage caused to these structures.

UAVs can carry various types of sensors, including cameras, video, infrared and ultra-violet sensors, radiation sensors, weather sensors, spectrum analyzers, and LIDAR reflectors. UAV imagery and videos can provide real-time, decision-relevant situational information for transportation planners who survey damaged roads, evacuation routes, and support for transport logistics [32].

UAVs are now recognized as a reliable data source for disaster information and for improving the estimation scale of damage [33]. With the combined use of UAV imagery, satellite, and aircraft data, it becomes possible to identify façade and roof damage to buildings by integrating geometrical transformation and environmental information [34]. Combining UAV imagery and crowd-sourcing enables annotation in the context of features of interest, such as damaged shelters and roads blocked by debris [30].

However, UAVs are still in the development stage and several issues need to be resolved; such as short battery life, which leads to a limited area of coverage; unforeseen behavior in different atmospheric conditions; limited scope of pilot training for users; and legislation that severely limits the use of UAVs in most countries. Due to the UAV technology, privacy issues arise. The community concern about exposing victims' locations needs to be addressed by officials and UAV developers. Public safety needs to be protected for privacy and individual rights during UAV operations in a complex disaster environment [35,36].

3.3. Wireless Sensor Web and Internet of Things

Chen et al. [37] utilized existing and available Wireless Sensor Network (WSN) technologies to develop an early warning system for natural disasters. The WSN technologies provide reliable data transmission and incorporate data fusion from heterogeneous sensors and minimizing energy consumption. Erdeji et al. [38] reviewed the overall structure of the WSN and UAV systems for natural disasters, and proposed the Inundation Monitoring and Alarm Technology in a System of Systems to incorporate smartphone data with WSN for enhanced situational awareness. Erman et al. [39] integrated low-cost embedded devices based on WSN and UAVs for improving the response time in critical situations, minimizing the latency, and maximizing the success ratio of delivery. WSN has also been extensively used in facilitating the intercommunication between a disaster impacted population and rescue teams when traditional infrastructure communication systems fail [40,41]. Tuna et al. [42] utilized a group of mobile robots to explore an unknown region after a disaster and used WSN to extend the range of communication for human existence detection.

While being widely deployed in disaster management, WSN is lacking in terms of the overall coordination of heterogeneous data sources and protocols from "socio-techno-economic perspectives" [43]. WSN contributes as an essential part of Internet of Things (IoT) technologies. Ray et al. [43] reviewed several IoT-enabled disaster management systems, including BRINCO (a notification system for earthquake and tsunami warning), BRCK (communication system under low connectivity areas), and GRILLO (earthquake alarming sensor network). The advantages of IoT include its capability to compensate for the poor infrastructure of a vulnerable population especially in developing countries, and being an alternative means of communication where IoT-enabled devices (battery powered wireless devices) can be used benefits data network resilience during disaster situations [44].

3.4. LiDAR

Airborne and terrestrial Light Detection and Ranging (LiDAR) is a method that provides the ability to extract high-quality elevation models and other features, providing reliable information about on-the-ground conditions during a disaster situation. LiDAR equipment is relatively expensive and collecting and processing of data with it can often prove to be time consuming. However, DEMs generated from LiDAR data can have a very high resolution and are very accurate. Unlike aerial photography, LiDAR scanners collect data at a very-fine (centimeter) scale resolution and can gather information about the ground surface below the vegetation. This ability is very useful for geological mapping and for measuring geological features, including monitoring volcano growth and predicting eruption patterns [45,46]. In addition, LiDAR data is highly sensitive to water and thus proves to be a suitable data source for flood prediction and assessment [47,48]. LiDAR has been useful in identifying and assessing elevation changes or structural damages after natural disasters [49]. For example, Kwan and Ransberger [50] demonstrated the use and analysis of LiDAR data before and after Hurricane Katrina in detecting transport network obstructions during a disaster response. They also demonstrated how the data helped in reducing the time taken by first responders to reach disaster sites. Similarly, Moya et al. [51] detected collapsed buildings after the earthquake that struck Kumamoto, Japan on 16 April 2016. They did so by using digital surface models (DSMs), taken before and after the earthquake during LiDAR flights.

3.5. Simulation Data

Numerical simulation or forecasting is one of the most significant contributions to the prediction of natural disasters occurring on account of meteorological phenomena, land surface phenomena, and various types of pollutions [52–54]. In addition, 3D modeling has been helpful in predicting the potential damage and in assessing the changes occurring after the disaster [55,56].

Large amounts and different types of observation data are generated during natural disasters, and these data can be used to produce, verify, validate, and improve models to represent the complexity for all facets of disaster management. Agent-based models are suitable for exploring human behavior and rapid low-level environmental changes using available high resolution data. Dou et al. [57] developed an agent-based framework for human rescue operations in a landslide disaster to evaluate a contingency plan. The framework extracts information from high resolution remote sensing images, simulates a landslide environment based on a three-dimensional landslide geological model, and uses a multi-agent simulation approach to provide individuals' behavior simulation under dynamic disaster scenarios. The simulations provide positive effects on the evacuation process, e.g., speeding up the process and reducing the number of casualties. Mas et al. [58] reviewed the agent-based models for tsunami mitigation and evacuation planning through case studies in Indonesia, Thailand, Japan, and Peru. Kureshi et al. [59] utilized advanced by combining agent-based models with physical sensors, and adaptively managed the heterogeneous collection of data resources and agent-based models to create what-if scenarios in order to deter- mine the best course of action.

3.6. Vector-Based Spatial Data

Vector-based spatial data provides fundamental support for disaster management, including disaster forecasts about the extent of a particular hazard or disaster, vulnerability analysis for critical facilities (hospital, school, shelter, fire station, etc.) and human beings (age, gender, socioeconomic status, etc.), damage assessment on the actual impact of a hazard, resource inventory (supplies, equipment, vehicles, etc.), and infrastructure (transportation networks and utility grids). Tomaszewski et al. [60] reviewed the importance of geographic information systems (GIS) and major GIS data sources for disaster response, including Federal Emergency Management Agency (FEMA) GIS Data Feeds, Geonames, US Census Bureau TIGER, the National Map, WorldBank Data, and Open Street Map. Herold and Sawada [61] reviewed the GIS applications of disaster management, emphasizing developing countries.

3.7. Crowdsourcing

It is known that both crowdsourcing and social media data are contributed by the public. While crowdsourcing data are actively contributed, social media data are mostly passively contributed, because the contributors are not aware that their activities on social media are leading to data collection [62].

Active crowdsourcing platforms have been developed for users to enthusiastically contribute the required information [63–65]. These platforms are usually developed and implemented by members of the affected public, or by non-governmental organizations (NGO's) such as Ushahidi (https://www.ushahidi.com/). The purpose of such platforms is to improve the disaster response and resource allocation based on real-time reports from disaster victims. Despite several success stories, challenges of using active crowdsourcing platforms are still worthy of consideration, particularly regarding the credibility and the value of integrating crowdsourced data into the decision-making process [66].

The analysis and processing of crowdsourced disaster data requires varied tools and automation processes in view of its noisy nature, large volume, and fast streaming speed. Significance of the crowdsourced data can be prioritized in keeping with the location of the disaster area and the analyzed data [67]. Online platforms and mobile applications have been established to collect and distribute crowdsourced data during and after disasters [68]. dos Santos Rocha et al. [12] have discussed the strategies that can be adopted by disaster managers in improving the performance of digital volunteers. They also assert that crowd sensing and collaborative mapping can be used in functions involving distributed intelligence, participatory engagement, and self-mobilization. They suggested the adoption of a feedback mechanism amongst the stakeholders to enhance the effectiveness of communication.

Government agencies have been slow in launching their own crowdsourcing platforms for disaster management. This might be due to concerns about the technological and human resources costs and risks of adopting and managing crowdsourcing for government operations. However, government agencies were criticized for the poor management of the 2005 Hurricane Katrina for inadequate information collection compared to the citizen efforts on response and recovery [69,70]. In recent years, government agencies have started to interact with citizens during disaster situations for real-time urgent information and better coordinate search and rescue operations [71,72]. FEMA is coordinating different types of crowdsourcing on image analysis, text messages, and mapathons (https://www.data.gov/event/fema-disaster-crowdsourcing-exchange/) for Hurricanes Harvey, Irma, and Maria, and the California Wildfires that happened in 2017.

3.8. Social Media

Social media platforms, including Twitter, Youtube, Foursquare, and Flickr have been contributing significantly to disaster management. Geotagged social media data can be collected by streaming harvest from the APIs provided by the social media companies.

Social media services have contributed significantly to disaster management as a tool to communicate information during disaster management [73]. Social media is increasingly being used by both NGO's and government disaster management agencies to determine public sentiment and reaction to an event [74]. It is evident that the multidirectional flows of communication and information that crisis crowdsourcing online platforms facilitate can make response and recovery efforts more efficient [70].

Even though social media provides implicit varieties of crowdsourced data, it is being effectively used in disaster management. Granell and Ostermann [75] discussed that Twitter is more effective for detection and prediction and less significant in recovery and response functions. Other data sources such as spatial video, UAV, and phone call data can provide strong evidence (e.g., in-situ images) of the current situation in the affected area during post-disaster stages.

There are varied ways of using social media in disaster management, including data collection, analytic workflow, narrative construction, disaster relevant information extraction, geolocation pattern/text/image analytics, and the broadcasting of information through social media platforms [76].

A number of authors have shown that real time analytics based on social media data provide good opportunities to detect and monitor events automatically [77,78]. Text messages are the basic source of analysis [79]. Visual analytics through social media data facilitate spatiotemporal analysis and create a spatial decision support environment that assists in evacuation planning and disaster management [80]. Given that social media does not rely only on text messages and provides more useful information through images and videos posted by users, visual analytics and image/video-based analysis are becoming more important in extracting the key information from social media posts [81].

3.9. Mobile GPS and Call Data Record

Mobile GPS (Global Positioning System) has emerged as an effective means of gathering mobile sensing data because it can be used to detect human mobility and behavior during large-scale natural disasters. Horanont et al. [82] made use of data collected after the 2011 Great Japan Earthquake and provided useful information on how humans react in disaster scenarios and how the evacuation process can be monitored on a real-time basis. With GPS, it becomes possible to determine the location, magnitude, and other details about an earthquake fault. This is done by using one or more of the three basic components of GPS, which are absolute location, relative movement, and time transfer. GPS data allows automatic calculation of the location, magnitude, and other details about the earthquake fault [83]. Similarly, Song et al. [84] developed a human mobility model to assess the disaster management behavior and mobility patterns of people after the occurrence of the Great East Japan Earthquake and Fukushima nuclear accident. They found that the behavior of human beings and their mobility following large-scale disasters sometimes correlate with their mobility patterns during normal times. In addition, they found that mobility patterns of human beings are also strongly impacted by factors such as their social relationships, intensity of disaster, extent of damage, availability of government shelters, news reporting, population flows, etc. GPS can also be used to determine the resilience of transportation systems during a natural disaster. Donovan and Work [85] conducted quantitative research in measuring transportation resilience by observing about 700 million taxi trips in New York City, which they used to analyze and assess the resilience of the transportation infrastructure to Hurricane Sandy.

It is known that call monitoring and recording applications used by telecommunication companies generate an extremely large amount of call detail records (CDRs) in real-time, and that companies constantly need to analyze this data and make required changes to boost productivity. It is noteworthy in this regard that the volume of the calls and data captured by the call monitoring applications is so large that it is impossible to manually analyze and reach conclusions on the behavior of the network.

CDR datasets are collections of spatiotemporal traces that can characterize individual mobility and social network behaviors at very fine scales. CDR datasets contain information about the location and time at which a communication (call/SMS) is made, along with unique identifiers for the sender and receiver. These data may be valuable for disaster management when they are used to estimate population size and density in a region or city. The information is based on the number of phone subscribers that are present in the coverage area of each cellular tower [86,87]. Information on population distribution before and after a disaster can be useful when assessing exposure risk and response needs. Ghurye et al. [88] utilized call records in using the granular behavior models to evaluate the similarities and differences between the normal and disaster patterns by using data and information relative to the floods that occurred in Rwanda in 2012. Results showed that disasters tend to disrupt both mobility patterns and communication behaviors, while recovery efforts can take several weeks.

4. Usage of Big Data in Disaster Management Phases

The disaster management cycle is a continuous process involving different phases of planning and preparing for expected disasters, including long-term mitigation (i.e., "mitigation/prevention"); short-term preparation and prevention (i.e., "preparedness"); reducing impacts of a disaster through response and rescue efforts (i.e., "response"); and restoration through clean-up and reconstruction (i.e., "recovery") [89]. In this section, we classify the 149 articles into the four phases and review the "mitigation/prevention" phase in two aspects: (1) long-term risk assessment and reduction; and (2) forecasting and prediction; the "preparedness" phase in two aspects: (1) monitoring and detection, and (2) early warning; two aspects for the "response" phase: (1) damage assessment; and (2) post-disaster coordination and response; and the post-disaster "recovery" phase (Figure 3).

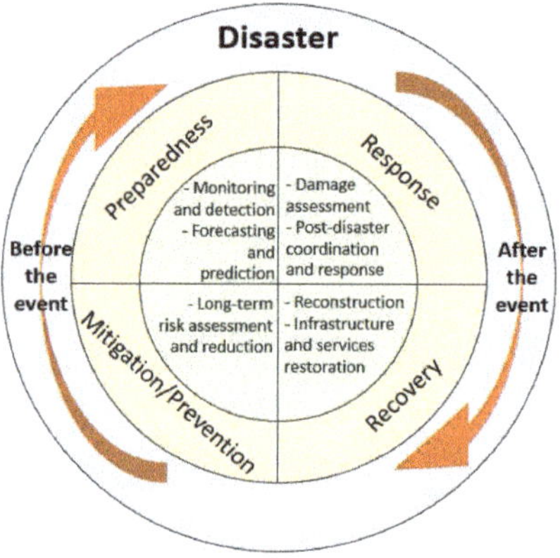

Figure 3. Disaster management cycle derived from Alexander [90].

Table 1 shows the disaster management phases, major data sources, and the application fields that were mapped based on the literature reviewed in this paper.

Table 1. Mapping disaster management phases with major data sources and application fields.

Disaster Management Phase	Data Source	Reviewed Application Fields
	1. Mitigation/Prevention	
Long-term risk assessment and reduction	Satellite, 33% Crowdsourcing, 17% Sensor web and IoT, 17% Social media, 13% Mobile GPS and CDR, 12% Simulation, 8%	General natural disaster [10] Earthquake [88,91] Oil spill [92] Flood [15,93,94]
Forecasting and Predicting	Simulation, 50% Satellite, 25% Sensor web and IoT, 13% Social media, 12%	Hurricane [52,54,95–100] Flood [101–103]
	2. Preparedness	
Monitoring and detection	Social Media, 31% Sensor web and IoT, 31% Satellite, 13% Combination of various data types, 9% Spatial data, 4% Lidar, 4% Mobile GPS and CDR, 4% Crowdsourcing, 4%	Wildfire [104] Flood [105–109] Earthquake [108,110] Landslide [111] Volcano [45,46]
Early warning	Social media, 29% Sensor web and IoT, 29% Simulation 14% Crowdsourcing 14% Satellite, 14%	Flood [112] Tsunami [76,112]
	3. Response	
Damage Assessment	Satellite, 53% UAV, 21% Social media, 16% Sensor web and IoT, 5% Crowdsourcing, 5%	Earthquake [19,20,113–115] Flood [116] Typhoon [117] Hurricane [118]
Post-disaster Coordination and Response	Social media, 25% Satellite, 16% Sensor web and IoT, 16% Crowdsourcing, 10% UAV, 9% Simulation, 6% Spatial data, 6% Lidar, 6% Mobile GPS and CDR, 3% Combination of various data types, 3%	General natural disaster [117–119] Flood [89,108] Earthquake [19,83–85,120]
	4. Recovery	
	Combination of various data types, 60% Crowdsourcing, 30% Satellite, 10%	Earthquake [121–123] Hurricane [124] Typhoon [125]

4.1. Mitigation/Prevention

4.1.1. Long-term Risk Assessment and Reduction

Satellite images are known to allow experts to identify geographical and infrastructure risks. Kwak [93] developed a multiple satellite-based flood mapping and monitoring system to assist in risk assessment during disaster risk management. In this system, the Moderate Resolution Imaging Spectroradiometer (MODIS)-derived synchronized floodwater index was used to detect the maximum extent of flood based on annual time-series data for the year 2015. Similarly, Skakun et al. [15] utilized Landsat-5/7 satellite images from 1989 to 2012 and produced a flood risk assessment study that included direct damage categories such as dwelling units, roads, health facilities, and schools. Among these facilities, buildings are one of the most important factors to assess for risk reduction, since they represent human settlements in villages and cities. Building collapse is one of the major sources of casualties in disaster scenarios such as hurricanes and earthquakes, primarily because population distributions can be estimated from building stocks [91].

User-generated data helps in filling past research gaps and provides new opportunities to conduct research on risk assessments. For example, McCormick [94] introduced a crowdsourcing system for oil spill risk assessment, which detected factors that were not found in official expert and government risk assessment reports. Residents observed exposures, such as smells, smoke, and other potential risk factors, that were not detected at harmful levels by expert risk assessors. After the 2015 Nepal Earthquake, Wilson et al. [88] used call detail records pertaining to phone metadata tracking numbers and times of calls to estimate population distribution and socioeconomic status for risk assessment. It became evident that the analysis of mobile data and online content can help policymakers understand the behavior of communities, which allows researchers to test their response to disaster management plans and training [10]. According to Horita et al. [94], combining sensor data with user-generated data proves to be an effective risk management strategy. They developed a spatial decision support system that integrated the information provided by WSN and VGI for flood risk management. This integration can support the decision-making of disaster management agencies by improving the maintainability and assessment of WSN, as well as providing data from poorly gauged or un-gauged areas through VGI.

4.1.2. Forecasting and Predicting

Predictive analytics can be powerful tools for natural disaster management. Natural disasters are difficult to forecast because of the complexity associated with the physical phenomena and variability of the parameters involved [96–98]. Existing natural disaster forecasting approaches mostly rely on the underlying physical models and parameters. Nevertheless, prediction capabilities of these approaches have been enhanced in recent years, enabling them to provide higher resolution results, create better physical schemes, and develop the ability to access new data more broadly.

Higher resolution prediction models have been developed for meteorological events, including tropical cyclones, hurricanes, winter storms, or general weather prediction. Murakami et al. [52] introduced the High-Resolution Forecast-Oriented Low Ocean Resolution (FLOR) model (HiFLOR), which has high resolution (~25-km mesh) atmosphere and land components and a more moderate-resolution (~100-km mesh) sea ice and ocean component. The decrease in horizontal resolution (from originally 50 km to 25 km) was made by directly reducing the horizontal grid spacing of the cubed sphere in the underlying model scheme. This resulted in a more realistic simulation of the structure, global distribution, and seasonal and inter-annual variations of tropical cyclones. In addition, it became possible for the higher resolution model to simulate and predict extremely intense tropical cyclones (Saffir–Simpson hurricane categories 4 and 5) and their inter-annual variations. This was the first time that a global coupled model was able to simulate such extremely intense tropical cyclones in situations such as multi-century simulation, sea surface temperature restoring simulations, and retrospective seasonal predictions. Another way to increase prediction resolution is through nested modeling. Goldenberg et al. [54] presented the 2012 version of the Hurricane Weather Research and Forecasting Model (H212). In being an enhanced version of the 2007 version, H212 has an enhanced horizontal resolution of up to 3 km, which makes it the first operational model that meets the condition of running on convection-permitting resolution. The foundational improvement was that the horizontal resolution for HWRF was increased to a 27-km outer domain with 9-km intermediate and 3-km innermost moving nests.

Better underlying physical schemes have been developed with the help of prediction models by integrating big earth observation data [95,96]. As argued by Zhang et al. [95], the most recent upgrades in boundary layer physics have benefited from analyses of in situ aircraft observations in the low-level eye-wall region of major hurricanes. The boundary layer height is an important parameter because it regulates the vertical distribution of turbulent fluxes and helps determine where turbulent fluxes are negligible. Zhang et al. [97] identified different height scales in the hurricane boundary layer in terms of the height of the maximum tangential wind speed, the inflow layer depth, and the mixed layer depth. By analyzing hundreds of GPS dropsondes released from aircraft, they found that there is a clear separation of the thermo-dynamical and dynamical boundary layer depths. They also found that both dynamical and thermo-dynamical boundary layer height scales tend to decrease as distance is reduced from the storm center. These findings were integrated into the operational Hurricane Weather Research and Forecasting (HWRF) Model [98]; and prediction results showed considerable improvements in the simulated storm size, surface inflow angle, near-surface wind profile, and kinematic boundary layer heights in simulations carried out with the improved use of physics [95].

The enhancement of prediction capability becomes partially better due to greater access to new data. A pertinent example in this regard is the Cyclone Global Navigation Satellite System (CYGNSS), which is a NASA earth science mission launched in 2016 that focuses on tropical cyclones and tropical convection [99]. The goal of CYGNSS is to support significant improvements in our ability to forecast tropical cyclone track, intensity, and storm surge through better observations and, ultimately, better understanding of inner-core processes. Another method of integrating observation data into prediction models is through data assimilation, which is defined as (https://www.ecmwf.int/en/research/data-assimilation) "typically a sequential time-stepping procedure, in which a previous model forecast is compared with newly received observations, the model state is then updated to reflect the observations, a new forecast is initiated, and so on." Zhang and Weng [100] assimilated high-resolution airborne radar observations into hurricane prediction. With this assimilation, four to five days before these storms made landfall, the system produced good deterministic and probabilistic forecasts of not only track and intensity, but also of the spatial distributions of surface wind and rainfall.

4.2. Preparedness

4.2.1. Monitoring and Detection

Effective monitoring is very helpful in improving the management of disasters. One of the major sources of big data for disaster monitoring and detection is remote sensing data, which are available in different spatial and temporal resolutions with an adequate level of accuracy. Satellite remote sensing capabilities have long been used to monitor for timely and near-real-time disaster detection [105]. Effective management of disasters such as fires or floods can be done through multitemporal remotely sensed imagery, which is imagery captured over the same location at several points in time, usually days apart. It can be used to monitor the ways in which the event is spreading and helps decision makers in developing and implementing mitigation strategies [104,106,107]. The major goal of the early monitoring phase is to define a boundary that delimits the affected area, so that preliminary information on the event can be generated. Large coverage remote sensing data with a low spatial resolution (>30 m) is appropriate for this phase, whereas high- and very-high-resolution data could be a waste of resources. Automatic extraction of data relative to the disaster impacted area facilitates quick availability of the required information.

The second major source of big data for disaster monitoring and detection is user-generated data available on various platforms, including social media websites and VGI. According to Arribas-Bel [110], each user is considered as a sensor that contributes data for filling the gaps in the availability of authentic data about disasters [126,127]. Earle et al. [108] investigated the capability of using Twitter for rapid earthquake detection. Their investigation was based on the observation that people start conveying messages on Twitter the moment they start feeling the tremors. Such messages or tweets on Twitter were found to be in correlation with the peaks of earthquake-related tweet-frequency time series and the times when the event started. Cervone et al. [127] conducted comprehensive research in analyzing the flood management efforts during the Boulder flood disaster in 2013. Real-time data from Twitter was monitored to identify hotspots and keywords for the disaster in prioritizing the collection of remote-sensing images for the different disasters.

The integration of traditional remote sensing data and newly emerged big social-sensing data has facilitated better understanding in the context of the location, timing, causes, and impacts of natural disasters. This helps a great deal in enhancing the speed and effectiveness of the disaster response. Musaev et al. [111] developed a landslide detection system that integrates multiple physical sensors (USGS seismic network, NASA TRMM rainfall network) and social media (Twitter, YouTube, and Instagram) to determine the varied origins and compositions of multi-hazards. The authors collected social media data that was systematically preprocessed along with keyword filtering, stop-words and phrases filtering, geotag filtering, classification filtering, and blacklist filtering. Earth surface was represented as fixed-size grids and the geotagged social media posts were grouped based on their coordinates to determine areas where landslides might have occurred. Data from the physical and social sensors were integrated by the fixed grid using a Bayesian-based relevance ranking strategy, while favoring landslide events detected from both social and physical sensors. Jongman et al. [116] integrated near-real-time satellite data and social media information for better flood monitoring. They found that satellite imagery is better suited for monitoring large floods, while Twitter can be used to monitor floods of any size, as long as the observations and discussions are shared by people on social media. Crowdsourced hazard detection techniques can enlist citizens to provide information by posting pictures of the disaster's activity.

4.2.2. Early Warning

Sensor data are usually the major data sources for early warning system development. Mandl et al. [109] demonstrated the "Namibia Flood SensorWeb" early warning system based on integrated satellite and ground sensor data for flood situational awareness and early warning. Different types of data are integrated in the system, including river gauge, TRMM rainfall, MODIS flood extent, Radarsat SAR derived flood extent, river flow estimated from the Terra satellite, and flood model data. The system is capable of a sensor planning service (query sensor and initiate data acquisition), web notification service (provide alerts for events such as data product availability), sensor observation service (provide access to raw data), and web processing service (provide data processing function to create high level data products). Poslad et al. [112] proposed a semantic early warning system based on IoT, in which metadata is used to enhance rich sensor data acquisition and ontology models describing multilevel knowledge-bases are used to support decision support and workflow orchestration.

Social media also benefits early warning in a certain way. Carley et al. [76] examined the use of Twitter in Indonesia for early warning and planning for disasters. The coverage, spatiotemporal patterns, and identification of opinion leaders were assessed for the suitability of Twitter for early warning. The paper concluded that with careful collection, assessment, and coordination with official disaster Twitter sites, Twitter is capable of supporting early warning; and that a local Twitter opinion leader will have a critical role in the early warning process.

4.3. Response

4.3.1. Damage Assessment

The most well-known data source for damage assessment is remote sensing imagery. Different types and resolutions of remote sensing data are in demand depending on the extent of details required. Low resolution but large scale remote sensing data provides a quick initial assessment of the impacted area when in situ observations are not yet available. The initial assessment also guides disaster responders to prioritize the areas to be inspected with higher resolution data. Detailed damage assessment, including buildings and roads, requires higher resolution and three-dimensional (3D) data to provide accurate information about the intensity of the damage. Using Formosat-2 and Satellite pour Observation de la Terre (SPOT)-5 satellite images, Liou et al. [19] were able to identify the structural deformation of land areas, changes in directions of rivers, creation of new lakes, and the water levels of rivers and lakes in the 2008 Wenchuan earthquake affected area. The method used by Liou et al. [19] for analyzing such satellite imagery was mainly visual interpretation, even though methods of automatic extraction were also developed [20]. Very-High-Resolution satellite imagery (with a spatial resolution finer than1 m) are mostly suited for detecting disruption to transportation networks and for identifying open spaces that can be used for locating shelters [113,128].

UAV-based aerial imagery is an emerging method that allows data collection for disaster management and provides a better resolution than that available through manned aircrafts. They provide more flexible data acquisition that improves the quality of the point clouds that can be derived for this purpose. UAV imagery and videos analyze the disaster-impacted area in a high spatiotemporal resolution. Using UAV data that has a resolution better than 1cm, Fernandez Galarreta et al. [114] identified a number of damage features for each building focusing on aspects such as total collapse, collapsed roof, rubble piles, and inclined façades. The analysis was first conducted through visual inspection by experts using a 3D point cloud (generated from the raw UAV imagery) for damaged features. Thereafter, a more detailed object-based damage feature extraction from imagery was done in the context of façade and roof analysis of the buildings that did not show any damage features in the previous step. Damage information acquired from the 3D point and the imagery was aggregated to identify the extent of major damages.

Crowdsourced platforms, where images and descriptions of the disaster impacted places can be uploaded from disaster impacted residents, have been found to be helpful for damage assessment. During the 2010 Haiti Earthquake, the Global Earth Observation Catastrophe Assessment Network (GEN-CAN) was established by ImageCat Inc. The firm trained over 600 image professionals to identify collapsed and heavily damaged structures over an affected area of 1000 square kilometers [115]. Similarly, in regard to the Super Typhoon Haiyan that swept the Philippines in 2013, volunteers conducted a remote damage assessment based on satellite imagery on an OpenStreetMap-based application, which was led by the Humanitarian OpenStreetMap Team [129]. Social media data was also collected for Tyhoon Haiyan from Chinese microblog "Weibo" to estimate the extent of damage in the typhoon impacted area in China. Twitter was used as a big data tool during and after Hurricane Sandy struck the East Coast of the United States in 2012 [130]. FEMA organized groups of public and private agencies to analyze the tweets to plot out locations where aid and resources were most needed.

4.3.2. Post-Disaster Coordination and Response

It is imperative for a post-disaster response such as search and rescue operations to be conducted quickly and efficiently. A major problem in this regard is the lack of communication and situational awareness during disasters that forces first responder teams to improvise and thus lessen the efficiency of the rescue mission. A major issue with disasters is that they are mostly characterized by limited resources and dynamic change in the environment, and under such circumstances, it is always problematic to use the limited resources effectively in providing the best communication services. Big data analytics provides possible solutions to understand the situations in disaster areas, because it has the potential of using the limited resources optimally.

Aerial ad-hoc networks are associated with the advantage of being deployed in critical situations where terrestrial mobile devices might not operate. However, their implementation is challenging from the point of view of mobility management and coverage lifetime. Di Felice et al. [117] investigated the utilization of low-altitude aerial mesh networks created by Small Unmanned Aerial Vehicles

(SUAVs). The objective was to re-establish connectivity among isolated end-user devices located on the ground. They proposed a distributed mobility algorithm based on the virtual spring model, through which the SUAV-based mesh node can self-organize into a mesh structure by guaranteeing Quality of Service over the aerial link and connecting the maximum number of devices. They also proposed a distributed charging scheduling scheme through which a constant coverage of sensing devices can be guaranteed over the disaster scenario. Mosterman et al. [118] proposed an automated disaster response system that integrates varied kinds of autonomous vehicles, including ground vehicles, for setting up local stations, fixed wing aircraft for assessing infrastructure damage, and rotorcraft for delivering disaster supplies. These vehicles are coordinated and controlled by the proposed cyberspace operation system that integrates information and assists disaster response automation. Lu et al. [119] designed and implemented a system called TeamPhone, which provides smart phones with the capabilities of communications during disaster recovery, including a messaging system and a self-rescue system. The messaging system integrates cellular networking, ad-hoc networking, and opportunistic seamless networking and enables effective communications between rescue workers. The self-rescue system efficiently groups the smart phones of trapped survivors and sends out disaster messages in facilitating rescue operations.

The different sources through which big data is used by entities involved in the assessment of disaster damage need to be effectively coordinated in order to confirm the damage, know of the destruction that may have been left undetected, and prevent the duplication of efforts by disaster responders. Cervone et al. [127] found that the integration of user generated data (crowdsourcing and social media) and physical sensing data (satellite and aerial imagery) can enhance the accuracy of the former and increase the temporal resolution of the latter, thus providing a more comprehensive compact dataset. For example, after the Haiti earthquake, coordinating systems such as the UN inter-agency OneResponse website, the Sahana Free and Open Source Disaster Management System, and the crowdsourcing platform Ushaidi were established to coordinate a massive amount of information [120]. These systems allow the public to provide information about missing persons or to track missing people's location. The comprehensive data proves to be valuable in designing and executing task response operations, which means it should be made available as early as possible so as to limit the death toll and additional loss of property [19].

4.4. Recovery

The post-disaster recovery phase relates to the period after which initial relief has been provided and is characterized by efforts directed at bringing back normalcy in people's lives and improving the overall circumstances. The major data source for post-disaster recovery monitoring is remote sensing data, including satellite and aerial imagery. Based on change detection from multi-temporal remote sensing data, needs for reconstruction around damaged areas can be detected and monitored. The methodology of change detection is similar in comparison to damage assessment. For example, de Alwis Pitts and So [121] utilized an object-based change detection mechanism to identify the changes before and after two earthquake events; the Van earthquake in eastern Turkey in 2011 and the Kashmir earthquake in northwest Pakistan in 2005. They used high resolution satellite imagery, including the WorldView-2 (0.46 m for panchromatic band and 1.85 m for multispectural band), the Geoeye-1 (0.41 m for panchromatic band and 1.65 m for multispectual band), and the Quickbird-2 (2.44 m for spectral and 0.61 for panchromatic band). Pre- and post-disaster imagery was acquired and road information was obtained from openStreetMap. The changes in edges, texture, and gradient of primary roads were calculated, and changes of open green spaces were also detected. It became apparent that the quantified information contributed to the observation of disaster recovery over time. Contreras et al. [122] reported the progress of recovery efforts after the earthquake at L'Aquila, Italy in 2009. The recovery evaluation was based on remote sensing and ground observations. They used QuickBird imagery to detect the progress during the recovery process from damaged buildings.

GPS and other ground observations were collected progressively (every two years). Results of the recovery revealed the socio-economic progress after the earthquake.

The literature has vastly cited the outcomes emanating from user generated data obtained from social media and VGI platforms during post-disaster recovery. A major challenge for disaster management in utilizing such data is their limited capability of handling big data. Afzalan et al. [124] examined how active and influential members of Facebook groups aided in disaster recovery after the occurrence of Hurricane Sandy. Network analysis was carried out to find influential members and a web-survey was conducted to learn about their background and volunteer activity inside and outside Facebook. In effect, disaster recovery organizations can approach these influential people and collaborate with them in conducting recovery procedures. Yan et al. [125] utilized geotagged Flickr photos to monitor and assess post-disaster tourism recovery after the Philippines earthquake and Typhoon Haiyan in 2013. Geotagged Flickr photos were analyzed through quality enhancement (both locational accuracy and thematic accuracy), and quantitative and qualitative investigation of the available visual contents. Results showed spatiotemporal patterns of the recovery status and trends.

5. Two Emerging Topics—Evolutionary Technologies

Figure 4 illustrates the increasing popularity of "machine learning" and a significant popularity of "cyberinfrastructure" applied in the research of "big data" in "disaster management". Here, we review the two major topics, machine learning and big data cyberinfrastructure, as evolutionary technologies which facilitate disaster management in various ways.

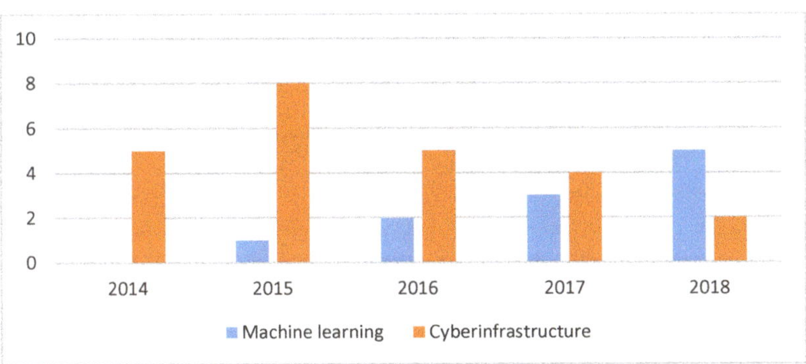

Figure 4. Distribution of the reviewed article with subtopic "machine learning" and "cyberinfrastructure" respectively over the recent years.

5.1. Machine Learning

A great deal of uncertainty is created when different sources of big data are integrated for the purposes of disaster awareness and response. Manual interpretation and analysis of the integrated data is no longer adequate, which is why sophisticated automatic analysis methods are required to make the process efficient and effective. Machine learning was introduced in the field of disaster management two decades back and has since evolved to become one of the most effective methods for eliminating unrelated data and speeding up the analysis in disaster situations, which helps in quick prediction analysis and identifying optimal response strategies.

Text classification of social media data classifies disaster response-related tweets and facilitates the rapid identification of disaster relief contents. Imran et al. [80] developed a system of Artificial Intelligence for Disaster Response (AIDR) to classify messages that people post during disasters into a set of user-defined categories of information (e.g., "needs", "damage", etc.). In meeting this purpose, the system continuously sources data from Twitter, processes it (i.e., using machine learning classification techniques), and leverages human-participation (through crowdsourcing) in real-time. AIDR has been successfully tested to classify informative vs. non-informative tweets posted during the 2013 Pakistan Earthquake. Based on AIDR, Bejiga et al. [131] proposed a hybrid crowdsourcing and real-time machine learning solution to rapidly process large volumes of aerial data for disaster response in a time-sensitive manner. Crowdsourcing was used to annotate features of interest in aerial images (such as damaged shelters and roads blocked by debris). These human-annotated features were then used to train a supervised machine learning system to learn to recognize such features in new unseen images.

Utilizing techniques through which machines are trained to detect damages helps in reducing human interaction and improves the performance of dynamic decision making. During the damage assessment stage, visual interpretation of satellite/aerial imagery and videos often proves to be time consuming, inaccurate, and costly. In addition, ancillary data such as UAV products, LiDAR, or GIS databases are mostly unavailable in regions where the death toll is highest [123]. Machine learning algorithms actively adapt with and learn the problem without relying on statistical assumptions about data distribution [132]. Machine learning algorithms have an overall improved accuracy compared to traditional classification and change detection methods. They work with nonlinear datasets [133], allow learning with limited training data [134], and successfully solve classification problems that are difficult to distinguish. For example, Cooner et al. [135] used high spatial resolution imagery acquired through WorldView-1 and Quickbird 2 to investigate the effectiveness of Artificial Neural Networks and Random Forests in detecting damages to buildings caused by the 2010 Haiti earthquake.

More recently, new learning emerged as one of the breakthroughs in machine learning that could be applied to disaster management. For example, Cha et al. [136] proposed a convolutional neural network approach to detect concrete cracks without the need to calculate a defect feature that is normally impacted by the noise included with the data. The proposed approach classified the detected cracks into different characteristics, including strong light spots, shadows, and very thin cracks and demonstrated robust performance in comparison to traditional edge detection methods (canny and sobel), which fail to provide meaningful crack information. Pouyanfar and Chen [137] proposed an ensemble deep learning framework to extract information from YouTube disaster videos in the context of flood, damage, fire, mud-rock, tornado, and lightning. Videos are firstly preprocessed into shots by boundary detection and key-frame selection, and then features are extracted through deep learning reference models on each key-frame. Well-known classifiers, such as Decision Tree and Support Vector Machine, were applied to the features for such classification.

In addition to real-time damage assessment and disaster monitoring, predictive damage assessment tools can now leverage Machine Learning and historical and Big Data analytics to improve damage forecasting modeling. Asim et al. [138] evaluated four machine learning techniques including the pattern recognition neural network, recurrent neural network, random forest, and linear programming boost ensemble classifier to model relationships between calculated seismic parameters and future earthquake occurrences. In addition, by automating such processes, it becomes possible to respond to the evolving disaster scenarios.

5.2. Big Data Cyberinfrastructure

In the context of data recovery and security, Chang [139] proposed a private cloud approach that allows data to be restored to multiple sites with multiple methods in facilitating the organization to recover close to 100 percent of the data. In doing so, data centers must adopt multi-purpose approaches in ensuring that all the data in the Big Data system can be recovered and retrieved without experiencing a prolonged downtime and complex recovery process.

Disaster management becomes more efficient if data is acquired from different sources in a higher spatial and temporal resolution. However, challenges emerge because of the constantly increasing quantum of image and video data. Emerging technological innovations including social media, location-based systems, radio frequency identification, and big data analytics are considered as powerful tools that may help during the disaster management cycle. Processing and analyzing the heterogeneous big disaster data requires efficient data collection, aggregation, information extraction, visualization, and efficient distribution. The growth of data and the need for an efficient distribution makes the development and operation of cyberinfrastructure very demanding. Wan et al. [140] developed a cloud-based flood cyber-infrastructure that collects, organizes, visualizes, and manages several global flood databases in facilitating location-based visualization and statistical analysis for authorities and the public. This system allows the collection of crowdsourcing data from smart phones or the internet to report new flood events.

Belaud et al. [141] proposed a cloud computing platform for scientific simulation that facilitates efficient natural hazards management with an analysis of big data and its exchange between distant locations. This platform provides pre- and post-processing visualization services, 3D large scientific data-set scalable compression and transmission methods, collaborative virtual environments, and 3D visualization. Puthal et al. [142] presented a framework that supports disaster event detection and the generation of alerts by analyzing the data stream, which includes efficient data collection, data aggregation, and efficient dissemination. One of the goals of such a framework is to support an end-to-end security architecture that protects the data stream from unauthorized manipulation, as well as the leakage of sensitive information. Hua et al. [143] proposed a near-real-time and cost-efficient scheme, called SmartEye, in the cloud-assisted disaster environment, which provides efficient image sharing for disaster detection and scene recognition by de-duplicating and aggregating similar features into one single flow. Wilson et al. [88] deployed a computational architecture and analytical capacity within nine days of the Nepal earthquake of 25 April 2015, to provide spatiotemporally detailed estimates of population displacements from call detail records based on the movements of 12 million de-identified mobile phones users. The analysis reveals the detailed mobility patterns of people after the earthquake and the patterns of their return to affected areas.

Bartoli et al. [144] proposed an efficient architecture of a smart public safety platform that performs a smart and functional integration of heterogeneous components as a smart data gathering and analysis system, a novel professional communication system, wireless sensor networks, and social networks. It integrates advanced infrastructures and analysis methods to coordinate the information flow between first responders, public authorities, and citizens. It allows responders and authorities to access the information available on the platform by using an advanced broadband communication system that can receive multimedia data for more effective and efficient operations in the field. Trono et al. [145] developed a Disruption Tolerant Network (DTN) distributed computing system for disaster map data generation and sharing. The system distributes computing tasks to multiple computing resources in minimizing individual computation loads. First responders and volunteers act as mobile sensing nodes as their GPS traces are collected for real-time disaster relief efforts.

6. Big Data Challenges in Disaster Management

6.1. Big Data Collection

In disaster management scenarios, a rapidly generated, big volume of data (e.g., 452,000 tweets/min, extreme weather simulation TB/day) needs to be analyzed in a real-time fashion for immediate action. Archives of historical data needs to be shared with researchers online and services of evaluation and validation of analysis methods or models need to be standardized and made publically available.

Disaster management requires a large variety of heterogeneous data from different data sources and might fill in the gaps between them, providing valuable information for all phases in the disaster management cycle. Sensors might generate different types of data, including time series, semi-structured data, and textural data; and these data might include noise and misinformation. Integrating multiple sources of data might contribute to the improvement of data quality and data completeness, but individual data validation needs to be conducted before data integration.

Data integration in the context of disaster management can benefit from the semantics or properties associated with the data itself.

Noises and misinformation from big data are almost inevitable as a lot of these are unintentional, especially from social media and crowdsourcing. In addition, data privacy and accuracy issues are still one of the main challenges associated with big data collection, even though protocols and analytical methods for dealing with these issues are crucially required during disaster management. Integrating multiple sources of data can help eliminate such noise and misinformation. Streaming data, such as sensor outputs and crowdsourcing data, needs to be enhanced with anomaly detection to identify incorrect data due to system failure or misguiding data collection methods. Machine learning can contribute to automating such data integrating and the filtering process, and ultimately increase the data quality.

6.2. Big Data Analytics

With the synthesis of multi-platform, multi-scale, and multi-discipline data, the capability of predictive modeling of natural disasters should become more efficient. Research and activities related to utilizing the synthesized information and predictive analysis results are expected to better enhance our ability to adaptively respond and plan. It has been found that crowdsourced data, especially the ones provided by the disaster affected people, have significant value during the disaster awareness phase. However, analytical methods are still needed to reliably and accurately integrate these crowdsourced data into the physical sensing data (e.g., satellite, UAV) and authoritative data (e.g., terrain data, census data). Only then can the disaster be effectively characterized in terms of situational awareness, spatiotemporal damage patterns, and community resilience. Consequently, decision-making processes can benefit from the analytical results and design rescue and response efforts in terms of both space and time, for the affected populations and communities.

6.3. Cyberinfrastructures

There is crucial need for the design and development of cyberinfrastructures in ways that big data is effectively integrated into disaster sensing, analysis, and response phases to support first responders and disaster management agencies for real-time decision making. Such capabilities of cyberinfrastructure provide decision makers and responders from different organizations and disciplines with shared knowledge and a communicating platform that allows for conducting the disaster response process in an effective manner. Research efforts and related activities are still required to look into the challenges emanating from big sensing data, particularly in the context of the emerging data volume of streaming videos, including efficient data management, fast data transfer, and intuitive data visualization.

7. Conclusions

In analyzing the recent achievements associated with leveraging Big Data to disaster management, this paper has presented the findings of several researchers on varied scientific and technological perspectives that have a bearing on the efficacy of big data in facilitating disaster management. It has become apparent that in the present age of information technology, a major objective of scientists is to analyze the varied aspects of big data and find ways of making the best of the available technologies in storing the available information in well-integrated structures and using it for the welfare of human societies, particularly in the context of using bid data to effectively deal with natural disasters. The paper has analyzed major big data sources that are valuable in disaster management. A detailed analysis has been conducted to highlight the significance of different big data sources in various disaster management phases. The main challenges pertain to effectively dealing with data collection and management technologies and developing efficient systems of mitigating the adversities associated with natural disasters and managing disasters in ways that result in minimum losses to human lives and property. Other challenges pertain to developing algorithms by way of systems that can be used in resolving operational issues and attaining greater accuracy in predicting disasters. The paper has highlighted the need for further research on big data applications in enhancing the efficiency of the public sector in further developing technology to mitigate the adverse effects of natural disasters. Overall, further research efforts need to be made to look into the challenges emerging from Big Sensing Data, particularly in the context of the emerging data volume of streaming videos, including efficient data management, fast data transfer, and intuitive data visualization.

Author Contributions: Manzhu Yu contributed in the study conception and design, acquisition of the reviewed paper, analysis of the literature review, and the writing. Chaowei Yang contributed in the initial idea of the paper. Yun Li revised an earlier version of the paper.

Funding: This work was supported by the NSF Cyber Polar, Innovation Center, EarthCube and Computer Network System Programs: [Grant NumbersPLR-1349259;IIP-1338925; CNS-1117300; ICER-1343759]; NASA: [Grant Number NNG12PP37I; NNX15AH51G].

Acknowledgments: We thank the anonymous reviewers for their insightful comments, which significantly improved the paper.

Conflicts of Interest: The authors declare no conflict of interest.

References

1. Blaikie, P.; Cannon, T.; Davis, I.; Wisner, B. *At Risk: Natural Hazards, People's Vulnerability and Disasters*; Routledge: London, UK, 2014.
2. Celik, S.; Corbacioglu, S. Role of information in collective action in dynamic disaster environments. *Disasters* **2010**, *34*, 137–154. [CrossRef] [PubMed]
3. Sutanta, H.; Bishop, I.D.B.; Rajabifard, A.R. Integrating spatial planning and disaster risk reduction at the local level in the context of spatially enabled government. In *Spatially Enabling Society Research, Emerging Trends and Critical Assessment*; Abbas, R., Joep, C., Kalantari, M., Kok, B., Eds.; Leuven University Press: Leuven, Belgium, 2010; pp. 55–68, ISBN 978-90-5867-851-5.
4. Penning-Rowsell, E.C.; Sultana, P.; Thompson, P.M. The 'last resort'? Population movement in response to climate-related hazards in Bangladesh. *Environ. Sci. Policy* **2013**, *27*, 445. [CrossRef]
5. Gaillard, J.C.; Mercer, J. From knowledge to action: Bridging gaps in disaster risk reduction. *Prog. Hum. Geogr.* **2013**, *37*, 93–114. [CrossRef]
6. Villars, R.L.; Olofson, C.W.; Eastwood, M. *Big Data: What It Is and Why You Should Care*; IDC: Framingham, MA, USA, 2011.
7. Hashem, I.A.T.; Yaqoob, I.; Anuar, N.B.; Mokhtar, S.; Gani, A.; Khan, S.U. The rise of "big data" on cloud computing: Review and open research issues. *Inf. Syst.* **2015**, *47*, 98–115. [CrossRef]
8. National Research Council. *Public Response to Alerts and Warnings on Mobile Devices: Summary of a Workshop on Current Knowledge and Research Gaps*; National Academies Press: Washington, DC, USA, 2011; ISBN 978-0-309-21162-8.
9. Castillo, C. *Big Crisis Data: Social Media in Disasters and Time-Critical Situations*; Cambridge University Press: Cambridge, UK, 2016; ISBN 9781107135765.
10. Cinnamon, J.; Jones, S.K.; Adger, W.N. Evidence and future potential of mobile phone data for disease disaster management. *Geoforum* **2016**, *75*, 253–264. [CrossRef]
11. Erdelj, M.; Natalizio, E. UAV-assisted disaster management: Applications and open issues. In Proceedings of the 2016 International Conference on Computing, Networking and Communications (ICNC 2016), Kauai, HI, USA, 15–18 February 2016; IEEE: Piscataway, NJ, USA, 2016; pp. 1–5.
12. Dos Santos Rocha, R.; Widera, A.; van den Berg, R.P.; de Albuquerque, J.P.; Helingrath, B. Improving the Involvement of Digital Volunteers in Disaster Management. In Proceedings of the International Conference on Information Technology in Disaster Risk Reduction, Sofia, Bulgaria, 16–18 November 2016; Murayama, Y., Velev, D., Zlateva, P., Gonzalez, J.J., Eds.; Springer: Cham, Switzerland, 2016; pp. 214–224.
13. Federal Geographic Data Committee. Emerging Technologies and the Geospatial Landscape. A Report of the National Geospatial Advisory Committee. Available online: https://www.fgdc.gov/ngac/meetings/dec-2016/ngac-paper-emerging-technologies-and-the.pdf (accessed on 22 April 2018).

14. United Nation. Data-Pop Alliance Synthesis Report Big Data for Climate Change and Disaster Resilience: Realizing the Benefits for Developing Countries. Available online: http://datapopalliance.org/wp-content/uploads/2015/11/Big-Data-for-Resilience-2015-Report.pdf (accessed on 22 April 2018).
15. Skakun, S.; Kussul, N.; Shelestov, A.; Kussul, O. Flood hazard and flood risk assessment using a time series of satellite images: A case study in Namibia. *Risk Anal.* **2014**, *34*, 1521–1537. [CrossRef] [PubMed]
16. Plank, S. Rapid damage assessment by means of multi-temporal SAR—A comprehensive review and outlook to Sentinel-1. *Remote Sens.* **2014**, *6*, 4870–4906. [CrossRef]
17. Yamazaki, F.; Liu, W. Remote sensing technologies for post-earthquake damage assessment: A case study on the 2016 Kumamoto earthquake. In Proceedings of the 6th Asia Conference on Earthquake Engineering (6ACEE), Cebu City, Philippines, 22–24 September 2016.
18. Pradhan, B.; Tehrany, M.S.; Jebur, M.N. A new semiautomated detection mapping of flood extent from TerraSAR-X satellite image using rule-based classification and taguchi optimization techniques. *IEEE Trans. Geosci. Remote Sens.* **2016**, *54*, 4331–4342. [CrossRef]
19. Liou, Y.A.; Kar, S.K.; Chang, L. Use of high-resolution FORMOSAT-2 satellite images for post-earthquake disaster assessment: A study following the 12 May 2008 Wenchuan Earthquake. *Int. J. Remote Sens.* **2010**, *31*, 3355–3368. [CrossRef]
20. Ehrlich, D.; Kemper, T.; Blaes, X.; Soille, P. Extracting building stock information from optical satellite imagery for mapping earthquake exposure and its vulnerability. *Nat. Hazards* **2013**, *68*, 79–95. [CrossRef]
21. Dini, G.R.; Jacobsen, K.; Rottensteiner, F.; Al Rajhi, M.; Heipke, C. 3D building change detection using high resolution stereo images and a GIS database. In Proceedings of the International Archives of the Photogrammetry, Remote Sensing and Spatial Information Sciences, Melbourne, Australia, 25 August–1 September 2012; Volume 39, pp. 299–304. [CrossRef]
22. Tian, J.; Nielsen, A.A.; Reinartz, P. Building damage assessment after the earthquake in Haiti using two post-event satellite stereo imagery and DSMs. *Int. J. Image Data Fusion* **2015**, *6*, 155–169. [CrossRef]
23. Tong, X.; Hong, Z.; Liu, S.; Zhang, X.; Xie, H.; Li, Z.; Yang, S.; Wang, W.; Bao, F. Building-damage detection using pre-and post-seismic high-resolution satellite stereo imagery: A case study of the May 2008 Wenchuan earthquake. *ISPRS J. Photogramm. Remote Sens.* **2012**, *68*, 13–27. [CrossRef]
24. Tian, J.; Cui, S.; Reinartz, P. Building Change Detection Based on Satellite Stereo Imagery and Digital Surface Models. *IEEE Trans. Geosci. Remote Sens.* **2014**, *52*, 406–417. [CrossRef]
25. Koyama, C.N.; Gokon, H.; Jimbo, M.; Koshimura, S.; Sato, M. Disaster debris estimation using high-resolution polarimetric stereo-SAR. *ISPRS J. Photogramm. Remote Sens.* **2016**, *120*, 84–98. [CrossRef]
26. Chini, M.; Piscini, A.; Cinti, F.R.; Amici, S.; Nappi, R.; DeMartini, P.M. The 2011 Tohoku (Japan) Tsunami inundation and liquefaction investigated through optical, thermal, and SAR data. *IEEE Geosci. Remote Sens. Lett.* **2013**, *10*, 347–351. [CrossRef]
27. Pesaresi, M.; Ehrlich, D.; Ferri, S.; Florczyk, A.; Freire, S.; Haag, F.; Halkia, M.; Julea, A.M.; Kemper, T.; Soille, P. Global human settlement analysis for disaster risk reduction. *Int. Arch. Photogramm. Remote Sen. Spat. Inf. Sci.* **2015**, *40*, 837. [CrossRef]
28. McCallum, I.; Liu, W.; See, L.; Mechler, R.; Keating, A.; Hochrainer-Stigler, S.; Mochizuki, J.; Fritz, S.; Dugar, S.; Arestegui, M.; et al. Technologies to support community flood disaster risk reduction. *Int. J. Disaster Risk Sci.* **2016**, *7*, 198–204. [CrossRef]
29. Raspini, F.; Bardi, F.; Bianchini, S.; Ciampalini, A.; Del Ventisette, C.; Farina, P.; Ferrigno, F.; Solari, L.; Casagli, N. The contribution of satellite SAR-derived displacement measurements in landslide risk management practices. *Nat. Hazards* **2017**, *86*, 327–351. [CrossRef]
30. Ofli, F.; Meier, P.; Imran, M.; Castillo, C.; Tuia, D.; Rey, N.; Briant, J.; Millet, P.; Reinhard, F.; Parkan, M.; et al. Combining human computing and machine learning to make sense of big (aerial) data for disaster response. *Big Data* **2016**, *4*, 47–59. [CrossRef] [PubMed]
31. Nonami, K.; Kendoul, F.; Suzuki, S.; Wang, W.; Nakazawa, D. Introduction. In *Autonomous Flying Robots*; Springer: Tokyo, Japan, 2010; pp. 1–29, ISBN 978-4-431-53855-4.
32. Kim, K.; Davidson, J. Unmanned aircraft systems used for disaster management. *Transp. Res. Rec. J. Transp. Res. Board* **2015**, *2532*, 83–90.
33. Foresti, G.L.; Farinosi, M.; Vernier, M. Situational awareness in smart environments: Socio-mobile and sensor data fusion for emergency response to disasters. *J. Ambient Intell. Humaniz. Comput.* **2015**, *6*, 239–257. [CrossRef]

34. Kakooei, M.; Baleghi, Y. Fusion of satellite, aircraft, and UAV data for automatic disaster damage assessment. *Int. J. Remote Sens.* **2017**, *38*, 2511–2534. [CrossRef]
35. Tanzi, T.J.; Chandra, M.; Isnard, J.; Camara, D.; Olivier, S.; Harivelo, F. Towards" drone-borne" disaster management: Future application scenarios. *ISPRS Ann. Photogramm. Remote Sens. Spat. Inf. Sci.* **2016**, 3, 181. [CrossRef]
36. Griffin, G.F. The use of unmanned aerial vehicles for disaster management. *Geomatica* **2014**, *68*, 265–281. [CrossRef]
37. Chen, D.; Liu, Z.; Wang, L.; Dou, M.; Chen, J.; Li, H. Natural disaster monitoring with wireless sensor networks: A case study of data-intensive applications upon low-cost scalable systems. *Mob. Netw. Appl.* **2013**, *18*, 651–663. [CrossRef]
38. Erdelj, M.; Natalizio, E.; Chowdhury, K.R.; Akyildiz, I.F. Help from the sky: Leveraging UAVs for disaster management. *IEEE Pervasive Comput.* **2017**, *16*, 24–32. [CrossRef]
39. Erman, A.T.; van Hoesel, L.; Havinga, P.; Wu, J. Enabling mobility in heterogeneous wireless sensor networks cooperating with UAVs for mission-critical management. *IEEE Wirel. Commun.* **2008**, *15*, 38–46. [CrossRef]
40. Carli, M.; Panzieri, S.; Pascucci, F. A joint routing and localization algorithm for emergency scenario. *Ad Hoc Netw.* **2014**, *13*, 19–33. [CrossRef]
41. Khalil, I.M.; Khreishah, A.; Ahmed, F.; Shuaib, K. Dependable wireless sensor networks for reliable and secure humanitarian relief applications. *Ad Hoc Networks* **2014**, *13*, 94–106. [CrossRef]
42. Tuna, G.; Gungor, V.C.; Gulez, K. An autonomous wireless sensor network deployment system using mobile robots for human existence detection in case of disasters. *Ad Hoc Netw.* **2014**, *13*, 54–68. [CrossRef]
43. Ray, P.P.; Mukherjee, M.; Shu, L. Internet of things for disaster management: State-of-the-art and prospects. *IEEE Access* **2017**, *5*, 18818–18835. [CrossRef]
44. Sakhardande, P.; Hanagal, S.; Kulkarni, S. Design of disaster management system using IoT based interconnected network with smart city monitoring. In Proceedings of the International Conference on Internet of Things and Applications (IOTA), Pune, India, 22–24 January 2016; IEEE: Piscataway, NJ, USA, 2016; pp. 185–190.
45. Bisson, M.; Spinetti, C.; Neri, M.; Bonforte, A. Mt. Etna volcano high-resolution topography: Airborne LiDAR modelling validated by GPS data. *Int. J. Dig. Earth* **2016**, *9*, 710–732. [CrossRef]
46. Nomikou, P.; Parks, M.M.; Papanikolaou, D.; Pyle, D.M.; Mather, T.A.; Carey, S.; Watts, A.B.; Paulatto, M.; Kalnins, M.L.; Livanos, I.; et al. The emergence and growth of a submarine volcano: The Kameni islands, Santorini (Greece). *GeoResJ* **2014**, *1*, 8–18. [CrossRef]
47. Costabile, P.; Macchione, F.; Natale, L.; Petaccia, G. Flood mapping using LIDAR DEM. Limitations of the 1-D modeling highlighted by the 2-D approach. *Nat. Hazards* **2015**, *77*, 181–204. [CrossRef]
48. Chen, B.; Krajewski, W.F.; Goska, R.; Young, N. Using LiDAR surveys to document floods: A case study of the 2008 Iowa flood. *J. Hydrol.* **2017**, *553*, 338–349. [CrossRef]
49. Gong, J. (Mobile lidar data collection and analysis for post-sandy disaster recovery. In Proceedings of the Computing in Civil Engineering, Los Angeles, CA, USA, 23–25 June 2013; American Society of Civil Engineers: Reston, VA, USA, 2013; pp. 677–684.
50. Kwan, M.P.; Ransberger, D.M. LiDAR assisted emergency response: Detection of transport network obstructions caused by major disasters. *Comput. Environ. Urban Syst.* **2010**, *34*, 179–188. [CrossRef]
51. Moya, L.; Yamazaki, F.; Liu, W.; Yamada, M. Detection of collapsed buildings due to the 2016 Kumamoto, Japan, earthquake from Lidar data. *Nat. Hazards Earth Syst. Sci.* **2017**, *18*, 65. [CrossRef]
52. Murakami, H.; Vecchi, G.A.; Underwood, S.; Delworth, T.L.; Wittenberg, A.T.; Anderson, W.G.; Chen, J.H.; Gudgel, R.G.; Harris, L.M.; Lin, S.J.; et al. Simulation and prediction of category 4 and 5 hurricanes in the high-resolution GFDL HiFLOR coupled climate model. *J. Clim.* **2015**, *28*, 9058–9079. [CrossRef]
53. MacLachlan, C.; Arribas, A.; Peterson, K.A.; Maidens, A.; Fereday, D.; Scaife, A.A.; Gordon, M.; Vellinga, M.; Williams, A.; Comer, R.E.; et al. Global Seasonal forecast system version 5 (GloSea5): A high-resolution seasonal forecast system. *Q. J. R. Meteorol. Soc.* **2015**, *141*, 1072–1084. [CrossRef]
54. Goldenberg, S.B.; Gopalakrishnan, S.G.; Tallapragada, V.; Quirino, T.; Marks, F., Jr.; Trahan, S.; Zhang, X.; Atlas, R. The 2012 triply nested, high-resolution operational version of the Hurricane Weather Research and Forecasting Model (HWRF): Track and intensity forecast verifications. *Weather Forecast.* **2015**, *30*, 710–729. [CrossRef]

55. Kim, J.C.; Jung, H.; Kim, S.; Chung, K. Slope based intelligent 3D disaster simulation using physics engine. *Wirel. Pers. Commun.* **2016**, *86*, 183–199. [CrossRef]
56. Yu, M.; Huang, Y.; Xu, Q.; Guo, P.; Dai, Z. Application of virtual earth in 3D terrain modeling to visual analysis of large-scale geological disasters in mountainous areas. *Environ. Earth Sci.* **2016**, *75*, 563. [CrossRef]
57. Dou, M.; Chen, J.; Chen, D.; Chen, X.; Deng, Z.; Zhang, X.; Xu, K.; Wang, J. Modeling and simulation for natural disaster contingency planning driven by high-resolution remote sensing images. *Future Gener. Comput. Syst.* **2014**, *37*, 367–377. [CrossRef]
58. Mas, E.; Koshimura, S.; Imamura, F.; Suppasri, A.; Muhari, A.; Adriano, B. Recent advances in agent-based tsunami evacuation simulations: Case studies in Indonesia, Thailand, Japan and Peru. *Pure Appl. Geophys.* **2015**, *172*, 3409–3424. [CrossRef]
59. Kureshi, I.; Theodoropoulos, G.; Mangina, E.; O'Hare, G.; Roche, J. Towards an info-symbiotic decision support system for disaster risk management. In Proceedings of the 19th International Symposium on Distributed Simulation and Real Time Applications, Chengdu, Sichuan, China, 14–16 October 2015; IEEE Press: Piscataway, NJ, USA, 2015; pp. 85–91.
60. Tomaszewski, B.; Judex, M.; Szarzynski, J.; Radestock, C.; Wirkus, L. Geographic information systems for disaster response: A review. *J. Homel. Secur. Emerg. Manag.* **2015**, *12*, 571–602. [CrossRef]
61. Herold, S.; Sawada, M.C. A review of geospatial information technology for natural disaster management in developing countries. *Int. J. Appl. Geospat. Res.* **2012**, *3*, 14–62. [CrossRef]
62. Stefanidis, A.; Crooks, A.; Radzikowski, J. Harvesting ambient geospatial information from social media feeds. *GeoJournal* **2013**, *78*, 319–338. [CrossRef]
63. Loukis, E.; Charalabidis, Y. Active and passive crowdsourcing in government. In *Policy Practice and Digital Science*; Janssen, M., Wimmer, M.A., Deljoo, A., Eds.; Springer: Cham, Switzerland, 2015; pp. 261–289, ISBN 978-3-319-12783-5.
64. Tong, Y.; Cao, C.C.; Chen, L. TCS: Efficient topic discovery over crowd-oriented service data. In Proceedings of the 20th ACM SIGKDD International Conference on Knowledge Discovery and Data Mining, New York, NY, USA, 24–27 August 2014; ACM: New York, NY, USA, 2014; pp. 861–870.
65. Qin, H.; Rice, R.M.; Fuhrmann, S.; Rice, M.T.; Curtin, K.M.; Ong, E. Geocrowdsourcing and accessibility for dynamic environments. *GeoJournal* **2016**, *81*, 699–716. [CrossRef]
66. Büscher, M.; Liegl, M.; Thomas, V. Collective intelligence in crises. In *Social Collective Intelligence*; Miorandi, D., Maltese, V., Rovatsos, M., Nijholt, A., Stewart, J., Eds.; Springer: Cham, Switzerland, 2014; pp. 243–265.
67. Tavakkol, S.; To, H.; Kim, S.H.; Lynett, P.; Shahabi, C. An entropy-based framework for efficient post-disaster assessment based on crowdsourced data. In Proceedings of the Second ACM SIGSPATIAL International Workshop on the Use of GIS in Emergency Management, San Jose, CA, USA, 3–5 November 2010; ACM: New York, NY, USA, 2010; p. 13.
68. Poblet, M.; García-Cuesta, E.; Casanovas, P. Crowdsourcing tools for disaster management: A review of platforms and methods. In *AI Approaches to the Complexity of Legal Systems*; Casanovas, P., Pagallo, U., Palmirani, M., Sartor, G., Eds.; Springer: Berlin/Heidelberg, Germany, 2014; pp. 261–274.
69. Palen, L.; Hiltz, S.R.; Liu, S. Online Forums Supporting Grassroots Participation in Emergency Preparedness and Response. *Commun. ACM* **2007**, *50*, 54–58. [CrossRef]
70. Roche, S.; Propeck-Zimmermann, E.; Mericskay, B. GeoWeb and crisis management: Issues and perspectives of volunteered geographic information. *GeoJournal* **2013**, *78*, 21–40. [CrossRef]
71. Sievers, J.A. Embracing Crowdsourcing: A strategy for state and local governments Approaching "Whole Community" Emergency Planning. *State Local Gov. Rev.* **2015**, *47*, 57–67. [CrossRef]
72. Nonnecke, B.M.; Mohanty, S.; Lee, A.; Lee, J.; Beckman, S.; Mi, J.; Krishnan, S.; Roxas, R.E.; Oco, N.; Crittenden, C.; et al. Malasakit 1.0: A participatory online platform for crowdsourcing disaster risk reduction strategies in the philippines. In Proceedings of the Global Humanitarian Technology Conference (GHTC), San Jose, CA, USA, 18–21 October 2017; IEEE: Piscataway, NJ, USA, 2017; pp. 1–6.
73. Charalabidis, Y.N.; Loukis, E.; Androutsopoulou, A.; Karkaletsis, V.; Triantafillou, A. Passive crowdsourcing in government using social media. *Transform. Gov. People Process Policy* **2014**, *8*, 283–308. [CrossRef]
74. Emergency Alerting Platforms Working Group. Social Media & Complementary Alerting Methods–Recommended Strategies & Best Practices. Available online: https://transition.fcc.gov/bureaus/pshs/advisory/csric5/WG2_FinalReport_091416.docx (accessed on 12 March 2018).

75. Granell, C.; Ostermann, F.O. Beyond data collection: Objectives and methods of research using VGI and geo-social media for disaster management. *Comput. Environ. Urban Syst.* **2016**, *59*, 231–243. [CrossRef]
76. Carley, K.M.; Malik, M.; Landwehr, P.M.; Pfeffer, J.; Kowalchuck, M. Crowd sourcing disaster management: The complex nature of Twitter usage in Padang Indonesia. *Saf. Sci.* **2016**, *90*, 48–61. [CrossRef]
77. Middleton, S.E.; Middleton, L.; Modafferi, S. Real-time crisis mapping of natural disasters using social media. *IEEE Intell. Syst.* **2014**, *29*, 9–17. [CrossRef]
78. Nguyen, D.T.; Jung, J.E. Real-time event detection for online behavioral analysis of big social data. *Future Gener. Comput. Syst.* **2017**, *66*, 137–145. [CrossRef]
79. Imran, M.; Castillo, C.; Lucas, J.; Meier, P.; Vieweg, S. AIDR: Artificial intelligence for disaster response. In Proceedings of the 23rd International Conference on World Wide Web, Seoul, Korea, 7–11 April 2014; ACM: New York, NY, USA, 2014; pp. 159–162.
80. Chae, J.; Thom, D.; Jang, Y.; Kim, S.; Ertl, T.; Ebert, D.S. Public behavior response analysis in disaster events utilizing visual analytics of microblog data. *Comput. Graph.* **2014**, *38*, 51–60. [CrossRef]
81. Landwehr, P.M.; Carley, K.M. Social media in disaster relief. In *Data Mining and Knowledge Discovery for Big Data*; Chu, W.W., Ed.; Springer: Berlin, Germany, 2014; pp. 225–257, ISBN 978-3-642-40836-6.
82. Horanont, T.; Witayangkurn, A.; Sekimoto, Y.; Shibasaki, R. Large-scale auto-GPS analysis for discerning behavior change during crisis. *IEEE Intell. Syst.* **2013**, *28*, 26–34. [CrossRef]
83. Li, M.; Li, W.; Fang, R.; Shi, C.; Zhao, Q. Real-time high-precision earthquake monitoring using single-frequency GPS receivers. *GPS Solut.* **2015**, *19*, 27–35. [CrossRef]
84. Song, X.; Zhang, Q.; Sekimoto, Y.; Shibasaki, R. Prediction of human emergency behavior and their mobility following large-scale disaster. In Proceedings of the 20th ACM SIGKDD International Conference on Knowledge Discovery and Data Mining, New York, NY, USA, 24–27 August 2014; ACM: New York, NY, USA, 2014; pp. 5–14.
85. Donovan, B.; Work, D.B. Using coarse GPS data to quantify city-scale transportation system resilience to extreme events. In Proceedings of the 2015 Transportation Research Board Annual Meeting, Washington, DC, USA, 11–15 January 2015; arXiv: Ithaca, NY, USA, 2015.
86. Bharti, N.; Lu, X.; Bengtsson, L.; Wetter, E.; Tatem, A.J. Remotely measuring populations during a crisis by overlaying two data sources. *Int. Health* **2015**, *7*, 90–98. [CrossRef] [PubMed]
87. Wilson, R.; zu Erbach-Schoenberg, E.; Albert, M.; Power, D.; Tudge, S.; Gonzalez, M.; Guthrie, S.; Chamberlain, H.; Brooks, C.; Hughes, C.; et al. Rapid and near real-time assessments of population displacement using mobile phone data following disasters: The 2015 Nepal Earthquake. *PLoS Curr.* **2016**, *8*. [CrossRef] [PubMed]
88. Ghurye, J.; Krings, G.; Frias-Martinez, V. A Framework to Model Human Behavior at Large Scale during Natural Disasters. In Proceedings of the 17th IEEE International Conference on Mobile Data Management (MDM), Porto, Portugal, 13–16 June 2016; Chow, C., Jayaraman, P., Wu, W., Eds.; IEEE: Piscataway, NJ, USA, 2016; Volume 1, pp. 18–27.
89. Poser, K.; Dransch, D. Volunteered geographic information for disaster management with application to rapid flood damage estimation. *Geomatica* **2010**, *64*, 89–98. [CrossRef]
90. Alexander, D.E. *Principles of Emergency Planning and Management*; Oxford University Press on Demand: Oxford, UK, 2002; ISBN 9780195218381.
91. Ehrlich, D.; Tenerelli, P. Optical satellite imagery for quantifying spatio-temporal dimension of physical exposure in disaster risk assessments. *Nat. Hazards* **2013**, *68*, 1271–1289. [CrossRef]
92. McCormick, S. After the cap: Risk assessment, citizen science and disaster recovery. *Ecol. Soc.* **2012**, *17*. [CrossRef]
93. Kwak, Y.J. Nationwide Flood Monitoring for Disaster Risk Reduction Using Multiple Satellite Data. *ISPRS Int. J. Geo-Inf.* **2017**, *6*, 203. [CrossRef]
94. Horita, F.E.; de Albuquerque, J.P.; Degrossi, L.C.; Mendiondo, E.M.; Ueyama, J. Development of a spatial decision support system for flood risk management in Brazil that combines volunteered geographic information with wireless sensor networks. *Comput. Geosci.* **2015**, *80*, 84–94. [CrossRef]
95. Zhang, J.A.; Nolan, D.S.; Rogers, R.F.; Tallapragada, V. Evaluating the impact of improvements in the boundary layer parameterization on hurricane intensity and structure forecasts in HWRF. *Mon. Weather Rev.* **2015**, *143*, 3136–3155. [CrossRef]

96. Yablonsky, R.M.; Ginis, I.; Thomas, B. Ocean modeling with flexible initialization for improved coupled tropical cyclone-ocean model prediction. *Environ. Model. Softw.* **2015**, *67*, 26–30. [CrossRef]
97. Zhang, J.A.; Marks, F.D.; Montgomery, M.T.; Lorsolo, S. An estimation of turbulent characteristics in the low-level region of intense Hurricanes Allen (1980) and Hugo (1989). *Mon. Weather Rev.* **2011**, *139*, 1447–1462. [CrossRef]
98. Zhang, J.A.; Gopalakrishnan, S.G.; Marks, F.D.; Rogers, R.F.; Tallapragada, V. A developmental framework for improving hurricane model physical parameterizations using aircraft observations. *Trop. Cycl. Res. Rev.* **2012**, *1*, 419–429. [CrossRef]
99. Ruf, C.S.; Atlas, R.; Chang, P.S.; Clarizia, M.P.; Garrison, J.L.; Gleason, S.; Katzberg, S.J.; Jelenak, Z.; Johnson, J.T.; Majumdar, S.J.; O'brien, A. New ocean winds satellite mission to probe hurricanes and tropical convection. *Bull. Am. Meteorol. Soc.* **2016**, *97*, 385–395. [CrossRef]
100. Zhang, F.; Weng, Y. Predicting hurricane intensity and associated hazards: A five-year real-time forecast experiment with assimilation of airborne Doppler radar observations. *Bull. Am. Meteorol. Soc.* **2015**, *96*, 25–33. [CrossRef]
101. Furquim, G.; Filho, G.P.R.; Jalali, R.; Pessin, G.; Pazzi, R.W.; Ueyama, J. How to Improve Fault Tolerance in Disaster Predictions: A Case Study about Flash Floods Using IoT, ML and Real Data. *Sensors* **2018**, *18*, 907. [CrossRef] [PubMed]
102. Kalyuzhnaya, A.V.; Nasonov, D.; Ivanov, S.V.; Kosukhin, S.S.; Boukhanovsky, A.V. Towards a scenario-based solution for extreme metocean event simulation applying urgent computing. *Future Gener. Comput. Syst.* **2018**, *79*, 604–617. [CrossRef]
103. Yan, J.; Jin, J.; Chen, F.; Yu, G.; Yin, H.; Wang, W. Urban flash flood forecast using support vector machine and numerical simulation. *J. Hydroinformatics* **2018**, *20*, 221–231. [CrossRef]
104. San-Miguel-Ayanz, J.; Schulte, E.; Schmuck, G.; Camia, A.; Strobl, P.; Liberta, G.; Giovando, C.; Boca, R.; Sedano, F.; Kempeneers, P.; et al. Comprehensive monitoring of wildfires in Europe: The European forest fire information system (EFFIS). In *Approaches to Managing Disaster-Assessing Hazards, Emergencies and Disaster Impacts*; Tiefenbacher, J., Ed.; InTech: Rijeka, Croatia, 2012; pp. 87–108, ISBN 978-953-51-0294-6.
105. Lacava, T.; Brocca, L.; Coviello, I.; Faruolo, M.; Pergola, N.; Tramutoli, V. Integration of optical and passive microwave satellite data for flooded area detection and monitoring. In *Engineering Geology for Society and Territory-Volume 3*; Lollino, G., Arattano, M., Rinaldi, M., Giustolisi, O., Marechal, J., Gordon, E.G., Eds.; Springer: Cham, Switzerland, 2014; pp. 631–635, ISBN 978-3-319-09053-5.
106. Kussul, N.; Shelestov, A.; Skakun, S. Flood monitoring from SAR data. In *Use of Satellite and In-Situ Data to Improve Sustainability*; Kogan, F., Powell, A., Fedorov, O., Eds.; Springer: Dordrecht, The Netherlands, 2011; pp. 19–29, ISBN 978-90-481-9617-3.
107. De Groeve, T. Flood monitoring and mapping using passive microwave remote sensing in Namibia. *Geomat. Nat. Hazards Risk* **2010**, *1*, 19–35. [CrossRef]
108. Earle, P.S.; Bowden, D.C.; Guy, M. Twitter earthquake detection: Earthquake monitoring in a social world. *Ann. Geophys.* **2012**, *54*. [CrossRef]
109. Mandl, D.; Frye, S.; Cappelaere, P.; Handy, M.; Policelli, F.; Katjizeu, M.; Van Langenhove, G.; Aube, G.; Saulnier, J.F.; Sohlberg, R.; et al. Use of the earth observing one (EO-1) satellite for the namibia sensorweb flood early warning pilot. *IEEE J. Sel. Top. Appl. Earth Obs. Remote Sens.* **2013**, *6*, 298–308. [CrossRef]
110. Arribas-Bel, D. Accidental, open and everywhere: Emerging data sources for the understanding of cities. *Appl. Geogr.* **2014**, *49*, 45–53. [CrossRef]
111. Musaev, A.; Wang, D.; Pu, C. LITMUS: A Multi-Service Composition System for Landslide Detection. *IEEE Trans. Serv. Comput.* **2015**, *8*, 715–726. [CrossRef]
112. Poslad, S.; Middleton, S.E.; Chaves, F.; Tao, R.; Necmioglu, O.; Bügel, U. A semantic IoT early warning system for natural environment crisis management. *IEEE Trans. Emerg. Top. Comput.* **2015**, *3*, 246–257. [CrossRef]
113. Vetrivel, A.; Gerke, M.; Kerle, N.; Vosselman, G. Identification of damage in buildings based on gaps in 3D point clouds from very high resolution oblique airborne images. *ISPRS J. Photogramm. Remote Sens.* **2015**, *105*, 61–78. [CrossRef]
114. Fernandez Galarreta, J.; Kerle, N.; Gerke, M. UAV-based urban structural damage assessment using object-based image analysis and semantic reasoning. *Nat. Hazards Earth Syst. Sci.* **2015**, *15*, 1087–1101. [CrossRef]

115. Corbane, C.; Saito, K.; Dell'Oro, L.; Bjorgo, E.; Gill, S.P.; Emmanuel Piard, B.; Huyck, C.K.; Kemper, T.; Lemoine, G.; Spence, R.J.; et al. A comprehensive analysis of building damage in the 12 January 2010 MW7 Haiti earthquake using high-resolution satellite and aerial imagery. *Photogramm. Eng. Remote Sens.* **2011**, *77*, 997–1009. [CrossRef]
116. Jongman, B.; Wagemaker, J.; Romero, B.R.; de Perez, E.C. Early flood detection for rapid humanitarian response: Harnessing near real-time satellite and Twitter signals. *ISPRS Int. J. Geo-Inf.* **2015**, *4*, 2246–2266. [CrossRef]
117. Di Felice, M.; Trotta, A.; Bedogni, L.; Chowdhury, K.R.; Bononi, L. Self-organizing aerial mesh networks for emergency communication. In Proceedings of the 2014 IEEE 25th Annual International Symposium on Personal, Indoor, and Mobile Radio Communication (PIMRC), Washington, DC, USA, 2–5 September 2014; IEEE: Piscataway, NJ, USA, 2014; pp. 1631–1636.
118. Mosterman, P.J.; Sanabria, D.E.; Bilgin, E.; Zhang, K.; Zander, J. A heterogeneous fleet of vehicles for automated humanitarian missions. *Comput. Sci. Eng.* **2014**, *16*, 90–95. [CrossRef]
119. Lu, Z.; Cao, G.; La Porta, T. Networking smartphones for disaster recovery. In Proceedings of the 2016 IEEE International Conference on Pervasive Computing and Communications (PerCom), Sydney, Australia, 14–18 March 2016; IEEE: Piscataway, NJ, USA, 2016; pp. 1–9.
120. Van de Walle, B.; Dugdale, J. Information management and humanitarian relief coordination: Findings from the Haiti earthquake response. *Int. J. Bus. Contin. Risk Manag.* **2012**, *3*, 278–305. [CrossRef]
121. De Alwis Pitts, D.A.; So, E. Enhanced change detection index for disaster response, recovery assessment and monitoring of accessibility and open spaces (camp sites). *Int. J. Appl. Earth Obs. Geoinform.* **2017**, *57*, 49–60. [CrossRef]
122. Contreras, D.; Forino, G.; Blaschke, T. Measuring the progress of a recovery process after an earthquake: The case of L'aquila, Italy. *Int. J. Disaster Risk Reduct.* **2017**. [CrossRef]
123. Kahn, M.E. The death toll from natural disasters: The role of income, geography, and institutions. *Rev. Econ. Stat.* **2005**, *87*, 271–284. [CrossRef]
124. Afzalan, N.; Evans-Cowley, J.; Mirzazad-Barijough, M. From big to little data for natural disaster recovery: How online and on-the-ground activities are connected. *I/S J. Law Policy Inf. Soc.* **2015**, *11*, 153. [CrossRef]
125. Yan, Y.; Eckle, M.; Kuo, C.L.; Herfort, B.; Fan, H.; Zipf, A. Monitoring and assessing post-disaster tourism recovery using geotagged social media data. *ISPRS Int. J. Geo-Inf.* **2017**, *6*, 144. [CrossRef]
126. Goodchild, M.F.; Glennon, J.A. Crowdsourcing geographic information for disaster response: A research frontier. *Int. J. Dig. Earth* **2010**, *3*, 231–241. [CrossRef]
127. Cervone, G.; Sava, E.; Huang, Q.; Schnebele, E.; Harrison, J.; Waters, N. Using Twitter for tasking remote-sensing data collection and damage assessment: 2013 Boulder flood case study. *Int. J. Remote Sens.* **2016**, *37*, 100–124. [CrossRef]
128. Byun, Y.; Han, Y.; Chae, T. Image fusion-based change detection for flood extent extraction using bi-temporal very high-resolution satellite images. *Remote Sens.* **2015**, *7*, 10347–10363. [CrossRef]
129. Westrope, C.; Banick, R.; Levine, M. Groundtruthing OpenStreetMap building damage assessment. *Procedia Eng.* **2014**, *78*, 29–39. [CrossRef]
130. Virtual Social Media Working Group and DHS First Responders Group. Lessons Learned: Social Media and Hurricane Sandy. Available online: https://www.dhs.gov/sites/default/files/publications/Lessons%20Learned%20Social%20Media%20and%20Hurricane%20Sandy.pdf (accessed on 12 March 2018).
131. Bejiga, M.B.; Zeggada, A.; Nouffidj, A.; Melgani, F. A convolutional neural network approach for assisting avalanche search and rescue operations with uav imagery. *Remote Sens.* **2017**, *9*, 100. [CrossRef]
132. Goldberg, D.; Holland, J. Genetic algorithms and machine learning. *Mach. Learn.* **1988**, *3*, 95–99. [CrossRef]
133. Benediktsson, J.; Swain, P.H.; Ersoy, O.K. Neural network approaches versus statistical methods in classification of multisource remote sensing data. *IEEE Trans. Geosci. Remote Sens.* **1990**, *28*, 540–552. [CrossRef]
134. Shao, Y.; Lunetta, R.S. Comparison of support vector machine, neural network, and CART algorithm for the land-cover classifcation using limited data points. *ISPRS J. Photogramm. Remote Sens.* **2012**, *70*, 78–87. [CrossRef]
135. Cooner, A.J.; Shao, Y.; Campbell, J.B. Detection of urban damage using remote sensing and machine learning algorithms: Revisiting the 2010 Haiti earthquake. *Remote Sens.* **2016**, *8*, 868. [CrossRef]

136. Cha, Y.J.; Choi, W.; Büyüköztürk, O. Deep Learning-Based Crack Damage Detection Using Convolutional Neural Networks. *Comput.-Aided Civil Infrastruct. Eng.* **2017**, *32*, 361–378. [CrossRef]
137. Pouyanfar, S.; Chen, S.C. Automatic video event detection for imbalance data using enhanced ensemble deep learning. *Int. J. Semant. Comput.* **2017**, *11*, 85–109. [CrossRef]
138. Asim, K.M.; Martínez-Álvarez, F.; Basit, A.; Iqbal, T. Earthquake magnitude prediction in Hindukush region using machine learning techniques. *Nat. Hazards* **2017**, *85*, 471–486. [CrossRef]
139. Chang, V. Towards a Big Data system disaster recovery in a Private Cloud. *Ad Hoc Netw.* **2015**, *35*, 65–82. [CrossRef]
140. Wan, Z.; Hong, Y.; Khan, S.; Gourley, J.; Flamig, Z.; Kirschbaum, D.; Tang, G. A cloud-based global flood disaster community cyber-infrastructure: Development and demonstration. *Environ. Model. Softw.* **2014**, *58*, 86–94. [CrossRef]
141. Belaud, J.P.; Negny, S.; Dupros, F.; Michéa, D.; Vautrin, B. Collaborative simulation and scientific big data analysis: Illustration for sustainability in natural hazards management and chemical process engineering. *Comput. Ind.* **2014**, *65*, 521–535. [CrossRef]
142. Puthal, D.; Nepal, S.; Ranjan, R.; Chen, J. A secure big data stream analytics framework for disaster management on the cloud. In Proceedings of the 2016 IEEE 18th International Conference on High Performance Computing and Communications; IEEE 14th International Conference on Smart City; IEEE 2nd International Conference on Data Science and Systems (HPCC/SmartCity/DSS), Sydney, Australia, 12–14 December 2016; IEEE: Piscataway, NJ, USA, 2016; pp. 1218–1225.
143. Hua, Y.; He, W.; Liu, X.; Feng, D. SmartEye: Real-time and efficient cloud image sharing for disaster environments. In Proceedings of the 2015 IEEE Conference on Computer Communications (INFOCOM), Hong Kong, China, 26 April–1 May 2015; IEEE: Piscataway, NJ, USA, 2015; pp. 1616–1624.
144. Bartoli, G.; Fantacci, R.; Gei, F.; Marabissi, D.; Micciullo, L. A novel emergency management platform for smart public safety. *Int. J. Commun. Syst.* **2015**, *28*, 928–943. [CrossRef]
145. Trono, E.M.; Arakawa, Y.; Tamai, M.; Yasumoto, K. Dtn mapex: Disaster area mapping through distributed computing over a delay tolerant network. In Proceedings of the 2015 Eighth International Conference on Mobile Computing and Ubiquitous Networking (ICMU), Hakodate, Japan, 20–25 January 2015; IEEE: Piscataway, NJ, USA, 2015; pp. 179–184.

Article

Shallow Shear-Wave Velocity Beneath Jakarta, Indonesia Revealed by Body-Wave Polarization Analysis

Rexha Verdhora Ry [1,2,*], Phil Cummins [1,3] and Sri Widiyantoro [2]

1 Research School of Earth Sciences, Australian National University, Canberra, Australian Capital Territory 0200, Australia
2 Global Geophysics Research Group, Institute Teknologi Bandung, Bandung 40116, Indonesia
3 Geoscience Australia, Canberra, ACT 2609, Australia
* Correspondence: rexha.ry@anu.edu.au

Received: 28 June 2019; Accepted: 30 August 2019; Published: 3 September 2019

Abstract: Noting the importance of evaluating near-surface geology in earthquake risk assessment, we explored the application to the Jakarta Basin of a relatively new and simple technique to map shallow seismic structure using body-wave polarization. The polarization directions of P-waves are sensitive to shear-wave velocities (V_s), while those of S-waves are sensitive to both body-wave velocities. Two dense, temporary broadband seismic networks covering Jakarta city and its vicinity were operated for several months, firstly, from October 2013 to February 2014 consisting of 96 stations, and secondly, between April and October 2018 consisting of 143 stations. By applying the polarization technique to earthquake signals recorded during these deployments, the apparent half-space shear-wave velocity (V_s^{ahs}) beneath each station is obtained, providing spatially dense coverage of the sedimentary deposits and the edge of the basin. The results showed that spatial variations in V_s^{ahs} obtained from polarization analysis are compatible with previous studies, and appear to reflect the average V_s of the top 150 m. The low V_s that characterizes sedimentary deposits dominates most of the area of Jakarta, and mainly reaches the outer part of its administrative margin to the southwest, more than 10 km away. Further study is required to obtain a complete geometry of the Jakarta Basin. In agreement with previous studies, we found that the polarization technique was indeed a simple and effective method for estimating near-surface V_s that can be implemented at very low-cost wherever three-component seismometers are operated, and it provides an alternative to the use of borehole and active source surveys for such measurements. However, we also found that for deep basins such as Jakarta, care must be taken in choosing window lengths to avoid contamination of basement converted phases.

Keywords: Jakarta basin; site effects; earthquake risk; shear-wave velocity

---

1. Introduction

The growth of the global population over the past century, combined with the accelerating pace of urbanization, has resulted in the explosive growth in the number of megacities (population over 10 million). Especially in the developing world, many of the buildings in these cities have not adhered to earthquake resilient construction practice with the result that "an epicentral hit on a megacity has the potential to cause 1 million fatalities" [1]. Also, many of these cities are concentrated in sedimentary basins, because of their flat topography and access to fertile soils, sources of water and maritime commerce. Improving earthquake risk assessment is one of the most important ways of alerting policymakers to the danger earthquakes pose. This requires not only knowledge of the population and building exposure and "hard rock" seismic hazard, but also an understanding of the potential for sedimentary basins to cause amplification and resonance of seismic wave motion. The observation of such effects in recent destructive earthquakes—the 2015 Kathmandu [2] and 2017 Mexico City earthquakes [3]—imparts some urgency to the evaluation of basin effects in other megacities. Such an evaluation can be conducted by profiling the shear-wave velocity (V_S) of sedimentary deposits down to bedrock.

Many approaches have been developed to estimate V_s depth profiles based on geotechnical and geophysical methods. Direct approaches such as borehole drilling, vertical seismic profiling (VSP), and standard penetration testing (SPT) can give accurate information on the velocity profile. However, these are often too expensive to cover the entire area of a large city. Other approaches exist such as seismic refraction or reflection surveys, but these use active sources like explosives and require large, regularly-spaced sensor arrays that are impractical in built-up urban areas. These problems are avoided using passive approaches such as microtremor measurements [4–7] and interferometry studies [8–11], which have become very popular for use in densely-populated areas.

Recently, Park and Ishii [12] introduced an alternative approach for estimating "near-surface" V_s by using measurements of body-wave polarization. The term polarization is used here to address the wave's particle motion (below we explain that "near-surface" V_s is actually apparent half-space V_S, which we denote as V_S^{ahs}). They utilized the well-known relationships between the V_s and compressional-wave velocity (V_P), and the polarization of body waves interacting with the free surface of a half-space. The incoming polarization directions of P-waves are sensitive to V_S, while those of S-waves are sensitive to both V_P and V_S. Therefore, it is possible to estimate near-surface V_s and V_P by observing the polarization directions of P-waves and S-waves generated by earthquakes at a seismic station. This approach requires no artificial source or other expensive equipment, and it can be applied with a minimum of computational effort.

Park and Ishii [12] concluded that their technique could be used very widely to study near-surface velocity structure wherever three-component seismometers are deployed. Since it is applicable to urban areas, we applied this technique based on P-waves polarization to map shallow V_s beneath Jakarta, the capital city of Indonesia. Because it lies on a thick young sedimentary basin [13–16], seismic risk in Jakarta is likely to be enhanced by amplification and resonance effects. In late 2013 and again in 2018, two temporary seismic networks consisting of three-component broadband seismometers were deployed for approximately 3 months and 5 months, respectively, covering most of the city and its vicinity. While Park and Ishii [12] presented their application using the Hi-net array in Japan, the combination of seismicity, dense seismic network, and sedimentary basin structure in Jakarta provides a good opportunity to evaluate this technique on a more local scale. The results are benchmarked against borehole data and shallow V_s estimated from other studies.

2. Geologic Setting and Seismometer Deployments

Located on the northern coast of the island of Java, the surface geology of Jakarta mainly consists of alluvial deposits. According to Turkandi et al. [17], the city's surface geology from the coastline to around 6 km southward to the center of the city consists of Holocene sand dunes and alluvial fan, while deposits of Pleistocene alluvial fan cover the southern part of the city with little trace of Tertiary volcanic deposits. The sedimentary deposits are estimated to thicken northward, with alluvial fan sediments reaching thicknesses of 300 m or more in the city center. Unfortunately, although the geological setting of Indonesia is described generally in Van-Bemmelen [18], there are only a few recent studies of the detailed geological setting of the Jakarta Basin (e.g., Turkandi et al. [17]).

Nevertheless, the existence of a thick sedimentary basin overlying tertiary bedrock strongly suggests that amplification and resonance could enhance the damaging effects of earthquakes. After seismic waves propagate through the transition between stiff rock and soft soil, their energy is trapped within the soil layer by internal reflections [19]. The reverberating seismic waves add constructively at certain frequencies but interfere destructively at others, with the former being resonant frequencies that are determined by the thickness and V_s of the sedimentary layer. Because of resonance, the energy and amplitude of seismic waves, which normally decay rapidly in intensity with distance from the earthquake, can instead increase dramatically in both amplitude and duration. Not accounting for such basin resonances may lead to an underrated earthquake risk assessment.

To better understand such basin effects, three-component broadband seismometers (Trillium Compact) and digitizers built by the Australian National University were installed at various sites (Figure 1) in two separate deployments. Stations spaced at 3–5 km, were deployed temporarily on a concrete slab floor in schools throughout the city. The first deployment comprised 96 stations and operated from October 2013 to February 2014, covering most of the area of Jakarta. Twenty-six stations were maintained for 3 months of recording as semi-permanent stations and the other 26 stations were redeployed in three phases, each of one-month duration. The second deployment, comprising 143 stations was deployed in 2018 and aims to include coverage just outside Jakarta in order to reveal the extent of the basin edge. The 30 stations were maintained and redeployed in five phases with at least one-month duration for every site.

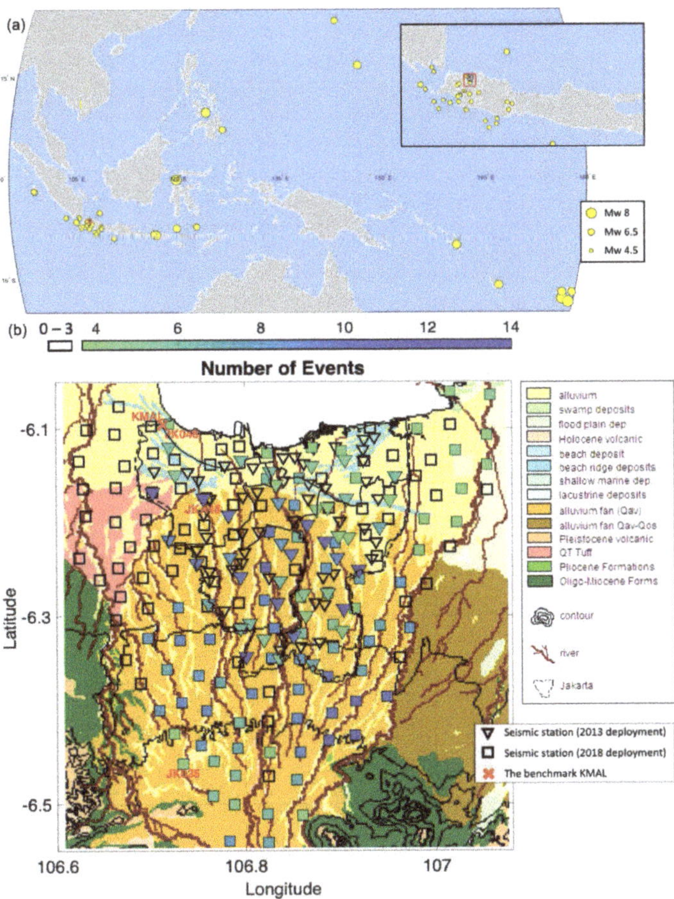

Figure 1. (**a**) Distribution of earthquakes (yellow circles) observed around the study area (red zone). (**b**) Map of the seismic deployments and surface geology around Jakarta (modified from [20]). The inverted triangles represent the seismic network deployed between October 2013 and February 2014. The squares represent the seismic network deployed between April 2018 and October 2018. The administrative boundary of Jakarta is shown with black lines. The blue cross is the benchmark site KMAL. The colors represent the number of events observed at each station, while the colorless stations did not record any useful earthquake signals.

Using data from the first deployment, Saygin et al. [15] extracted Rayleigh wave Green's functions from cross-correlograms of ambient noise at the different station pairs and imaged the basin structure using Ambient Noise Tomography. They found that the sedimentary basin covers most of the area of the city with a thickness up to 1500 m below central Jakarta [15]. Considering this evidence, they concluded that basin effects in Jakarta are likely to enhance the damaging effects of earthquake-generated seismic waves. This conclusion was also reached by Cipta et al. [16,20], who used the same data to invert Horizontal-to-Vertical Spectral Ratio (HVSR) curves to achieve better resolution of the basement architecture. However, neither of these models revealed the basin edges, which extend outside the city of Jakarta beyond the extent of the 2013–2014 seismometer deployment. For this reason, the 2018 seismometer deployment was undertaken to extend the coverage beyond the city of Jakarta itself, and hopefully resolve the basin edges. In addition, the extent to which either of these studies resolves the very shallow (<100 m depth) V_s structure is unclear.

Both the 2013–2014 and 2018 seismometer deployments were intended to make use of noise interferometry to study the basin structure, especially the first seismic network [15]. Nevertheless, earthquake signals were recorded during both deployments. We have evaluated recorded signals from 56 earthquakes with a good signal-to-noise ratio (SNR), varying from local to regional and teleseismic earthquakes. Applying body-wave polarization analysis to these signals seems feasible and worthwhile, and this is what we report on in this study. However, not all stations record the same events due to the different phases of deployment. The numbers of observed data at every station are summarized by color representation in Figure 1b.

3. Method

3.1. Apparent Incident Angles of Body-Waves

When a body wave arrives at Earth's free surface, the incident wave is both reflected and converted. In particular, the incident P-wave generates a reflected P-wave and a converted SV wave. This means the particle motion of P-waves recorded by a three-component seismometer on the free surface is determined by the combination of the incoming and the two outgoing waves, which is defined by the apparent incidence angle ($\overline{\theta}$).

Considering a P-wave incident on the free surface of a uniform half-space with P(S)-velocity $V_P(V_S)$, the apparent incidence angle is different from the true incidence angle (θ), with their relationship derived in Wiechert [21] as:

$$p = \frac{\sin\theta}{V_p} = \frac{\sin\left(\frac{1}{2}\overline{\theta}\right)}{V_s} \tag{1}$$

where p is the ray parameter (or horizontal slowness) of the P-wave.

The derivation of their relationship is also shown by [12] using the free surface boundary conditions. Equation (1) can be rewritten as:

$$V_s = \frac{\sin\left(\frac{1}{2}\overline{\theta}\right)}{p} \tag{2}$$

$$\overline{\theta} = 2\arcsin(p\,V_s) \tag{3}$$

which defines the half-space V_s if the ray parameter of the P-wave and the apparent angle are known [22].

In this study, Equation (2) is used to estimate a "near-surface" V_s for the Jakarta Basin. However, since the true V_s profile in the basin is not that of a half-space but has V_s increasing with depth, our estimates of "near-surface" V_s are actually an estimate of apparent half-space V_s, which we denote as V_s^{ahs} in what follows. V_s^{ahs} should be representative of the actual V_s averaged over some depth range. This closely follows Svenningsen and Jacobsen [22], who use a different notation, $V_{s,app}$ to denote the apparent half-space V_s.

3.2. Calculating Polarization

The apparent incidence angles of P-waves are measured from the particle motion of the observed body-waves, with particle motion in the direction of apparent incidence for P-waves. This particle motion can be observed using a "particle motion" plot of the curve connecting particle position in the vertical and radial plane at successive times (see Figures 3c and 4c). Therefore, it is natural to project the recorded three-component seismograms (vertical, north-south, and east-west signal amplitudes) onto the vertical-radial-transversal plane using a priori information of the earthquake source and seismic station location. Herein, the discretized vertical and radial component time series data are defined by the column vectors z and r, respectively.

Polarization of particle motion can be measured in the time domain using principal component analysis (PCA) [23]. For the selected signals, a data covariance matrix is arranged as follows:

$$z = [z_1, z_2, \ldots, z_M]^T; \; r = [r_1, r_2, \ldots, r_M]^T$$

$$C = \frac{1}{M}\begin{bmatrix} z^T z & z^T r \\ r^T z & r^T r \end{bmatrix} \quad (4)$$

where M is the number of data points for each component. The particle motion is then specified by the eigenvectors of this covariance matrix. The eigenvalues λ and eigenvectors v of the covariance matrix C are calculated by solving:

$$(C - \lambda I)v = 0 \quad (5)$$

where I is the 2×2 identity matrix. From the data comprising vectors z and r, solving Equation (5) yields two eigenvalues λ_1 and λ_2, which have respective eigenvectors v_1 and v_2. In contrast to Park and Ishii [12], we only utilized eigenvector v_1 related to λ_1 ($\lambda_1 > \lambda_2$), which defines the maximum energy in the data. The principal polarization of the selected signals in the vertical and radial component is given by the eigenvector $v_1 = [v_z, v_r]^T$.

In estimating the polarization, first of all, the data are selected by windowing to isolate the direct P-wave signal, with window length chosen to include as much of the respective waveform as possible without including coda that is contaminated by arrivals of different wave type or incidence angle (see below). For every window, Equations (6) and (7) are applied to estimate the principal P-wave polarization $v^p{}_1$. Then, the apparent incidence angles are defined by:

$$\overline{\theta} = arctan\left(\frac{v^p{}_r}{v^p{}_z}\right) \quad (6)$$

3.3. Estimating Apparent Half-Space Velocities

The procedure to invert the observed apparent incidence angles $\overline{\theta}$ for estimates of V_s^{ahs} is described as follows. For the observed data, Equation (6) describes the observed apparent incidence angle of the ray from a particular earthquake to the seismic station. Equation (3) is used for forward modeling the apparent incidence angles for given V_s^{ahs} and p. Arranged as an objective function for a single station, we use the misfit modified from [12]:

$$f(V_s) = \frac{\sum_{i=1}^{N}\left[w_i^p\left(\overline{\theta}_i^{obs} - \overline{\theta}_i^{cal}\left(V_S^{ahs}\right)\right)^2\right]}{\sum_i^N w_i^p} \quad (7)$$

where superscripts *obs* and *cal* denote observed and calculated apparent angles for the *i*-th earthquake, respectively, and the summation is over N earthquakes. The weighting values w_i^p are given based on the quality of measured data. In this case, we use the total variance in the measured data of particle motion, which is:

$$w = \frac{\lambda_1}{\lambda_1 + \lambda_2} \tag{8}$$

According to Equation (3), we require the ray parameters of P-waves for a certain station and earthquake geometry. The program Travel Time Toolbox (TTBox) [24] was utilized to compute seismic ray paths and travel times using a 1-D spherical velocity model. Then, $\overline{\theta_i}^{cal}$ can be computed as a function of V_s^{ahs}. A grid-search over V_s^{ahs} was used to determine the values that minimize the misfit in Equation (7).

4. Application in Jakarta

4.1. Window Selection

After the 56 earthquakes have been identified during the recording, the time information for the P-waves is required to calculate the time windows used for the polarization calculations.

At each station and for each of the 56 earthquakes, the arrival times (or onsets) of P-waves are automatically picked using a kurtosis based algorithm [25]. Unfortunately, this automatic step does not always work well for low SNR signals. Therefore, we manually checked the low SNR data and the P-waves onsets were refined manually to get more reliable times, while no manual re-picking was needed for signals with high SNR.

In order to calculate the polarization using PCA as described in Equations (4) and (5), the signal is selected and windowed to isolate the direct P-wave, starting from its onset. Band-pass filtering can be applied to strengthen the signals, as long as it still preserves the source frequency content. However, a problem arises when choosing the ideal length of the time window. The aim is to include as much of the waveform of the direct P-wave as possible without including contamination from arrivals of different wave type or horizontal slowness, as might be expected, for example, from phases converted at the basement of the sedimentary basin. Such contamination may bias or increase the uncertainty in estimates of seismic velocities. In addition, teleseismic earthquakes will require longer time windows compared to regional and local earthquakes.

Figure 2 shows examples of a regional and a teleseismic earthquake recorded at one of the stations. The band-pass filters of 0.1–2 Hz and of 0.1–1 Hz are used for the regional earthquake and the teleseismic earthquake, respectively. Their P-waves can be clearly distinguished. Obviously, the duration of the teleseismic earthquake is much longer. Then, the calculation of their polarizations using PCA is illustrated by Figures 3 and 4. The onsets are used as the starting point of the time window.

The length of the time window that can be exploited in selecting the signals may vary from very short to significantly long, and they affect the estimation. Depending on the time window length, the result of polarization direction can vary significantly, as shown in Figures 3b and 4b. These differences may result in bias and/or high uncertainty in the estimation of seismic velocities. Choosing a stable and consistent polarization becomes crucial at this point.

A shorter window achieves a purely direct P-wave signal that is uncontaminated by phases of different wave type or incidence angle, but is not necessarily stable owing to its small number of data points. A longer window may contain phases of different wave type or incidence angle, which could bias the result. The synthetic tests (Appendix A) shows that window lengths greater than 1–2 seconds are unreliable due to the contamination of converted waves from the basin basement. Therefore, we analyzed each record to judge whether a mean estimate of P-wave incidence angle could be made, while removing outliers at the beginning and end of the window that might represent instability due to limited data or contamination by converted phases, respectively.

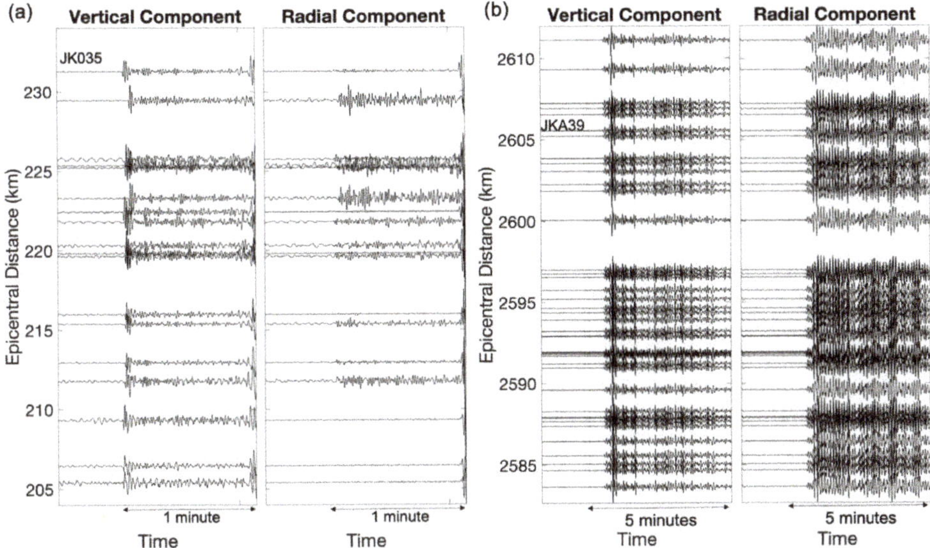

Figure 2. Examples of recorded earthquakes signals at stations shown in vertical and radial components. (**a**) Regional M_W 5.3 earthquake which occurred on 22 June 2018 in the Java Sea; (**b**) Teleseismic M_W 7.2 earthquake which occurred on 15 October 2013 in Bohol, the Philippines.

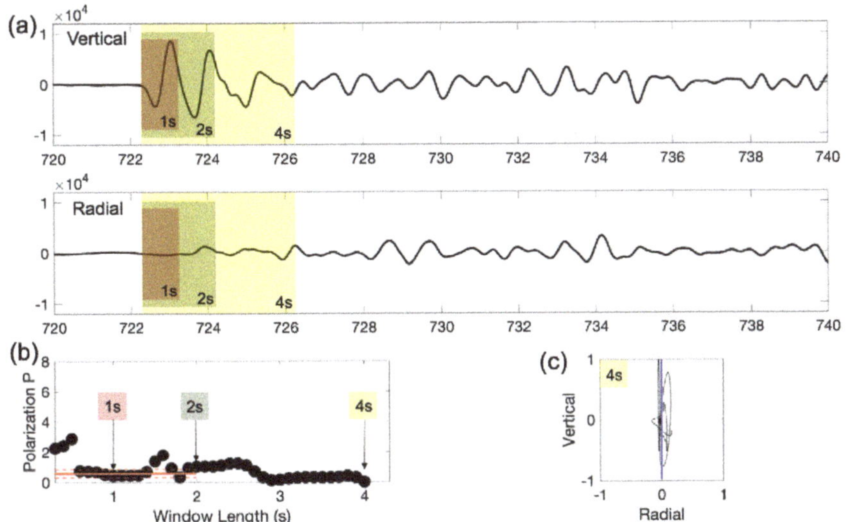

Figure 3. Polarization analysis of P-waves at station JK035 for the M_W 5.3 regional earthquake. The azimuth is 230° and the epicentral distance is 231 km. (**a**) Time windowing used for principal component analysis (PCA), with window lengths of 1, 2, and 4 seconds indicated by red, green, and yellow shades, respectively. (**b**) Distribution of P-wave polarizations calculated from different lengths of the time window, in increments of 0.1 s. The red line represents the best polarization after removing outliers. (**c**) The particle motion during the 4-s window. The blue line highlights its principal component.

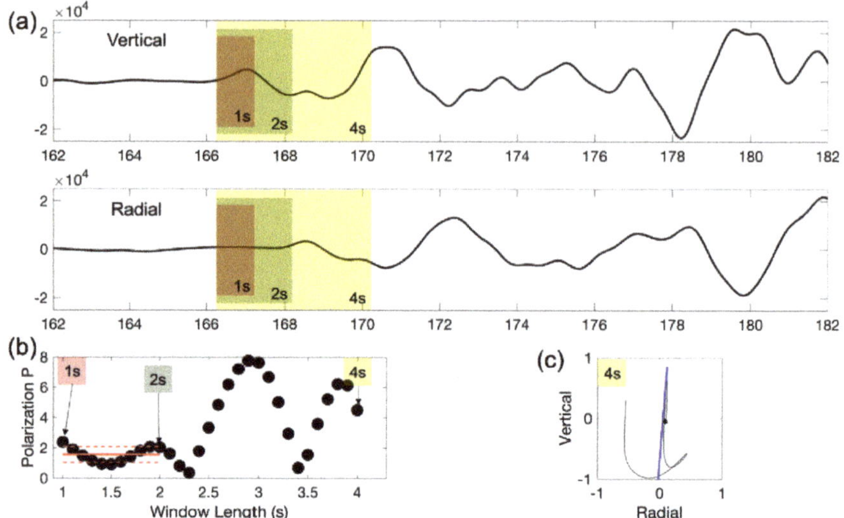

Figure 4. Polarization analysis of P-wave at station JKA39 for the M_w 7.2 teleseismic earthquake. The azimuth is 227° and the epicentral distance is 2605 km. (**a**) Time windowing used for PCA, with window lengths of 1, 2, and 4 seconds, indicated by red, green, and yellow shades, respectively. (**b**) Distribution of P-wave polarizations calculated from different lengths of the time window, in increments of 0.1 s. The red line represents the best polarization after removing outliers. (**c**) The particle motion during the 4 s. The blue line highlights its principal component.

4.2. Apparent Half-Space Velocity Estimates

A station may observe several earthquakes during its recording duration, each of which contributes measurements of P-wave polarizations. These body-wave polarizations are then used to estimate apparent half-space velocities (V_s^{ahs}) beneath the station. A grid-search optimization was undertaken using Equation (7) for every station, while ray parameters were calculated from the AK135 1-D earth model [26]. Figure 5 shows plots of the apparent incidence angles as a function of ray parameter for station JKA12. Stations that record only a few earthquakes will have poorly constrained apparent half-space velocities, so we only utilized stations that recorded at least four clear earthquakes. Aiming to get the best fit model, V_s^{ahs} were searched from 50 m/s to 2000 and 4000 m/s using increments of 10 m/s.

The maps of apparent half-space velocities V_s^{ahs} within Jakarta and its vicinity are shown in Figure 6. Note that stations that lack data were not included in the results. We focus on the V_s^{ahs} results derived mainly from the relatively high-quality P-wave polarization measurements.

The estimated values of V_s^{ahs} range from 200 to 800 m/s. Within the boundary of Jakarta, most of the area is dominated by low V_s^{ahs} between 200 and 400 m/s. The area characterized by low V_s^{ahs} extends outward beyond the city, only to the southwest. On the other hand, the eastern edge of the city is characterized by higher V_s^{ahs}, especially outside the city's eastern boundary. Unexpectedly, given the results of previous studies, higher V_s^{ahs} can also be observed in the northern part of the city very near to the coastline.

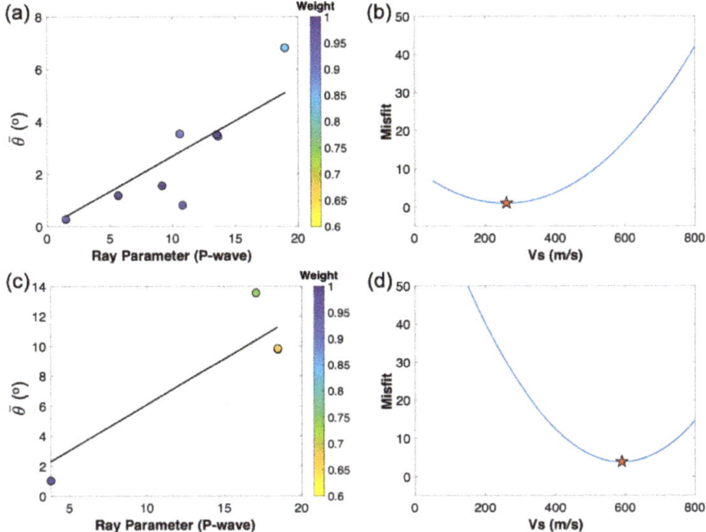

Figure 5. Observed (**a**) P-wave incidence angles (colored circles) at station JKA39, as a function of theoretical ray parameters computed using the AK135 model. The colors represent the weighting values. Black lines represent the best fitting calculated angles. (**b**) Grid-search results for station JKA39 along its misfit, with the best fit marked by the red star. (**c**) and (**d**) Same as (**a**) and (**b**), respectively, but at station JK035.

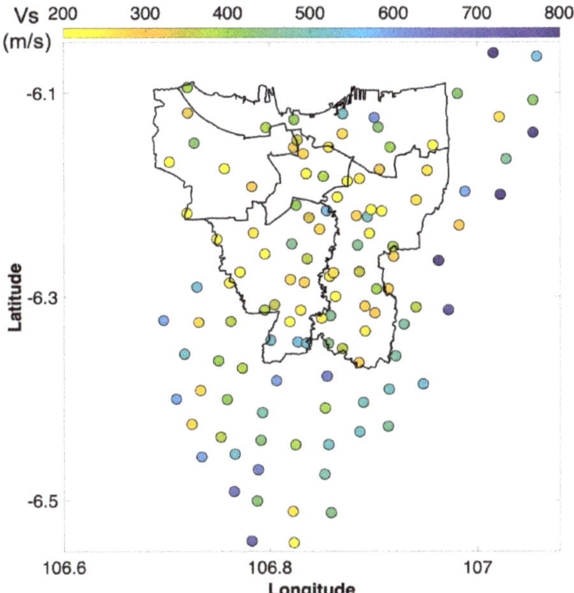

Figure 6. Map of near-surface shear-wave velocities represented by colored circles. Yellow is low and blue is high, ranging from 200 to 800 m/s.

4.3. Comparison and Depth Estimation

The polarization technique appears to provide considerable information on the variation in the shallow structure within the Jakarta basin, as reflected in the measured V_s^{ahs}. Unfortunately, it is difficult to know what depth range these apparent half-space values correspond to. In their use of a similar technique for measuring apparent half-space V_s from receiver functions, it is noted by Svenningsen and Jacobsen [22] that the results depend on the frequency content of the signals used. They advocate narrow-band filtering centered on inverse period f = 1/T to estimate a curve $V_s^{ahs}(T)$ that can be inverted for the shear-wave velocity profile. Given the complications that the basin structure may pose for receiver function computation, as well as the use of local and teleseismic events with varying frequency content, we adopted the simpler approach of Park and Ishii [12]. Comparison of their estimates for V_p^{ahs} and V_s^{ahs} at stations of Japan's broadband network Hi-net, allowed Park and Ishii [12] to conclude that their results were representative of the top ~100 m of the V_p and V_s profiles, respectively.

Therefore, in order to assess whether our half-space velocity measurements are representative of shallow V_s structure and if so, to suggest a depth range to which they correspond, we compared our results to previous studies. Ridwan et al. [14] used the spatial autocorrelation (SPAC) method to estimate depth profiles of V_s for depths less than 1 km at 55 sites throughout Jakarta. They provide a good set of benchmark V_s profiles at their site KMAL, located in northwest Jakarta, where they compared their SPAC result with cone penetrometer (SPT) and downhole seismic measurements (Figure 7, modified from Figure 10c in Ridwan et al. [14]). From Ridwan et al. [14], the shear-wave velocity at a depth of 100 m beneath the site KMAL is 350 m/s. Our closest station to this site, which is 700 m away, is JK046, where we used the polarization technique to obtain a very similar estimate for V_s^{ahs} of 390 m/s. The small difference between these values could reflect either measurement error or be a genuine difference in V_s due to the slight difference in the locations of measurement.

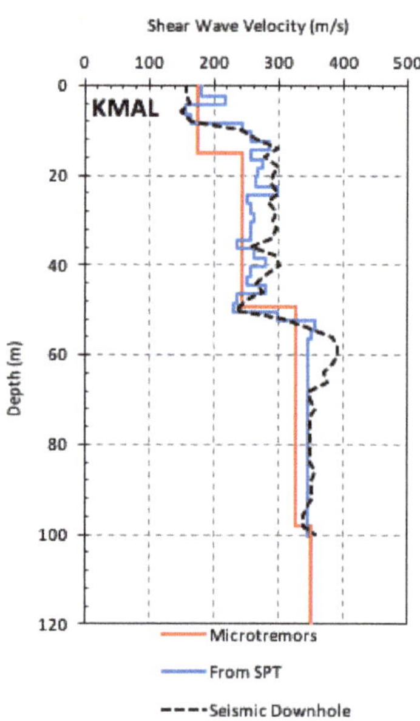

Figure 7. V_s profiles at the site KMAL obtained from microtremor arrays (red line), standard penetration testing (SPT) (blue line) and seismic downhole (dashed line). Modified from [14].

As discussed above, Cipta et al. [16] applied HVSR analysis to the 2013–2104 deployment data to obtain the V_S structure beneath Jakarta. We compared our results with average values of Cipta et al.'s [16] V_S profiles taken over different depths from the surface. We found that our results best match Cipta et al.'s [16] V_S profiles when the latter are averaged over the top 150 m, as shown in Figure 8. Note that the different samples shown are due to the lack of observed earthquakes at some stations in our study.

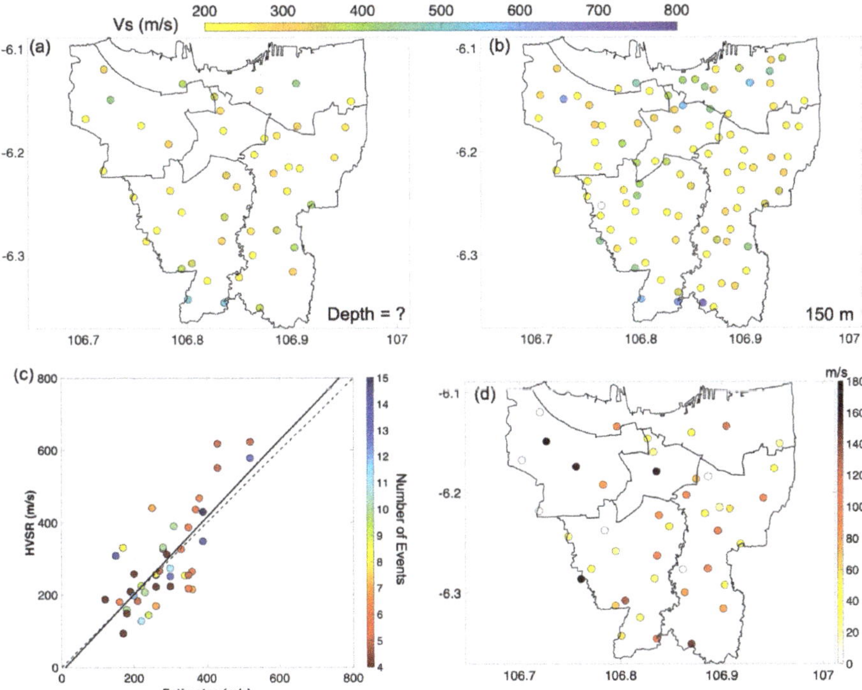

Figure 8. Map of near-surface shear-wave velocities represented by colored circles. Yellow is low and blue is high, ranging from 200 to 800 m/s. (**a**) V_S^{ahs} obtained in this study based on body-wave polarization. (**b**) Average V_S over depths in the top 150 m obtained from the HVSR study [16]. (**c**) The scatter plot between the V_S^{ahs} estimates and the HVSR study; black solid line is the linear regression line and the grey dashed line is the 1:1 line. (**d**) Absolute difference between (**a**) and (**b**).

By comparison to the benchmark site KMAL [14], our V_S^{ahs} matches the average of the V_S profile over the top 100 m reasonably well. Unfortunately, no data deeper than that is available and one site is insufficient to indicate an overall trend. Meanwhile, our V_S^{ahs} estimates agree with many of the 150 m depth values of the much wider dataset of average V_S profiles obtained in [16]. A few inconsistencies do exist, but these may be due to anomalies in the measurements such as misalignment or miscalibration of sensors. In any case, our comparisons with previous studies suggests that the V_S^{ahs} estimates obtained in this study approximately represent average V_S at around the top 150 m depth.

4.4. Correlation with Surface Geology

The distribution of low V_S^{ahs} agrees with the geologic mapping that shows sedimentary deposits comprise almost all areas of Jakarta's surface geology. Although the contrast between Holocene and Pleistocene deposits is small and their boundary is not well resolved in our V_S^{ahs} map, their deposits obviously fill the Jakarta basin to a depth of at least 150 m. These deposits reach the outer part of the Jakarta administrative margin mostly to the southwest, more than 10 km away. The result is in accordance with Cipta et al. [20] who reported the estimation of the basin extension based on extrapolation of the basement depth.

The higher V_s^{ahs} found in the southern part of the basin may relate to Tertiary volcanic deposits, or even the edges of the basin, which is Tertiary rock. This feature gets more distinct in the eastern part of Jakarta. The gradation from low V_s^{ahs} to higher V_s^{ahs} at the eastern administrative boundary underlines the contrast between sedimentary deposits and denser rock. The sedimentary deposits diminish eastward, while the thinnest layer lies at the southeastern corner of our study area. We suggest that the basin edge emerges near the surface in this area.

5. Conclusions

We applied the simple body-wave polarization technique of Park and Ishii [12] to obtain the variation in apparent half-space S-wave velocity (V_s^{ahs}) over the Jakarta Basin. By measuring the apparent incidence angles of earthquake-generated P-waves using principal component analysis, we obtained estimates of V_s^{ahs}. Although care had to be taken in choosing window lengths to avoid contamination of the direct P-wave by basement converted phases, we found we were able to obtain stable estimates of polarization by choosing windows of 1–2 sec duration.

The spatial variations we observed in V_s^{ahs} estimated using the body-wave polarization technique seem sensible when compared to other studies. In particular, when comparing our V_s^{ahs} with the V_s profiles in [14] and [16], the mapping of V_s^{ahs} appears to be correlated to the average of V_s profiles over the top 150 meters. In further studies, it might be useful to investigate the frequency-dependence of body-wave polarization in an attempt to reveal further details of the shallow V_s profiles.

Our estimates of V_s^{ahs} reflect the shallow Vs structure obtained within the Jakarta city limits in earlier studies [15,16], but extended this information beyond the city limits of Jakarta to what is thought to be the basin edge [20]. Although the surface geology of the entire study area is composed of quaternary sediments (Figure 1b), we found that on average, V_s^{ahs} increases towards the outer edge of the study area (Figure 6). This may indicate that V_s^{ahs} is sensitive to a reduction in basement depth that indicates the effective edge of the basement. In future studies we hope to obtain a more complete geometry of the Jakarta sedimentary basin that will enable a more accurate ground-motion simulation for hypothetical earthquake scenarios that can characterize seismic risk in Jakarta.

Author Contributions: R.V.R., P.C., S.W. conceived the study and contributed to the writing of the manuscript. All authors contributed to the preparation of the manuscript. The data and material that support the findings of this study are available on request from the corresponding author, R.V.R.

Funding: This study was partly supported through the ITB Research Grant 2018 awarded to SW and RVR and partly supported through the ANU Research Grant 2018 awarded to PC.

Acknowledgments: We gratefully acknowledge the Institute of Technology Bandung (ITB) and the Australian National University (ANU) for funding this research. We also thank the Australia Awards for the scholarship awarded to RVR and conducting this research at RSES, ANU.

Conflicts of Interest: We declare that we have no significant competing financial, professional or personal interests that might have influenced the performance or presentation of the work described in this manuscript. The funders had no role in the design of the study; in the collection, analyses, or interpretation of data; in the writing of the manuscript, or in the decision to publish the results.

Appendix A. Synthetic Test

As mentioned above, the idea of estimating the P-waves polarization relies greatly on the windowing to isolate P-wave signals, with window length chosen to include as much of the direct P waveform as possible without including coda that is contaminated by arrivals of different wave type or incidence angle. A complication arises in sedimentary basins due to the possible contamination from arrivals of phases converted at the basement which will arrive with a different incidence angle.

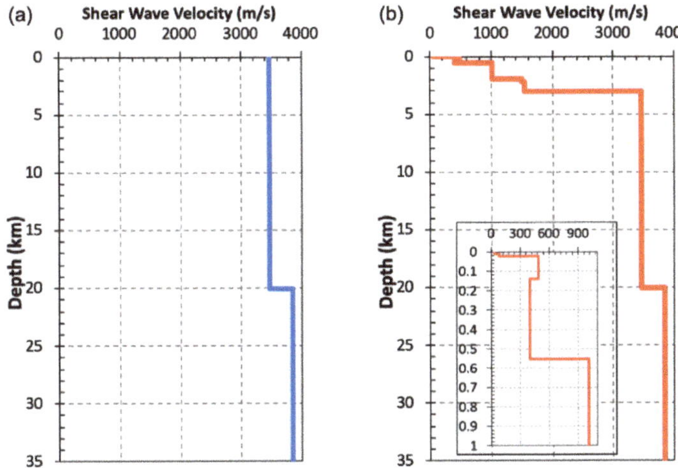

Figure A1. V_s profiles of (**a**) AK135 [26] and (**b**) AK135 adding sedimentary basin layers [16].

We performed synthetic tests to examine the effect of the converted wave in the Jakarta Basin. We generated a seismogram for the incoming P-wave based on a layered half-space model [27] for the Bohol, Philippines Mw 7.2 earthquake, which occurred on 2013 October 15. The ray parameter of the direct P-waves is 9.14 s/deg^{-1}. Seismograms for two different velocity models were generated, one for the AK135 1-D earth model [26] (Figure A1a) and the second for the AK135 model with the insertion of layers in the top 3 km representing the sedimentary fill of the Jakarta Basin [16] (Figure A1a).

As shown in Figure A2, the waveforms of the direct P-wave on vertical and radial components correlate very well. There are no obvious differences in the vertical- vs. radial-component waveforms that might indicate the presence of converted waves. Using window lengths varying from 1 to 8 seconds, the calculated P-wave incidence angle varied between 32 and 36 degrees. Applying Equation (2), the calculated Vs^{ahs} are 3.35 km/s and 3.75 km/s, respectively. These values seem reasonably consistent with the V_s = 3.46 km/s in the top layer of the AK135 model.

As shown in Figure A3, on the other hand, there is poor correlation between the radial and vertical components following the arrival of the direct P-wave. After 1–2 seconds following the arrival of the direct P-wave, a different phase arrives which we interpret as the arrival Ps wave converted at the basement, although it is difficult to identify clearly because of the basin's complex velocity structure. Due to the appearance of this basement-converted wave, the use of more than 1–2 seconds for the length of window is questionable. Using time window lengths of 1 to 8 seconds resulted in a variation of measured angle of incidence between 0.1 and 10 degrees. The average Vs of the basin model in the top 100 m is 300 m/s. Applying Equation (3), then the expected polarization should be 2.6 degree. While this lies within the range of incidence angles for windows shorter than 2 seconds, it is difficult to pick the optimal window length with certainty.

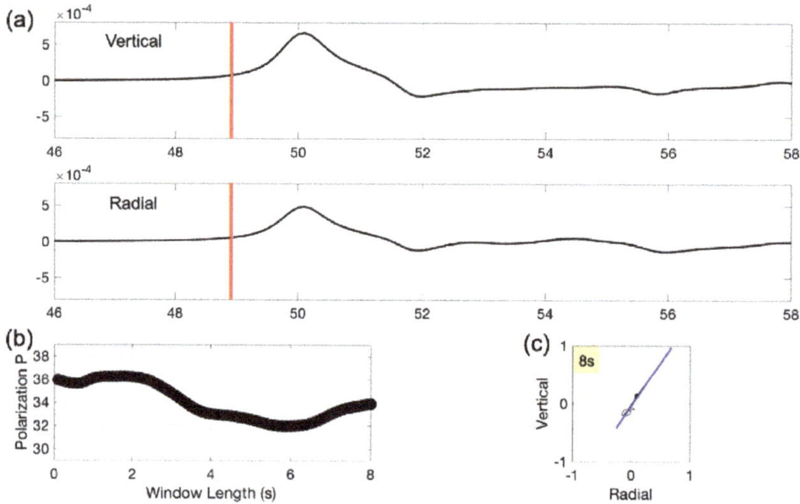

Figure A2. Seismograms of P-wave arrival in Jakarta for the M_w 7.2 Bohol earthquake at 2600 km distance, calculated for a velocity model with no sedimentary basin layers. (**a**) Waveforms of vertical and radial components. Red lines indicate the P-onset. (**b**) The variation in P-wave apparent incidence angle calculated for different time window lengths, in increments of 0.1 s, starting from the P-onset. (**c**) The particle motion during the 8 s window. The blue line highlights its principal component.

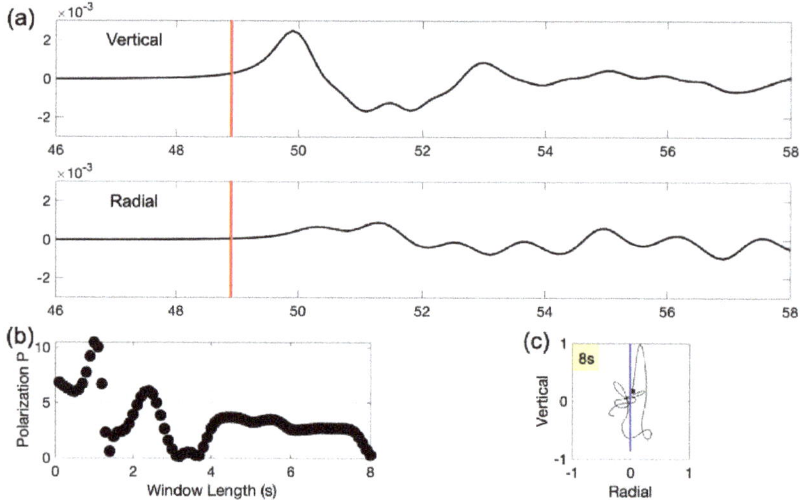

Figure A3. Seismograms of P-wave arrival in Jakarta for the M_w 7.2 Bohol earthquake at 2600 km distance, calculated for a velocity model that includes sedimentary basin layers. (**a**) Waveforms of vertical and radial components. Red lines are P-onset. (**b**) Variation in P-wave apparent incidence angle calculated for different time window lengths, in increments of 0.1 s, starting from P-onset. (**c**) The particle motion during the 8 s window. The blue line highlights its principal component.

References

1. Bilham, R. The seismic future of cities. *Bull. Earthq. Eng.* **2009**, *7*, 839–887. [CrossRef]
2. Galetzka, J.; Melgar, D.; Genrich, J.F.; Geng, J.; Owen, S.; Lindsey, E.O.; Xu, X.; Bock, Y.; Avouac, J.-P.; Adhikari, L.B.; et al. Slip pulse and resonance of the Kathmandu basin during the 2015 Gorkha earthquake, Nepal. *Science* **2015**, *349*, 1091–1095. [CrossRef]
3. Sahakian, V.J.; Melgar, D.; Quintanar, L.; Ramírez-Guzmán, L.; Pérez-Campos, X.; Baltay, A. Ground Motions from the 7 and 19 September 2017 Tehuantepec and Puebla-Morelos, Mexico, Earthquakes. *Bull. Seismol. Soc. Am.* **2018**, *108*, 3300–3312. [CrossRef]
4. Nakamura, Y. A method for dynamic characteristics estimation of subsurface using microtremor on the ground surface. *Railw. Tech. Res. Inst. Q. Rep.* **1989**, *30*, 25–33.
5. Behrou, R.; Haghpanah, F.; Foroughi, H. Empirical seismic site effect analysis for the city of Tehran using H/V and methods. *Int. J. Geotech. Eng.* **2018**, *6362*, 1–9. [CrossRef]
6. Fäh, D.; Kind, F.; Giardini, D. A theoretical investigation of average H/V ratios. *Geophys. J. Int.* **2001**, *145*, 535–549. [CrossRef]
7. Arai, H.; Tokimatsu, K. S-wave velocity profiling by joint inversion of microtremor dispersion curve and horizontal-to-vertical (H/V) spectrum. *Bull. Seismol. Soc. Am.* **2005**, *95*, 1766–1778. [CrossRef]
8. Young, M.K.; Rawlinson, N.; Bodin, T. Transdimensional inversion of ambient seismic noise for 3D shear velocity structure of the Tasmanian crust. *Geophysics* **2013**, *78*, WB49–WB62. [CrossRef]
9. Shapiro, N.; Campillo, M. Emergence of broadband Rayleigh waves from correlations of the ambient seismic noise. *Geophys. Res. Lett.* **2004**, *31*, 8–11. [CrossRef]
10. Denolle, M.A.; Miyake, H.; Nakagawa, S.; Hirata, N.; Beroza, G.C. Long-period seismic amplification in the Kanto Basin from the ambient seismic field. *Geophys. Res. Lett.* **2014**, *41*, 2319–2325. [CrossRef]
11. Zheng, D.; Saygin, E.; Cummins, P.; Ge, Z.; Min, Z.; Cipta, A.; Yang, R. Transdimensional Bayesian seismic ambient noise tomography across SE Tibet. *J. Asian Earth Sci.* **2017**, *134*, 86–93. [CrossRef]
12. Park, S.; Ishii, M. Near-surface compressional and shear wave speeds constrained by body-wave polarization analysis. *Geophys. J. Int.* **2018**, *213*, 1559–1571. [CrossRef]
13. Lubis, R.F.; Sakura, Y.; Delinom, R. Groundwater recharge and discharge processes in the Jakarta groundwater basin, Indonesia. *Hydrogeol. J.* **2008**, *16*, 927–938. [CrossRef]
14. Ridwan, M.; Widiyantoro, S.; Irsyam, M.; Yamanaka, H. Development of an engineering bedrock map beneath Jakarta based on microtremor array measurements. *Geol. Soc. London Spec. Publ.* **2016**, *441*, 153–165. [CrossRef]
15. Saygin, E.; Cummins, P.R.; Cipta, A.; Hawkins, R.; Pandhu, R.; Murjaya, J.; Masturyono, J.; Irsyam, M.; Widiyantoro, S.; Kennett, B.L.N. Imaging architecture of the Jakarta Basin, Indonesia with transdimensional inversion of seismic noise. *Geophys. J. Int.* **2016**, *204*, 918–931. [CrossRef]
16. Cipta, A.; Cummins, P.; Dettmer, J.; Saygin, E.; Irsyam, M.; Rudyanto, A.; Murjaya, J. Seismic velocity structure of the Jakarta Basin, Indonesia, using trans-dimensional Bayesian inversion of horizontal-to-vertical spectral ratios. *Geophys. J. Int.* **2018**, *215*, 431–449. [CrossRef]
17. Turkandi, T.; Sidarto, A.D.; Purbohadiwidjojo, M. *Geologic Map of the Jakarta and Kepulauan Seribu Quadrangle Java*; Geological Agency: Bandung, Indonesia, 1992.
18. Van-Bemmelen, R. *The Geology of Indonesia, Vol 1A, General Geology of Indonesia and Adjacent Archipelagoes*; Government Printing House: Hague, The Netherlands, 1949.
19. Anderson, J.G.; Bodin, P.; Brune, J.N.; Prince, J.; Singh, S.K.; Quaas, R.; Onate, M. Strong ground motion from the Michoacan, Mexico, earthquake. *Science* **1986**, *233*, 1043–1049. [CrossRef]
20. Cipta, A.; Cummins, P.; Irsyam, M.; Hidayati, S. Basin Resonance and Seismic Hazard in Jakarta, Indonesia. *Geosciences* **2018**, *8*, 128. [CrossRef]
21. Wiechert, E. Über Erdbebenwellen. Part I: Theoretisches über die. Ausbreitung der Erdbebenwellen, Nachrichten von der Königlichen. Gesellschaft der Wissenschaften zu Göttingen. *Math. Klasse* **1907**, 415–549.
22. Svenningsen, L.; Jacobsen, B.H. AbsoluteS-velocity estimation from receiver functions. *Geophys. J. Int.* **2007**, *170*, 1089–1094. [CrossRef]
23. Jolliffe, I.T. *Principal Component Analysis*, 2nd ed.; Springer: New York, NY, USA, 2002; Volume 98, ISBN 0-387-95442-2.

24. Knapmeyer, M. TTBox: A MatLab Toolbox for the Computation of 1D Teleseismic Travel Times. *Seism. Res. Lett.* **2004**, *75*, 726–733. [CrossRef]
25. Baillard, C.; Crawford, W.C.; Ballu, V.; Hibert, C.; Mangeney, A. An automatic kurtosis-based P-and S-phase picker designed for local seismic networks. *Bull. Seismol. Soc. Am.* **2013**, *104*, 394–409. [CrossRef]
26. Kennett, B.L.N.; Engdahl, E.R.; Buland, R. Constraints on seismic velocities in the Earth from traveltimes. *Geophys. J. Int.* **1995**, *122*, 108–124. [CrossRef]
27. Wang, R. A simple orthonormalization method for stable and efficient computation of Green's functions. *Bull. Seismol. Soc. Am.* **1999**, *89*, 733–741.

Article

Seismic Vulnerability and Old Towns. A Cost-Based Programming Model

Salvatore Giuffrida [1,*], Maria Rosa Trovato [1,*], Chiara Circo [1,*], Vittoria Ventura [2], Margherita Giuffrè [3] and Valentina Macca [2]

1. Department of Civil Engineering and Architecture, University of Catania, 54, 95125 Catania CT, Italy
2. University of Catania, 54, 95125 Catania CT, Italy; vittoriaventura01@gmail.com (V.V.); valentina.macca26@gmail.com (V.M.)
3. Institute of Environmental Geology and Geoengineering (IGAG) of the National Research Council (NRC), 00185 Rome, Italy; mar.giuffre@gmail.com

* Correspondence: sgiuffrida@dica.unict.it (S.G.); mrtrovato@dica.unict.it (M.R.T.); chiaracirco@virgilio.it (C.C.)

Received: 26 August 2019; Accepted: 7 September 2019; Published: 2 October 2019

Abstract: Vulnerability is a big issue for small inland urban centres, which are exposed to the risk of depopulation. In the climate of the centre-northern part of Italy, and in the context of the recent concentration of a high number of earthquakes in that area, seismic vulnerability can become the determinant cause of the final abandonment of a small town. In some Italian regions, as well as in Emilia Romagna, municipalities are implementing seismic vulnerability reduction policies based on the Emergency Limit Condition, which has become a basic point of reference for ordinary land planning. This study proposes an approach to seismic vulnerability reduction based on valuation planning for implementation within the general planning framework of the Faentina Union, a group of five small towns located in the southwestern part of the Province of Ravenna, Italy. This approach consists of three main stages: knowledge—the typological, constructive, and technological descriptions of the buildings, specifically concerning their degree of vulnerability; interpretation—analysis with the aim of outlining a range of hypotheses with respect to damage in case of a prospective earthquake; and planning—the identification of the courses of action intended to meaningfully reduce the vulnerability of buildings. This stage includes a cost modelling tool aimed at defining the trade-off between the extension and the intensity of the vulnerability reduction works, given the budget.

Keywords: urban fabrics; seismic vulnerability; critic analysis; cost modelling; urban preservation programming; building works programming

---

1. Introduction

The proposal contained within this contribution addresses the issue of reducing the seismic vulnerability of historic urban buildings with reference to the case of the city of Brisighella, Italy.

This research was conducted within the context of an agreement stipulated by the Union of Municipalities of the Romagna Faentina and the Department of Civil Engineering and Architecture of the University of Catania, Italy, the objective of which is a joint study on the seismic vulnerability of building aggregates in the historic centres of Brisighella, Casola Valsenio, Castel Bolognese, Riolo Terme, and Solarolo.

The research was multidisciplinary work that involved the disciplines of restoration, urban planning, and economic valuation assessment.

Economic-estimative evaluations deal with the issue of the seismic vulnerability of historical cities [1] in terms of a fundamental interest in the economic category of territorial capital and in its two forms: urban capital [2] and human capital. This scientific and methodological interest corresponds to the original civil commitment of economic-estimative evaluation with respect to distributive justice.

In the case of the redevelopment of minor historical centres, references to the notions of capital and distributive justice aim to answer one preliminary question: Is it worth it?

The answer to this dilemma involves questions concerning the way in which urban policies combine the need for capital in its urban form and the needs of capital in its economic essence.

With reference to these two needs, capital becomes an important interpretative filter in terms of the vulnerability of small urban centres [3]; the consistency of urban capital in terms of both volume and value is the main reason for the resilience of such centres. Marginal historic centres, especially those located in the mountain hinterland and characterised by poor accessibility, are condemned to the vulnerability trap. Seismic vulnerability, specifically, is not added to other vulnerabilities (social, economic, etc.), but rather constitutes a determining cause.

In the field of small urban centre vulnerability, the sciences of restoration and evaluation have common interests with respect to certain characteristics of architectural and urban heritage:

- The ability to endure, which is denoted as the resilience of the social-urban system; it is necessary to preserve its ability to continue to perform its essential functions or to survive during and after catastrophic events [4];
- Self-referentiality, or iconicity [5,6], which is ascribable to the material aspects (masonry and construction systems) that must be conserved in their own right, rather than on the basis of their functions or their appearance. Self-referentiality does not concern the present value of an artefact but, rather, the maintenance of its testimonial value, starting from its primary essence. Self-reference is a typical feature of the capital category under conditions of high tension in prices.

The resistance of a social-urban system to catastrophic events—that is, its resilience—depends on how much surplus social product have been set aside to maintain its security, that is, on the production or "renovation" fund designated for its structural consolidation.

A variety of different seismic vulnerability and risk analysis methods for unreinforced masonry structures indicate that having a high degree of vulnerability to earthquakes can also have remarkable socio-economic implications. Several studies have proposed an innovative holistic approach based on indicators related to physical exposure, social fragility, and the lack of resilience of urban areas; this could also involve geomatic tools (GIS), which would be able to describe the seismic risk results on the basis of scenarios of expected losses, but also on the basis of the probabilities of occurrence of predefined damage states [7,8].

Concerning the physical vulnerability of the built environment, many studies carried out on historic centres have presented relevant methodologies for assessing the vulnerability of the urban fabrics, both on an architectural scale and on an urban scale. In addition to the present study, some study cases have measured vulnerability by combining empirical and mechanical methodologies in order to structurally and typologically identify the buildings. In addition, an index-based method for masonry building aggregates has been applied [9]. The obtained vulnerability characteristics and the corresponding assessments provide relevant and consistent information in the form of typological capacity and fragility curves, which can be applied to urban areas presenting similar building classes [10].

This study exposes the methodology and the findings of the vulnerability assessment of the building heritage in the old town of Brisighella, referring both to the building aggregates and to the architectural units; starting from these results, it proposes an integrated model of analysis [11], evaluation and project initiation [12–16], aimed at outlining a variety of strategies for programming interventions to reduce vulnerability and, therefore, to optimize the urban policy choices.

2. Materials

2.1. Mitigation of Urban Vulnerability

The issue of urban vulnerability is studied in the specific context of the Emilia-Romagna Region with a wide range of approaches and tools that were only recently introduced to the regional planning aimed at reducing the urban seismic risk. The vulnerability is analysed according to two scales: that of the entire urban centre and that of the historic centre.

The aim is to ensure that, during a seismic event, an urban centre can persist, regarding both the efficiency of the main strategic activities for the recovery and the identity characteristics that distinguish it.

For the five municipalities of the "Unione Faentina", two studies were carried out: an evaluation of the seismic vulnerability of the historic centres agreement between the Union of Municipalities of Romagna Faentina and the University of Catania (approved with the official act N.132/2016 – scientific directors DICAR: Caterina F. Carocci, Salvatore Giuffrida; research team: Chiara Circo, Margherita Giuffrè, Luciano A. Scuderi, and Vittoria Ventura) and specific urban studies to define the Emergency Limit Condition (ELC), an urban scale analysis aimed at managing the behaviour of the settlement in the post-earthquake phase, carried out by the Technical Office of the Union of Municipalities [17].

The overlaying of the studies allowed an integrated project of intervention to be developed through which the economic evaluation illustrated in this paper was tested.

The study on the five historic centres of the "Unione Faentina" (Brisighella, Casola Valsenio, Castel Bolognese, Riolo Terme, Solarolo) was carried out in two phases.

In the first phase, homogeneous areas of the entire municipal territory with regard to the seismic vulnerability were identified, following a method established by the Department of Civil Protection and tested on Faenza and Solarolo in 2011 [18]. The second phase included the qualitative assessment of the seismic vulnerability of historical centres—identified as the most vulnerable areas of the urban fabric—following a procedure already tested on the historic centre of Faenza between 2011 and 2013. The aim was to define the criteria for the mitigation of the seismic vulnerability of the building aggregate with regard to the specific characteristics of each building fabric [19].

At the same time, the Technical Office carried out an analysis of the Emergency Limit Condition (ELC), introduced by the Italian Government Ordinance (OPCM n. 4007/2012), in accordance with Law n. 77/2009 art.11 "National plan for the prevention of seismic risk" for each of the five municipalities. The ELC is a municipal scale analysis set up based on the Civil Protection Plans and aimed at guaranteeing the functioning of the emergency management system in the post-earthquake phase. By definition, the ELC represents that limit condition for which, after the seismic event, the urban settlement loses all its functions (including residence) and preserves only the operation of most of the strategic functions for emergency management, their accessibility and connection with the territorial context. The ELC is, in fact, made up of strategic buildings, emergency areas, and the main links between the elements and the territorial context, as well as their interactions with the interfering elements [20].

The analysed emergency management systems, like the settlements they belong to, have a rather simple configuration. The connection infrastructures (roads) run across the urban centre, reaching strategic buildings inside the historic centre, or to its border. The small size of the centres, and the choice to place the strategic buildings in newly built areas outside the historic centres, imply a low presence of interfering structural aggregates.

2.2. The Historic Centre of Brisighella

The settlement of Brisighella rises from the slopes of the Tuscan-Romagna Apennines in the lower valley of the Lamone river. The first settlement, dating back to the end of the XIII century, consisted of a fortified nucleus, the current Rocca. In the 14th century, the fortification works were extended to the settlement of the "Borgo", creating an elevated arcaded path integrated into the houses for defensive purposes, (now Via degli Asini). During the 1400s, the nucleus expanded towards the valley, creating a new fortification wall beyond which, starting from 1500, the city developed.

The historical evolutionary process of an urban centre and the orographic peculiarities of its territory have greatly influenced the definition of the urban form, characterised by aggregates of townhouses built against the slope. The residential buildings have incorporated the ancient walls defining the unique configuration of the urban fabric.

3. Methods

3.1. The Analysis of the Seismic Vulnerability of the Historic Centre

The methodology used for the analysis of the vulnerability of the historic centres is based on the direct knowledge of the building aggregates and has been used often in contexts damaged by earthquakes [21] as well as under ordinary conditions [22]. The activities to be carried out are organized in three strictly connected phases: knowledge, interpretation, and project initiation.

The knowledge phase includes preliminary bibliographical research of the studies already carried out by the of Municipal Technical Office (MTO), which outlined the main evolutionary phases of the historic centre, and an on-site survey aimed at detecting all of the (constructional, typological, evolutionary) factors which may significantly affect, positively or negatively, the seismic behaviour of the urban fabric in its current configuration [23,24]. The elements that positively influence the seismic response (called resistance factors or strengths), such as the presence of anti-seismic devices, the good quality of the constructional technique, etc., and those that play negative roles (called vulnerability factors) are identified, with particular reference to the development of important damage mechanisms, such as the overturning of the façade. It should be noted that all of the information is collected within a direct survey of the urban fabric through observations from the outside of the building façades and the accessible courtyards.

Moreover, the survey highlights the specificities of the aggregates in terms of their use and construction technique, distinguishing from masonry residential buildings those constructed with reinforced concrete or another construction technique, and buildings with specialized functions, such as churches and historical palaces. The aim is to identify possible points of constructive discontinuity and the relationships of contiguity between buildings with different geometrical-structural characteristics.

This type of analysis conducted in the whole historic centre allows a map of the recurrent vulnerability and strength factors of the urban fabric to be obtained, constituting indispensable background knowledge for the definition of intervention criteria aimed at reducing vulnerabilities.

In the interpretative phase, the data collected on vulnerability and strength factors are critically selected with the aim of formulating a judgment on the mechanical quality of the urban fabric and therefore prefiguring the expected damage related to the precariousness observed.

In the project phase, the intervention criteria for the mitigation of vulnerabilities are established and, in the case of Brisighella, the economic evaluation of interventions with the aim of managing the public financial resources was carried out.

3.2. Vulnerability Reduction Assessment and Programming Model

The coherence between observations, assessments, and decisions in the planning of interventions for the reduction of vulnerability consists of the correspondence between (even quantitative) heterogeneous aspects and is therefore difficult to compare. On the one hand, there is public expenditure, which improves the resilience of the urban centre; on the other, there are benefits in terms of an increase in the value (private and public) of buildings and the urban fabric, as supposed by the program.

It is possible to distinguish between direct benefits, such as the seismic improvement of buildings and the increase in the overall resilience of the entire city and externalities [25–27], such as the increase in real estate market value, the perception of a greater sense of individual security, and so on.

From an economic point of view, the coherence between the value of investments and the value of security concerns two components of the calculation of seismic risk: hazard and exposure.

The measure of the hazard is affected by the randomness of the seismic event and by how this possibility is perceived by the public. This measure depends on two elements: the first, probabilistic, refers to the natural and geotechnical sphere (the probability of the seismic event); the second, instead, has a psycho-sociological and political nature. The political component of the dangerousness measure concerns the way in which the political-administrative system incorporates the geological evidence with the Urban Plan [28,29]. In this case, the Urban-Building Regulation of the Union of Municipalities of Romagna Faentina takes the ELC into account as a reference for ordinary planning.

The extent of the exposure varies according to the different "qualities of value" associated with the vulnerable buildings and their monetary measure, which allows planners and decision makers to compare the value of the security improvement with the value of the investment. This difficulty calls into question the effectiveness and completeness of the monetary measure, suggesting additional and alternative measures.

The economic-monetary measure of vulnerability can be carried out indirectly, starting from the critical observations and evaluations of the Restoration [30,31]. In such an inter-disciplinary context, and with reference to the urban centre as a whole, the works necessary to improve its resilience are selected, not only to guarantee the perfect integrity of the individual buildings, but also to prevent the urban centre from interrupting its basic functions. The convergence of two disciplines, restoration and urban planning, changes the objectives of both so that they become more specific and less ambitious.

The economic evaluation expands the possibilities of the plan: once the cost of the works has been calculated (see Section 4.3), it is possible to provide a coherent multiplicity of budget allocation hypotheses.

In this case, given some conflicting variables, such as the completeness of the interventions and the extension of the area involved, it is possible to define the trade-off relationship to maximize the cost-effectiveness function.

This greater integration between observations, evaluation, and decision-making [32], and the expansion of the contents taken into consideration, consolidates the consensus and the success of the project [33].

In this sense, as we will propose in a further study, the public consensus is on the awareness "of the value of the overall life safety of the building's occupants endangered, the direct economic losses due to the cost of the building restoration, in addition to the anticipated downtime and other indirect sources of loss associated with the recovery of the building to its full functionality" [34].

The evaluation model, as mentioned, consists of a database that carries out all of the calculations and logical operations necessary to transform the dataset into information based on which the assessments select the best project options. This integration of cognitive functions allows scenario analyses at the building and urban scale to be carried out.

The model coordinates the structural, material, geometric, technological, typological, and maintenance characteristics of the units of study [35], Relevant for (a) characterising their static attributes (seismic vulnerability); (b) hypothesising the design aspects by selecting the interventions corresponding to each degree of vulnerability; (c) calculating the costs based on the definition of typical bundles of works for each of the façade units facing the public areas [36]; (d) adjusting the intensity and extent of the interventions; (e) mapping the interventions corresponding to each combination of intensity and extension of the interventions [37,38]; and (f) calculating the total cost for each hypothesis by defining the trade-off functions [39] between the intensity and extension of the interventions.

The proposed model aims to integrate the way in which the ELC is formed and proposes its possible extension, according to the advantages resulting for the urban centre as a whole.

3.2.1. Calculation of Vulnerability

The calculation of vulnerability consists of a measure of the risk that the façades of the buildings interfering with the evacuation and rescue routes may overturn and collapse, obstructing them and/or affecting the safety of the fleeing people and rescuers. This measure is particularly important for buildings included in the ELC that must withstand the earthquake without collapsing to guarantee the functionality of the paths that connect the strategic nodes.

Vulnerability is calculated for each individual Façade Unit (FU) facing the public spaces. An FU is the vertical portion of masonry of an external façade located between two orthogonal structural walls, whose behaviour is assumed to be independent of the others.

For each of them, by applying the dynamic structural analysis model developed by C. Tocci [40], a "numerical indicator of the ground acceleration level capable of triggering elementary overturning kinematic mechanisms" was calculated. This indicator is defined, consistently with the conceptual layout of the Technical Standards for Constructions (NTC 2008), as "the triggering multiplier of the motion due to overturning (α_0) of the wall", taking into account: (i) the presence and extent of tapers, (ii) the direction of the floor main beams (parallel or perpendicular to the façade wall), (iii) the presence of tie-rods, (iv) the effectiveness of bonding between the façade and the (orthogonal) shear walls" [41]. The vulnerability index depends on the following significant geometric and typological parameters:

S_1: The thickness of the ground floor wall;
H: The total height of the wall;
L: The distance between the shear walls;
N: The total number of floors;
p: The number of floors without tie-rods (counted from the top);
k: The direction of the main floor beams: (k is 1 or 3 if parallel or perpendicular to the façade); and
r: The bonding of the façade with the shear walls (r is 0 in cases where no bonding occurs at all).

The overturning load multiplier is expressed by different equations that we define as the basic configuration and the varied configuration, the former being characterised by the simultaneous occurrence of two circumstances: (i) the absence of tie rods $(N = p)$ and (ii) floor beams parallel to the façade $(k = 1)$, and the latter being characterised by the lack of one or both of the above circumstances: $(N > p)$ and/or $(k = 3)$. For both configurations, the contribution of façade bonding introduces an additional term.

The equations are the following:

$$\alpha_0 \approx (1 + r) \cdot \frac{S_1}{H}.$$

For the basic configuration, $(N = p)$ and $(k = 1)$.

$$\alpha_0 \approx (1 + r) \cdot 0.3 \cdot \left(\frac{S_1}{H}\right)^{\left(1 - \frac{n}{100}\right)}.$$

For the varied configuration, $(N > p)$ and/or $(k = 3)$, where r and n are defined as

$$r = 0.001 \cdot (9 - L) \cdot \frac{(p + 1)^2}{k}$$

$$\begin{cases} n = 72 & if\, N = p \\ n = 83 - 21p + 13 \cdot (p + 1) \cdot \frac{(k-1)}{2} & if\, N > p \end{cases}.$$

Taking into account that the r factor (which quantifies the influence of bonding) applies only for $L < 9$ m, any form of bonding between the façade and the shear walls beyond this is substantially ineffective, and it can be assumed that $r = 0$.

The above equations show that the overturning load multiplier α_0 for each external wall is strongly dependent on the ratio $\alpha = \frac{S_1}{H}$ between the ground floor thickness of the wall and its total height, parameters that are both easily obtainable just by the external survey of the street façades.

The acceleration coefficient is assumed to act as an index of vulnerability [42] of each UF, to which the works directly and indirectly connected are associated: the former are aimed at avoiding their overturn; the latter are joint works, such as those for securing elements soaring above the roofs, e.g., the chimneys and the external and internal finishing works. This distinction is important for decisions [43,44] regarding the level of completeness of the program of interventions, as discussed below.

In the rows, the database contains in the rows all the FU, u_i, and in the columns all the characteristics necessary to calculate the acceleration that the ground must give to the building so that the façade crumbles down. A low acceleration coefficient indicates a high vulnerability, and vice versa [45].

As highlighted by relevant studies, the methodology used to characterise the structural vulnerability is well advanced nowadays and overcomes the acceleration coefficient [46–48].

3.2.2. Cost Calculation

Once the acceleration coefficient has been calculated and therefore the degree of vulnerability of each UF has been determined, logical and research functions associate it with safety measures, starting from the most common, up to the most consistent or invasive ones, such as inserting chains, filling of superficial lesions, integration of masonry damaged by passing lesions, introduction of reinforced masonry, securing of jutting and towing elements, and external and internal finishing works related to both walls and ceilings, articulated in a total of 36 items on a price list.

The elementary costs associated with the interventions on each façade unit are then aggregated to calculate the total cost of each hypothesis related to the ELC.

Each hypothesis related to the ELC is defined by varying the intensity of the interventions and/or their extension. The intensity depends on the degree of completeness of the interventions with the same extension; the extension is the number of FUs involved with the same degree of safety. The result is two cost functions, one intensive, $C(j)$, the other extensive $C(k)$ [49].

The intensive cost function relates the total cost of each FU and the type of intervention, which depends on the bundle of works b_{jk} associated with a single u_i. Each bundle includes works corresponding to the entries of the Emilia Romagna Region Official Bill of Quantities for the public works, $b_{jk} \in B$, where B is the set of all works referable to the activities done to reduce the vulnerability of the buildings included in the ELC hypothesis.

The b_{jk} package can contain more or less works, according to their different relevance levels. In fact, we can distinguish between those that are strictly necessary, j_n; those of primary public interest, j_p; the less invasive ones, j_v; and those that are more or less adequate, j_a.

By combining the five degrees of completeness j with the five safety degrees k, 25 different hypothetical strategies with increasing costs have been defined [50].

The extensive cost function relates to the number of FUs included in the ELC according to their vulnerability level, measured by the acceleration coefficient defined above. The FUs are grouped according to five thresholds, $k_{60\%}, k_{70\%}, \ldots, k_{100\%}$, delimitating five corresponding sub-ranges of the acceleration coefficient associated with each FU: $k_{60\%}$ defines the sub-range of the façades whose acceleration coefficient is lower than the minimum one (α_{min}) which corresponds to the highest vulnerability degree and the minimum number of FUs included; vice versa, $k_{100\%}$ corresponds to the largest number of façades. The intermediate degrees are defined by progressively adding up to α_{min} a quarter of the range $\alpha_{max} - \alpha_{min}$. Then, $k_{70\%} = \alpha_{min} + (\alpha_{max} - \alpha_{min}) \cdot 0.25$; $k_{80\%} = \alpha_{min} + (\alpha_{max} - \alpha_{min}) \cdot 0.50$; $k_{90\%} = \alpha_{min} + (\alpha_{max} - \alpha_{min}) \cdot 0.75$.

It should be noted that the evaluation of the degree of safety of the interventions proposed here is part of an "ex-ante" programming model. Checks on the effectiveness of the planned operations will be carried out in the future (ex-post) during the implementation of the program according to Ministerial Decree 58/2017 [51].

4. Application and Results

4.1. Map of Vulnerabilities and Strengths, and the ELC of Brisighella

The elements identified in the context of the field survey are represented by icons in the "Map of vulnerabilities and strengths of the urban fabric". Depending on the method used to acquire knowledge (observations from the outside), the vulnerability discussed here is related to the possible overturning mechanisms of the walls facing the street.

From the analysis emerged a general good state of conservation for the historic centre of Brisighella, which does not present cases of buildings with different masonry structures, because it was not subject to post-war reconstruction.

The map of Brisighella shows the recurring vulnerabilities and strengths in all five centres, such as the presence of tall buildings, which are more vulnerable to the overturning mechanism due to the number of uncontained outer walls, and volumes jutting out from the external fronts, which are more frequent in the rear façades of the buildings. Furthermore, the map illustrates some specific vulnerability factors, such as the constructional irregularities of some parts of the urban fabric due to the integration in the ancient city walls. The contiguity between very different geometrical and structural configurations (in terms of storey height, wall thickness, etc.) can constitute a weakness from a seismic perspective, and this should be taken into consideration in the context of a possible intervention.

The strength factors that characterise the urban fabric of Brisighella concern the widespread use of historic anti-seismic devices (e.g., metallic tie-rods and buttresses) and a good construction technique—as concerns its visibility from the outside.

The ELC of Brisighella, unlike the other municipalities analysed, is included in the historical urban fabric, since the main connection infrastructures cross the historical centre in a rather extensive way. For this reason, a high number of interfering structural aggregates was noted, which are those that, following an earthquake, could collapse on the escape routes identified in the ELC.

4.2. Intervention Criteria for Vulnerability Reduction

The interpretation of information related to vulnerabilities and strengths allows the prefiguration of the seismic damage mechanisms that can affect the analysed urban fabric (Figure 1).

The recurring issues to be faced in seismic risk reduction are clarified and the essential features of a mitigation strategy that is respectful of the constructional and urban peculiarities of the historic centres are specified. The intervention criteria are not expressed by technical details but rather by the objectives to which the intervention must aim, which allows a design freedom with only one indispensable restriction: respect for the constructive logic of the masonry technique as a guarantee of effectiveness and compatibility of the intervention with the historical building.

With reference to the vulnerabilities observed in the historic centre of Brisighella, the improvement of the seismic response is pursued by means of interventions aimed at control of the thrusts in the roofs and reduction of the thrusts of vaulted elements as well as improvement of the connections between walls and floors with particular regard to the containment of the façade walls.

These indications are valid for the entire wall structure of the historic centre of Brisighella and allow a general framework of actions to be implemented in order to define the preventive mitigation of the seismic vulnerability [52].

It follows that the overlapping of the levels of knowledge in terms of vulnerability, strengths and intervention criteria with the previsions of the ELC of Brisighella can help to identify the strategic interventions to be promoted by the public authorities within the historic centre—favouring coordinated management of the financial resources—and to define reward mechanisms to promote the implementation of private interventions.

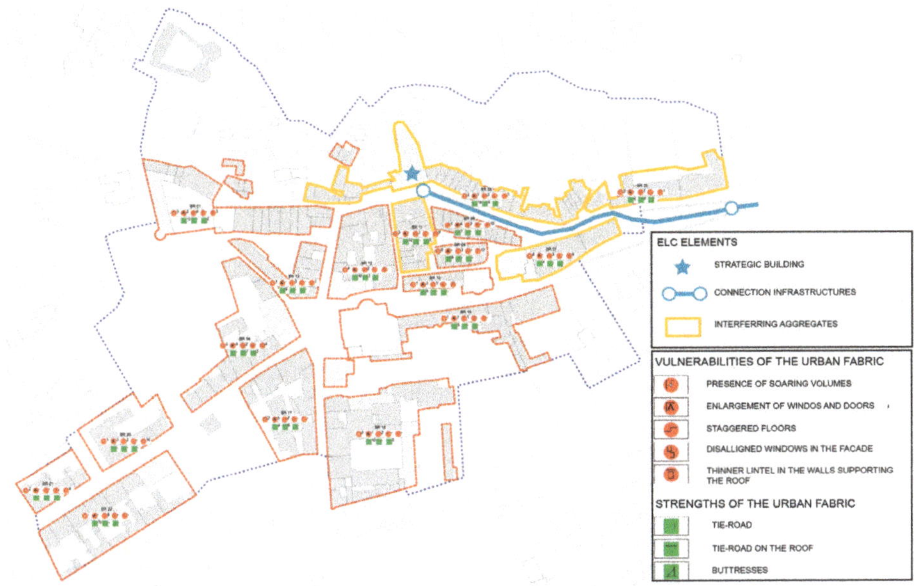

Figure 1. The Emergency Limit Condition of the historic Centre of Brisighella.

4.3. Valuation and Programming of the Interventions for Implementing the ELC

The information base used to calculate and select the interventions includes identification of the UF in the building complex to which it belongs (Block and Architectural Unit); land registry identification; type; façade units per aggregate; number of façades for each room; horizontal extension of the façade; wall thickness; room depth; surface of the room; number of floors above ground; gross surface area of the façade; heights of the different floors; average height of each floor; construction system; wall type; orientation of the structure of the floor with respect to the direction of the façade; soaring elements; braces, hypothesised to be required if the width of the front façade is greater than 6.50 m and the number of floors is greater than 1; the presence of chains; and the presence of injuries. In this study, the data were obtained from the documentation provided by the Union's Technical Office (Municipality of Brisighella) and by means of the quick inspections carried out on site.

The tendency to overturn of the façade was calculated according to (a) a pessimistic prudential scenario referred to as the basic configuration, quantified by the coefficient α_{0b}, and (b) an optimistic scenario referred to as the configuration changed and quantified by the coefficient α_{0v} (Table 1, Figure 2).

Depending on the degree of vulnerability of each of the 749 façade units analysed (only 685 need to be secured), the model identifies the interventions necessary for securing them. It should be noted that the interventions are not activated automatically and unambiguously, but based on the type of strategy that the decision maker chooses.

Table 2 lists the securing works according to the proposed vulnerability of the façade and to the Architectural Unit characteristics.

Table 1. Portion of the database displaying the calculation of the vulnerability indexes.

Calculation of the Acceleration Coefficient												
AU	FU	*S1*	*H*	*L*	*N*	*p*	*k*	*r*	*n*	Base (0) Varied (1)	Alpha0 b	Alpha0 v
1	1	0.4	6.3	8.1	2	2	1	0.0810	72	0	0.068	0.068
1	2	0.4	6.3	6.2	2	2	3	0.0828	72	1	0.068	0.150
1	3	0.4	3.2	6.3	1	1	1	0.1088	72	0	0.141	0.141
1	4	0.4	3.2	7.2	1	1	3	0.0244	72	1	0.130	0.172
1	5	0.5	3.2	6.2	1	1	3	0.0373	72	1	0.164	0.186
1	6	0.4	3.2	4.5	1	1	3	0.0595	72	1	0.134	0.178
1	7	0.4	6.3	4.9	2	2	3	0.1227	72	1	0.071	0.155
2	1	0.4	6.5	4.4	2	2	3	0.1377	72	1	0.070	0.156
2	2	0.4	6.5	3.0	2	2	3	0.1806	72	1	0.073	0.162
2	3	0.4	6.5	6.5	2	1	3	0.0339	88	1	0.064	0.222
2	4	0.4	6.2	5.7	2	2	3	0.1005	72	1	0.071	0.153
2	5	0.6	6.5	4.4	2	2	3	0.1371	72	1	0.106	0.175
2	6	0.5	9.1	8.2	2	2	3	0.0234	72	1	0.056	0.136
2	7	0.4	9.3	7.3	3	3	1	0.2672	72	0	0.055	0.055
2	8	0.4	10.1	5.6	3	3	3	0.1813	72	1	0.047	0.144
2	9	0.4	10.1	7.0	3	3	1	0.3200	72	0	0.052	0.052
2	10	0.4	9.2	6.3	3	3	3	0.1445	72	1	0.050	0.143
2	11	0.6	9.2	6.5	3	3	3	0.1333	72	1	0.074	0.159
2	12	0.6	9.2	5.2	3	3	3	0.2027	72	1	0.079	0.168
2	13	0.6	9.2	7.0	3	3	1	0.3232	72	0	0.087	0.087
2	14	0.6	12.8	7.0	4	4	1	0.5000	72	0	0.070	0.070
2	15	0.4	12.8	5.8	4	4	3	0.2683	72	1	0.040	0.144
2	16	0.6	12.9	7.8	4	4	1	0.2925	72	0	0.060	0.060
2	17	0.6	13.1	5.9	4	1	3	0.0420	88	1	0.048	0.216
2	18	0.4	13.1	8.9	4	1	1	0.0032	62	1	0.031	0.080
2	19	0.4	12.7	5.0	4	3	3	0.2160	72	1	0.038	0.139
2	20	0.4	12.8	7.8	4	3	1	0.1920	20	1	0.037	0.022
2	21	0.4	10.1	7.0	3	3	3	0.1077	72	1	0.044	0.135
2	22	0.6	8.9	6.2	3	3	3	0.1504	72	1	0.078	0.162
2	23	0.4	12.2	6.3	4	3	3	0.1445	72	1	0.038	0.132
2	24	0.4	13.9	6.5	4	3	1	0.4032	20	1	0.040	0.025
2	25	0.4	13.9	3.2	4	4	3	0.4850	72	1	0.043	0.165
2	26	0.4	13.9	3.9	4	4	3	0.4217	72	1	0.041	0.158

Figure 2. Map of the vulnerability of the Façade Units of the old town of Brisighella. The table classifies the Façade Units by vulnerability degree.

Table 2. List of the proposed vulnerability of reduction works based on the vulnerability degree and the Architectural Unit characteristics.

Emilia Romagna Region Official Bill of Quantities for the Public Works			
Item Cod.	**Short Description**	**Unit of Measure**	**Unit Price**
F01100a	scaffolding assembly	sq. m.	€9.06
F01100b	higher freight scaffolding	sq. m.	€1.33
F01100c	scaffolding disassembly	sq. m.	€3.09
B02018b	tie rods: masonry perforations	m.	€36.59
B02021	tie rods: plates niches	sq. m.	€484.04
B02022	tie rods: plates	kg	€6.26
B02024	tie rods: implementation	kg	€9.24
B02025	tie rods: stake	kg	€7.79
B02026	tie rods: re-stringing	cad	€133.53
B02028a	tie rods: injection pressure drilling	m.	€18.74
A09002b	tie rods: countertop	sq. m.	€25.13
A20001	tie rods: countertop: painting preparation	sq. m.	€1.82
A20012c	tie rods: countertop: painting	sq. m.	€8.67
A20001	tie rods: wall: painting preparation	sq. m.	€1.82
A20002	tie rods: wall: grouting	sq. m.	€1.24
A20012c	tie rods: wall: painting	sq. m.	€8.67
B02002b	shear walls: reinforcement of existing masonries	sq. m.	€170.15
A05004a	shear walls: new masonry building	c. m.	€323.49
A20001	shear walls: painting: wall preparation	sq. m.	€1.82
A20002	shear walls: wall grouting painting	sq. m.	€1.24
A20012c	shear walls: painting: wall preparation	sq. m.	€8.67
B02006a	deep crack masonry integration	c. m.	€576.10
A08005d	deep crack masonry integration: external plaster	sq. m.	€24.23
A20007	deep crack masonry integration: external painting: wall preparation	sq. m.	€10.69
A20015b	deep crack masonry integration: external painting	sq. m.	€14.20
A08004d	deep crack masonry integration: internal plaster	sq. m.	€23.64
A20001	deep crack masonry integration: internal painting; wall preparation	sq. m.	€1.82
A20002	deep crack masonry integration: internal painting; wall grouting	sq. m.	€1.24
A20012c	deep crack masonry integration: internal painting	sq. m.	€8.67
B02009	injections	c. m.	€150.09
A08005d	injections: external plasters	sq. m.	€24.23
A20007	injections: external painting; wall preparation	sq. m.	€10.69
A20015b	injections: external painting	sq. m.	€14.20
A20001	injections: internal painting: wall preparation	sq. m.	€1.82
A20002	injections: internal painting: grouting	sq. m.	€1.24
A20012c	injections: internal painting	sq. m.	€8.67
market survey	chimney securing	each	€600

Figure 3 summarises the model of selection of the works according to the strategy and indicates the costs of the 25 strategies according to the degree of completeness and security; the graphs below show the trade-off functions between the degree of security (the extension of the ELC) and the completeness of the interventions on each FU and the related AU, for each amount of the total cost.

The completeness of the interventions is described in Figure 3a:

- Completeness degree 1 includes only the works that can be considered necessary, of public interest, of minimum security, and non-invasive;
- Completeness degree 2 includes the works that are necessary, of public interest, of minimum security, of maximum security for 70% out the total amount of the FUs, and non-invasive;
- Completeness degree 3 includes the works that are necessary, of public interest and of private interest for 50% out the total amount of the FUs, of minimum and maximum security, non-invasive and invasive;
- Completeness degree 4 includes the works that are necessary and unnecessary for 30% out the total amount of the FUs, of public and private interest, of minimum and maximum security, non-invasive and invasive;
- Completeness degree 5 includes all of the works.

Kind of works	Completeness degree 1	2	3	4	5
necessary	1	1	1	1	1
unnecessary				0.3	1
public	1	1	1	1	1
private			0.5	1	1
min security	1	1	1	1	1
max security		0.7	1	1	1
not invasive	1	1	1	1	1
invasive		1	1	1	1

(a)

Number of fixed Façade Units	Completeness degree 1	2	3	4	5
196	0.71	1.11	1.29	1.65	2.44
232	0.83	1.30	1.52	1.95	2.89
374	1.39	2.04	2.33	3.03	4.57
577	2.10	2.96	3.33	4.34	6.60
685	2.46	3.42	3.84	4.99	7.57

(b)

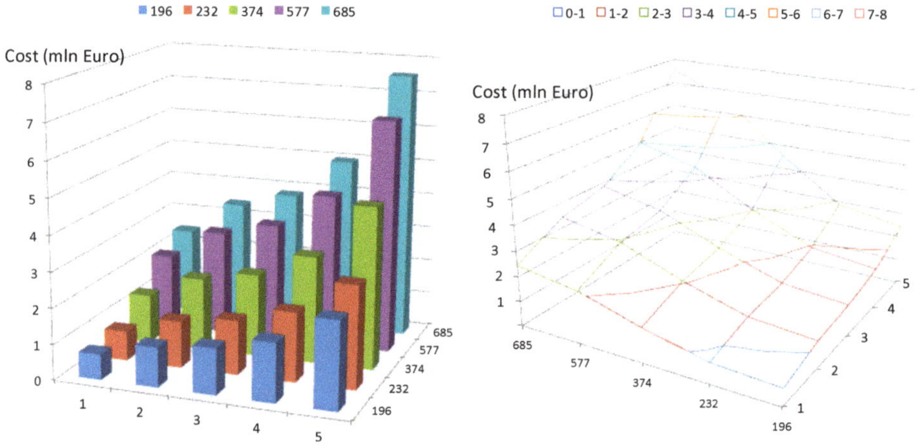

(c)

(d)

Figure 3. (**a**) Combination of security and completeness; (**b**) table of total cost for each strategy given by the combination of the two above mentioned performances; (**c**) graph of the total costs of each of the 25 strategies displayed in table (**b**); (**d**) 3D isocost functions displaying the trade-off between completeness and the number of façades involved for each level of cost.

A further function of the model is mapping the 25 different strategies of vulnerability reduction providing information on the FU for which the intervention is necessary (given by the position of the bubbles in the map) and a graphic representation of the cost (given by the dimension of the bubbles), as sampled in Figure 4 displaying four of the 25 strategies. In this figure, the position of the strategy on the isocost graph is shown. The sequence represents four strategies with increasing costs due to the simultaneous improvement in completeness and security.

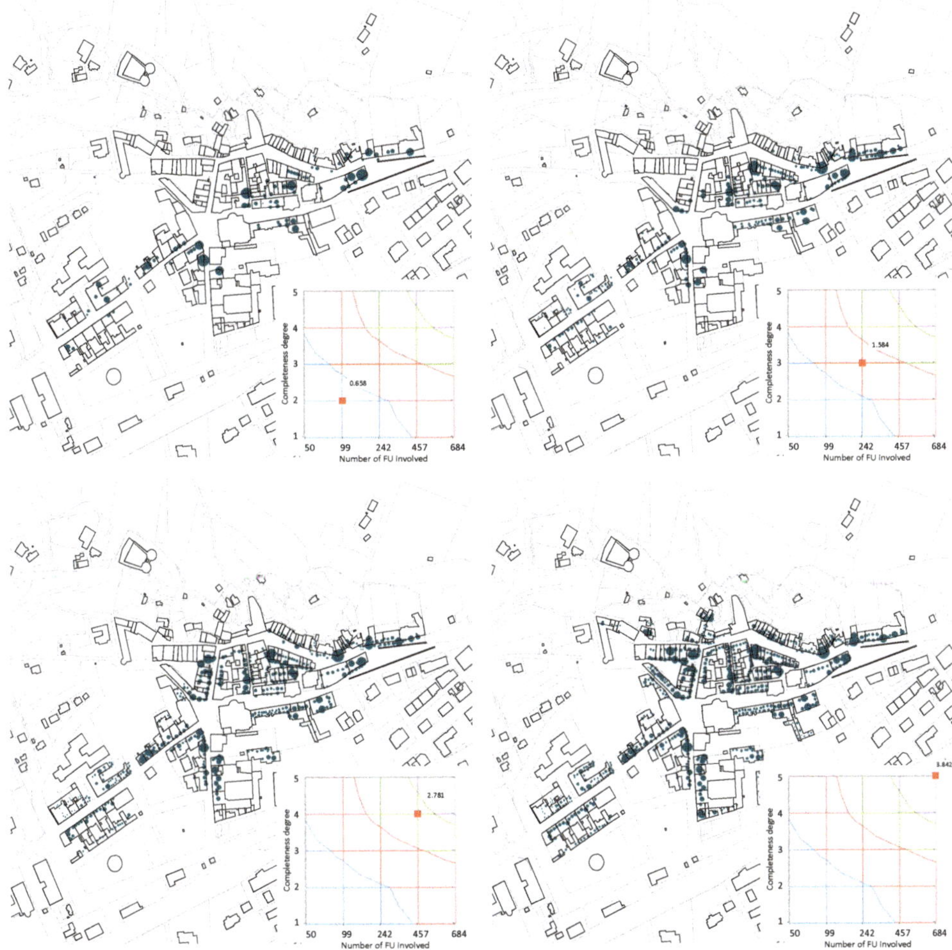

Figure 4. Mapping of the different layouts of the strategies having an increasing total cost.

In Figure 5, instead, the strategies displayed are those having approximately the same cost so that they differ regarding the completeness and security degree.

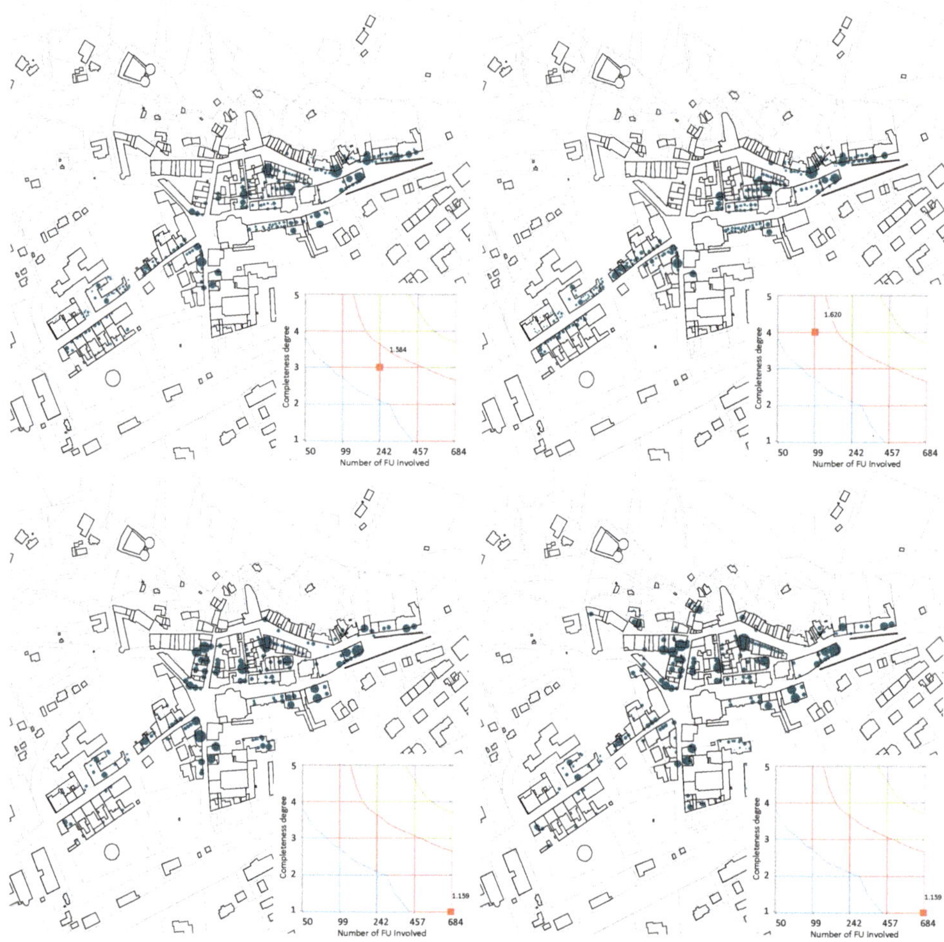

Figure 5. Mapping of the different layouts of the strategies with (approximately) the same total cost.

5. Discussion and Conclusions

The proposed model outlined a wide range of possible options concerning how to combine the overall degree of security corresponding to the number of buildings secured (from 196 to 685 FU out of 749) and the budget to cover the total costs (from 0.71 to €7.57 million). The different possible mixes of values contained between these extremes correspond to precise "statements" on the degree of resilience of the urban organism, on the access to safety, which depends on the social subjects to whom the advantage of safety will be granted, and to the extent and value of the resilience of the urban organism acknowledged by the administration.

The central scenario envisages that 374 FUs can be secured by means of average completeness level interventions, having a total cost of €2.33 million.

The table and the diagram, taken along the main diagonal, measure how the cost increases as the resilience and completeness of the interventions increase. If, on the other hand, the table and the diagram are traversed along the isocost function, the trade-off relationships between the completeness of the interventions and the degree of resilience for the same budget (that means for each different isocost function) and for each amount are shown.

The combination, integration, and consequentiality of factual, axiological and decisional aspects shows how this study considers ELC as an opportunity to go beyond its original purpose and its immediate significance. The ELC is an informative and normative device intended to provide a minimal ability to adapt the urban fabric so that it does not lose its identity. The case of Abruzzo, in which most of the cities comprised in the seismic crater have been evacuated for a long time, is an example of the different ways in which responsibility can be distributed among the spheres of proactive and reactive policies [53,54].

The two variables—the costs and degree of completeness of the interventions—in fact, allow the political-decisional profile of the ELC regarding the allocation of advantages and responsibilities between private and public actors to be identified.

The urban centre in its entirety achieves the requirement of resilience only when the ELC is fully realized—therefore, from the moment when all the buildings included in it are secured. For this reason, it is necessary to coordinate the interests and motivations of all owners of the assets involved.

Furthermore, there is no doubt about the unequal distribution of the positive externalities associated with the ELC. In particular, the sudden succession of significant catastrophic events in central Italy has made the seismic risk a piece of evidence with significant symbolic implications, since it involves the overall landscape dimensions of a settled community [55,56].

The presence of these externalities allows the local administration to start negotiations on the works to be subsided, and, as a consequence, on the dimensions of the incentives [57] according to the general trend of the urban policies focused on the trade-off between efficiency and fairness [58].

Therefore, the natural completion of this study will concern the measurement of the exposure, that is, the evaluation of the vulnerable assets, in terms of the joint value of the human and urban capital. This evaluation allows the costs of the seismic retrofit to be compared with the advantages of safety and to provide further evidence to formalize the equalization model.

The planning of the seismic retrofit at the urban scale, in fact, requires a well-structured public–private partnership, capable of capturing all benefits coming from both the avoided reconstruction costs [59,60], and any real estate externalities [61–66], to be taken into account in order to redistribute the added value generated in terms of resilience of the settled community as a result of the seismic retrofit program coordinated by the public.

This aspect is relevant, especially in historical urban contexts characterised by settlement complexity, structural fragility, typological and formal inertia, and a low population density. These characteristics can influence the answer to the original question: "Is it worth it?"

In this study, we tried to understand how the involvement and coordination of measures, judgments, and decision profiles allows the typical object-based approach implied by the ELC to be overcome. This, in fact, is attributable to a "prescriptive grammar" (it says what needs to be done), while the proposed approach is attributable to the logic of "generative grammar" (it says what can be done).

The proposed pattern, instead, implements an "axiological approach" that integrates natural, environmental, and technological aspects, with cultural, landscape, and political decision issues. Such an approach expands the way in which the original question can be answered.

Author Contributions: Conceptualization, S.G.; methodology, S.G., M.R.T., V.V.; software, S.G. and V.V.; validation, M.R.T., V.V.; formal analysis, S.G., M.R.T., V.V.; investigation, C.C., M.G., V.M.; resources, C.C., M.G.; data curation, V.V.; writing—original draft preparation, all these authors; writing—review and editing, S.G. and M.R.T.; visualization, all these authors; supervision, S.G., M.R.T.; project administration, S.G., M.R.T.; funding acquisition, S.G.

Funding: This research was funded by the Unione della Romagna Faentina, within the Agreement signed with the Department of Civil Engineering and Architecture of the University of Catania in 03/03/2016 based on the Decision n. 132/2016; scientific responsible C. Carocci e S. Giuffrida.

Conflicts of Interest: The authors declare no conflict of interest.

References

1. Giuffrida, S.; Ferluga, G.; Ventura, V. Planning Seismic Damage Prevention in the Old Towns Value and Evaluation Matters. In *Proceedings of the International Conference on Computational Science and Its Applications (ICCSA 2015), Banff, AB, Canada, 22–25 June 2015*; Gervasi, O., Murgante, B., Misra, S., Gavrilova, M.L., Rocha, A.M.A.C., Torre, C.M., Taniar, D., Apduhan, B.O., Eds.; Springer: Berlin, Germany, 2015; Part III; pp. 253–268. [CrossRef]
2. Napoli, G.; Giuffrida, S.; Trovato, M.R.; Valenti, A. Cap Rate as the Interpretative Variable of the Urban Real Estate Capital Asset: A Comparison of Different Sub-Market Definitions in Palermo, Italy. *Buildings* **2018**, *7*, 80. [CrossRef]
3. Fuentes, D.D.; Laterza, M.; D'Amato, M. *Seismic Vulnerability and Risk Assessment of Historic Constructions: The Case of Masonry and Adobe Churches in Italy and Chile*; Springer: Cham, Switzerland, 2019; Volume 18, pp. 1127–1137. [CrossRef]
4. Sawada, Y. The Economic Impact of Earthquake on Households. In *The Economic Impact of Natural Disasters*; Guha-Sapir, D., Santos, I., Eds.; Oxford Scholarship: New York, NY, USA, 2013. [CrossRef]
5. Giannelli, A.; Giuffrida, S.; Trovato, M.R. *Il Parco Madrid Rio. Valori simbolici e valutazione contingente, Valori e valutazioni*; E-Flow Dei Tipografia del Genio Civile: Rome, Italy, 2018; pp. 75–86, ISSN 2036-2404. Valori e Valutazioni.
6. Giuffrida, S.; Casamassima, G.; Trovato, M.R. Le norme EMAS-ISO nella valutazione della qualità del servizio idrico integrato. *Aestimum* **2017**, *70*, 109–134. [CrossRef]
7. Barbat, A.H.; Carreño, M.L.; Pujades, L.G.; Lantada, N.; Marulanda, M.C. Seismic vulnerability and risk evaluation methods for urban areas. A review with application to a pilot area. *Struct. Infrastruct. Eng.* **2010**, *6*, 17–38. [CrossRef]
8. Formisano, A. Local- and global-scale seismic analyses of historical masonry compounds in San Pio delle Camere (L'Aquila, Italy). *Nat. Hazards* **2017**, *86*, 465–487. [CrossRef]
9. Chieffo, N.; Clementi, F.; Formisano, A.; Lenci, S. Comparative fragility methods for seismic assessment of masonry buildings located in Muccia (Italy). *J. Build. Eng.* **2019**, *25*, 100813. [CrossRef]
10. Lamego, P.; Lourenço, P.B.; Sousa, M.L.; Marques, R. Seismic vulnerability and risk analysis of the old building stock at urban scale: Application to a neighbourhood in Lisbon. *Bull. Earthq. Eng.* **2017**, *15*, 2901–2937. [CrossRef]
11. Calabrò, F. Local Communities and Management of Cultural Heritage of the Inner Areas. An Application of Break-Even Analysis. In *Proceedings of the International Conference on Computational Science and Its Applications (ICCSA 2017), Trieste, Italy, 3–6 July 2017*; Gervasi, O., Murgante, B., Misra, S., Borruso, G., Torre, C.M., Rocha, A.M.A.C., Taniar, D., Apduhan, B.O., Stankova, E., Cuzzocrea, A., Eds.; Springer: London, UK, 2017; Part III; pp. 542–557. [CrossRef]
12. Giuffrida, S.; Napoli, G.; Trovato, M.R. Industrial Areas and the City. Equalization and Compensation in a Value-Oriented Allocation Pattern. In *Proceedings of the International Conference on Computational Science and Its Applications (ICCSA 2016), Beijing, China, 4–7 July 2016*; Gervasi, O., Murgante, B., Misra, S., Rocha, A.M.A.C., Torre, C.M., Taniar, D., Apduhan, B.O., Stankova, E., Wang, S., Eds.; Springer International Publishing: Cham, Switzerland, 2016; Part IV, pp. 79–94. [CrossRef]
13. Giuffrida, S.; Gagliano, F.; Nocera, F.; Trovato, M.R. Landscape assessment and Economic Accounting in wind farm Programming: Two Cases in Sicily. *Land* **2018**, *7*, 120. [CrossRef]
14. Giuffrida, S.; Trovato, M.R.; Giannelli, A. Semiotic-Sociological Textures of Landscape Values. Assessments in Urban-Coastal Areas. In *Proceedings of the International Conference on Information and Communication Technologies in Agricultural Food and Environment (HAICTA 2017), Chania, Crete, Greece, 21–24 September 2017*; Salampasis, M., Bournaris, T., Eds.; Springer: Cham, Switzerland, 2019; Volume 953, pp. 35–50. [CrossRef]
15. Naselli, F.; Trovato, M.R.; Castello, G. An evaluation model for the actions in supporting of the environmental and landscaping rehabilitation of the Pasquasia's site mining (EN). In *Proceedings of the International Conference on Computational Science and Its Applications (ICCSA 2014), Guimarães, Portugal, 30 June–3 July 2014*; Murgante, B., Misra, S., Rocha, A.M.A.C., Torre, C.M., Rocha, J.G., Falcão, M.I., Taniar, D., Apduhan, B.O., Gervasi, O., Eds.; Springer: Cham, Switzerland, 2014; Part III; pp. 26–41. [CrossRef]

16. Giuffrida, S.; Trovato, M.R. A Semiotic Approach to the Landscape Accounting and Assessment. An Application to the Urban-Coastal Areas. In *Proceedings of the 8th International Conference on Information and Communication Technologies in Agriculture, Food and Environment (HAICTA 2017), Chania, Crete Island, Greece, 21–24 September 2017*; Salampasis, M., Bournaris, T., Eds.; Springer International Publishing: Cham, Switzerland, 2017; pp. 696–708, ISSN 16130073.
17. Conti, T.; Salvini, V.; Galeati, S. *Comune Di Brisighella, Piano Di Recupero Del CS di Brisighella*; Comune Di Brisighella: Ravenna, Italy, 1994. Available online: http://www.romagnafaentina.it/I-servizi/Edilizia-e-Urbanistica/Tutela-e-governo-del-territorio/Regolamento-Urbanistico-ed-Edilizio-RUE/Regolamento-Urbanistico-ed-Edilizio-RUE-Intercomunale-Approvato (accessed on 29 September 2019).
18. Dolce, M.; Di Pasquale, G.; Speranza, E. A multipurpose method for seismic vulnerability assessment of urban areas. In Proceedings of the 15th World Conference on Earthquake Engineering (WCEE), Lisboa, Portugal, 24–28 September 2012.
19. Circo, C.; Giuffrè, M. La pianificazione urbanistica come possibile sovrapposizione di strategie per la riduzione del rischio sismico. Considerazioni sul Piano Regolatore della Sismicità dei Comuni dell'Unione della Romagna Faentina. *Urban. Inf.* **2018**, *278*, 49–51.
20. Carocci, C.F. Small centres damaged by 2009 L'Aquila earthquake: On site analyses of historical masonry aggregates. *Bull. Earthq. Eng.* **2012**, *10*, 45–71. [CrossRef]
21. Carocci, C.F. Conservazione del tessuto murario e mitigazione della vulnerabilità sismica. Introduzione allo studio degli edifici in aggregato. In *Architettura Storica e Terremoti. Protocolli Operativi per La Conoscenza e La Tutela*; Blasi, C., Ed.; Wolters Kluwer: Lucca, Italy, 2013; pp. 138–153.
22. Torsello, P. *La Materia Del Restauro*; Marsilio: Venezia, Italy, 1988.
23. Brando, G.; De Matteis, G.; Spacone, E. Predictive model for the seismic vulnerability assessment of small historic centres: Application to the inner Abruzzi Region in Italy. *Eng. Struct.* **2017**, *153*, 81–96. [CrossRef]
24. Rapone, D.; Brando, G.; Spacone, E.; De Matteis, G. Seismic vulnerability assessment of historic centers: Description of a predictive method and application to the case study of Scanno (Abruzzi, Italy). *Int. J. Arch. Herit.* **2018**, *12*, 1171–1195. [CrossRef]
25. Trovato, M.R.; Giuffrida, S. A DSS to Assess and Manage the Urban Performances in the Regeneration Plan: The Case Study of Pachino. In *Proceedings of the International Conference on Computational Science and Its Application (ICCSA 2014), Guimarães, Portugal, 30 June–3 July 2014*; Murgante, B., Misra, S., Rocha, A.M.A.C., Torre, C.M., Rocha, J.G., Falcão, M.I., Taniar, D., Apduhan, B.O., Gervasi, O., Eds.; Springer International Publishing: Cham, Switzerland, 2014; Part III; pp. 224–239. [CrossRef]
26. Trovato, M.R.; Giuffrida, S. The choice problem of the urban performances to support the Pachino's redevelopment plan. *IJBIDM* **2014**, *9*, 330–355. [CrossRef]
27. Trovato, M.R. A multi-criteria approach to support the retraining plan of the Biancavilla's old town. In *Proceedings of the International Symposium on New Metropolitan Perspectives, Reggio Calabria, Italy, 22–25 May 2018*; Calabrò, F., Della Spina, L., Bevilacqua, C., Eds.; Springer International Publishing: Cham, Switzerland, 2019; Volume 101, pp. 434–441. [CrossRef]
28. Pelling, M. *The Vulnerability of Cities: Natural Disasters and Social Resilience*; Earthscan Publishing: London, UK, 2003.
29. Dolce, M.; Manfredi, G. *Libro Bianco Sulla Ricostruzione Privata Fuori Dai Centri Storici Nei Comuni Colpiti Dal Sisma Dell'abruzzo Del 6 Aprile 2009*; Doppiavoce: Napoli, Italy, 2015.
30. Carreno, M.L.; Cardona, O.; Barbat, A.H. Urban Seismic Risk Evaluation: A Holistic Approach. *Nat. Hazard* **2007**, *40*, 137–172. [CrossRef]
31. Hung, H.C.; Ho, M.C.; Chien, C.Y. Integrating long-term seismic risk changes into improving emergency response and land use planning: A case study for the Hsinchu City, Taiwan. *Nat. Hazard* **2013**, *69*, 491–508. [CrossRef]
32. Della Spina, L. Integrated Evaluation and Multi-methodological Approaches for the Enhancement of the Cultural Landscape. In *Proceedings of the International Conference on Computational Science and Its Applications (ICCSA 2017), Trieste, Italy, 3–6 July 2017*; Gervasi, O., Murgante, B., Misra, S., Borruso, G., Torre, C.M., Rocha, A.M.A.C., Taniar, D., Apduhan, B.O., Stankova, E., Cuzzocrea, A., Eds.; Springer: London, UK, 2017; Part III; pp. 478–493. [CrossRef]
33. Della Spina, L.; Lorè, I.; Scrivo, R.; Viglianisi, A. An Integrated Assessment Approach as a Decision Support System for Urban Planning and Urban Regeneration Policies. *Buildings* **2017**, *7*, 85. [CrossRef]

34. O'Reilly, G.J.; Sullivan, T.J. Probabilistic seismic assessment and retrofit considerations for Italian RC frame buildings. *Bull. Earthq. Eng.* **2018**, *16*, 1447–1485. [CrossRef]
35. D'Ayala, D. Assessing the seismic vulnerability of masonry buildings. In *Handbook of Seismic Risk Analysis and Management of Civil Infrastructure Systems*; Tesfamariam, S., Goda, K., Eds.; Woodhead Publishing: Cambridge, UK, 2013; pp. 334–365.
36. Duzgun, H.S.B.; Yucemen, M.S.; Kalaycioglu, H.S.; Celik, K.; Kemec, S.; Ertugay, K.; Deniz, A. An integrated earthquake vulnerability assessment framework for urban areas. *Nat. Hazards* **2011**, *59*, 917. [CrossRef]
37. Christodoulou, S.E.; Fragiadakis, M.; Agathokleous, A.; Xanthos, S. From Historical and Seismic Performance to City-Wide Risk Maps. In *Handbook of Seismic Risk Analysis and Management of Civil Infrastructure Systems*; Tesfamariam, S., Goda, K., Eds.; Science Direct: Amsterdam, The Netherlands, 2013; pp. 247–267. [CrossRef]
38. Ma, X.; Ohno, R. Examination of Vulnerability of Various Residential Areas in China for Earthquake Disaster Mitigation. *Procedia Soc. Behav. Sci.* **2012**, *35*, 369–377. [CrossRef]
39. Manganelli, B.; Vona, M.; De Paola, P. Evaluating the cost and benefits of earthquake protection of buildings. *J. Eur. Real Estate Res.* **2018**, *11*, 263–278. [CrossRef]
40. Tocci, C. Vulnerabilità sismica e scenari di danno: Analisi speditiva delle catene di danno. In *Architettura Storica e Terremoti. Protocolli Operativi per La Conoscenza e La Tutela*; Blasi, C., Ed.; Wolkers Kluwer: Lucca, Italy, 2014; pp. 113–119.
41. Dolce, M.; Kappos, A.; Masi, A.; Penelis, G.; Vona, M. Vulnerability assessment and earthquake damage scenarios of the building stock of Potenza (Southern Italy) using Italian and Greek methodologies. *Eng. Struct.* **2005**, *28*, 357–371. [CrossRef]
42. Speranza, E.; Dolce, M.; Di Pascquale, G.; Tiberi, P. Una metodologia per la formulazione di scenari di danno a scala comunale: Applicazione pilota su 24 centri urbani della Valdaso. In Proceedings of the 14th National Congress L'Ingegneria Sismica in Italia, Bologna, Italy, 27–29 June 2009.
43. Banica, A.; Rosu, L.; Muntele, I.; Grozavu, A. Towards Urban Resilience: A Multi-Criteria Analysis of Seismic Vulnerability in Iasi City (Romania). *Sustainability* **2017**, *9*, 270. [CrossRef]
44. Dan, M.B. Decision making based on benefit-costs analysis: Costs of preventive retrofit versus costs of repair after earthquake hazards. *Sustainability* **2018**, *10*, 1537. [CrossRef]
45. Chieffo, N.; Formisano, A. Geo-Hazard-Based Approach for the Estimation of Seismic Vulnerability and Damage Scenarios of the Old City of Senerchia (Avellino, Italy). *Geosciences* **2019**, *9*, 59. [CrossRef]
46. Tiedemann, H. Earthquake and Volcanic Eruptions. In *A Handbook on Risk Assessment*; Swiss Reinsurance Company: Zurich, Switzerland, 1992.
47. Tomassetti, U.; Correia, A.A.; Graziotti, F.; Penna, A. Seismic vulnerability of roof systems combining URM gable walls and timber diaphragms. *Earthq. Eng. Struct. Dyn.* **2019**, *48*, 1297–1318. [CrossRef]
48. O'Reilly, G.J.; Sullivan, T.J. Quantification of modelling uncertainty in existing Italian RC frames. *Earthq. Eng. Struct. Dyn.* **2018**, *47*, 1054–1074. [CrossRef]
49. O'Reilly, G.J.; Perrone, D.; Fox, M.; Monteiro, R.; Filiatrault, A. Seismic assessment and loss estimation of existing school buildings in Italy. *Eng. Struct.* **2018**, *168*, 142–162. [CrossRef]
50. Giuffrida, S.; Gagliano, F.; Tocci, C. Emergency Limit Condition in the Old Towns. Axiology of seismic vulnerability and design. *Valori Valutazioni* **2015**, *15*, 55–67.
51. Ministerial Decree. *Guidelines for the Seismic Risk Classification of Construction—58/2017*; Ministerial Decree: Rome, Italy, 2017.
52. Grant, D.N.; Bommer, J.J.; Pinho, R.; Calvi, G.M.; Goretti, A.; Meroni, F.A. Schema di prioritizzazione per l'intervento sismico negli edifici scolastici in Italia. *Earthq. Spectra* **2012**, *23*, 291–314. [CrossRef]
53. Trovato, M.R.; Giuffrida, S. The Monetary Measurement of Flood Damage and the Valuation of the Proactive Policies in Sicily. *Geosciences* **2018**, *8*, 141. [CrossRef]
54. Trovato, M.R.; Giuffrida, S. The protection of territory in the perspective of the intergenerational equity. In *Green Energy and Technology*; Springer: Berlin, Germany, 2018; pp. 469–485. [CrossRef]
55. Giuffrida, S.; Napoli, S.; Trovato, M.R. The urban being between environment and landscape. On the old town as an emerging subject. In *New Metropolitan Perspectives*; Calabrò, F., Della Spina, L., Bevilacqua, C., Eds.; Springer: Cham, Switzerland, 2018; pp. 378–386. [CrossRef]

56. Giuffrida, S.; Trovato, M.R. From the Object to Land. Architectural Design and Economic Valuation in the Multiple Dimensions of the Industrial Estates. In *Proceedings of the International Conference on Computational Science and Its Applications (ICCSA 2017), Trieste, Italy, 3–6 July 2017*; Gervasi, O., Murgante, B., Misra, S., Borruso, G., Torre, C.M., Rocha, A.M.A.C., Taniar, D., Apduhan, B.O., Stankova, E., Cuzzocrea, A., Eds.; Springer: London, UK, 2017; Volume 10406, pp. 591–606. [CrossRef]
57. Napoli, G.; Giuffrida, S.; Trovato, M.R. Fair planning and affordability housing in urban policy. The case of Syracuse (Italy). In *Proceedings of the International Conference on Computational Science and Its Applications (ICCSA 2016), Beijing, China, 4–7 July 2016*; Gervasi, O., Murgante, B., Misra, S., Rocha, A.M.A.C., Torre, C.M., Taniar, D., Apduhan, B.O., Stankova, E., Wang, S., Eds.; Springer International Publishing: Cham, Switzerland, 2016; Part III; pp. 46–62. [CrossRef]
58. Napoli, G.; Giuffrida, S.; Trovato, M.R. Efficiency versus fairness in the management of public housing assets in Palermo (Italy). *Sustainability* **2019**, *11*, 1199. [CrossRef]
59. Carbonara, S. Il recupero dell'edilizia privata nell'Abruzzo post-sisma: Un'analisi delle procedure di stima. *Territorio* **2014**, *70*, 119–125. [CrossRef]
60. Carbonara, S.; Cerasa, D.; Spacone, E. Una proposta per la stima sommaria dei costi nella ricostruzione post-sismica. *Territorio* **2014**, *17*, 119–125. [CrossRef]
61. Gabrielli, L.; Giuffrida, S.; Trovato, M.R. From Surface to Core: A Multi-layer Approach for the Real Estate Market Analysis of a Central Area in Catania. In *Proceedings of the International Conference on Computational Science and Its Applications (ICCSA 2015), Banff, AB, Canada, 22–25 June 2015*; Gervasi, O., Murgante, B., Misra, S., Gavrilova, M.L., Rocha, A.M.A.C., Torre, C.M., Taniar, D., Apduhan, B.O., Eds.; Springer: Berlin, Germany, 2015; Part III; pp. 284–300. [CrossRef]
62. Gabrielli, L.; Giuffrida, S.; Trovato, M.R. Functions and Perspectives of Public Real Estate in the Urban Policies: The Sustainable Development Plan of Syracuse. In *Proceedings of the International Conference on Computational Science and Its Applications (ICCSA 2016), Beijing, China, 4–7 July 2016*; Gervasi, O., Murgante, B., Misra, S., Rocha, A.M.A.C., Torre, C.M., Taniar, D., Apduhan, B.O., Stankova, E., Wang, S., Eds.; Springer International Publishing: Cham, Switzerland, 2016; Volume IV, pp. 13–28. [CrossRef]
63. Gabrielli, L.; Giuffrida, S.; Trovato, M.R. Gaps and overlaps of urban housing sub market: A fuzzy clustering approach. In *Green Energy and Technology*; Springer: Berlin, Germany, 2017; pp. 203–219, Issue 9783319496757. [CrossRef]
64. Napoli, G.; Giuffrida, S.; Valenti, A. Forms and Functions of the Real Estate Market of Palermo (Italy). Science and Knowledge in the Cluster Analysis Approach. In *Appraisal: From Theory to Practice*; Stanghellini, S., Morano, P., Bottero, M., Oppio, A., Eds.; Springer: Cham, Switzerland, 2017; pp. 191–202, ISBN 978-3-319-49675-7. [CrossRef]
65. Valenti, A.; Giuffrida, S.; Linguanti, F. Decision Trees Analysis in a Low Tension Real Estate Market: The Case of Troina (Italy). In *Proceedings of the International Conference on Computational Science and Its Applications (ICCSA 2015), Banff, AB, Canada, 22–25 June 2015*; Gervasi, O., Murgante, B., Misra, S., Gavrilova, M.L., Rocha, A.M.A.C., Torre, C.M., Taniar, D., Apduhan, B.O., Eds.; Springer: Berlin, Germany, 2015; Volume 9157, pp. 237–252, ISBN 978-3-319-21469-6. [CrossRef]
66. Giuffrida, S.; Ferluga, G.; Valenti, A. Clustering Analysis in a Complex Real Estate Market the Case of Ortigia (Italy). In *Proceedings of the International Conference on Computational Science and Its Applications (ICCSA 2014), Guimarães, Portugal, 30 June–3 July 2014*; Murgante, B., Misra, S., Rocha, A.M.A.C., Torre, C.M., Rocha, J.G., Falcão, M.I., Taniar, D., Apduhan, B.O., Gervasi, O., Eds.; Springer: Cham, Switzerland, 2014; Volume 8581, pp. 106–121, ISBN 9783319091495.

Article

Geo-Hazard-Based Approach for the Estimation of Seismic Vulnerability and Damage Scenarios of the Old City of Senerchia (Avellino, Italy)

Nicola Chieffo [1] and Antonio Formisano [2,*]

1 Department of Civil Engineering, Faculty of Engineering, Politehnica University of Timisoara, Traian Lalescu Street, Timisoara 300223, jud. Timis, Romania; nicola.chieffo@student.upt.ro

2 Department of Structures for Engineering and Architecture, School of Polytechnic and Basic Sciences, University of Naples "Federico II", P. le V. Tecchio 80, 80125 Napoli, Italy

* Correspondence: antoform@unina.it

Received: 30 December 2018; Accepted: 22 January 2019; Published: 26 January 2019

Abstract: The large-scale seismic risk assessment is a crucial point for safeguarding people and planning adequate mitigation plans in urban areas. The current research work aims at analysing a sector of the historic centre of Senerchia, located in the province of Avellino, in order to assess the seismic vulnerability and damage of old masonry building compounds. First, the typological classification of the inspected building aggregates is developed using the CARTIS form developed by the PLINIVS research centre in collaboration with the Italian Civil Protection Department. The global seismic vulnerability assessment of the building sample is carried out using the macroseismic method according to the EMS-98 scale in order to identify the buildings most susceptible to seismic damage. Furthermore, 12 damage scenarios are developed by means of an appropriate seismic attenuation law. Finally, the expected damage scenarios considering the local hazard effects induced are developed in order to evaluate the damage increment, averagely equal to 50%, due to the seismic amplification of different soil categories.

Keywords: masonry aggregates; vulnerability assessment; vulnerability curves; damage scenarios; local hazard effect

1. Introduction

The concept of seismic risk, in general, can be understood as a multi-factorial combination of three main parameters, Vulnerability (V), Exposure (E) and Hazard (H), which directly or indirectly influence a specific area. Some of the most recent Italian seismic disasters (L'Aquila 2009, Emilia Romagna 2012, Central Italy 2016) [1–3], have shown again the inadequate structural performance of many existing buildings against seismic actions. The main reason is related to the widely accepted fact that seismic risk scenarios and the estimation of the economic and human losses produced by earthquakes are useful tools for seismic risk mitigation based on adequate prevention plans [4,5].

The formulation of an earthquake loss model into a given region is not only of interest for the economic impact of future earthquakes, but it is also important for risk mitigation. A specific loss model allows us to predict the damage to the built environment under a given data scenario (deterministic approach model) and may be particularly important for responding to emergencies and disaster planning by national governments [6].

Thus, focusing at the urban scale, risk and age of the buildings are dependent on each other, since the site hazard induces significant consequences in term of expected losses. In fact, the seismic risk assessment on an urban scale implies the management of a large number of variables, such as different types of buildings, people, roads, etc., in order to know the expected damage probability due to earthquakes [7].

The vulnerability to disasters in areas exposed at risk is one of the most underestimated and impressive problems for urban development.

Rapid urbanization has dramatically increased the vulnerabilities and risks of urban dwellers in densely populated areas. In fact, the high population, combined with the presence of numerous buildings that do not comply with seismic regulations, significantly increase the problem of seismic safety in urban areas [8]. The identification of the most vulnerable buildings in an urban context is not a simple task due to their heterogeneity and complexity. However, in order to both cope with such inefficiencies and to achieve the overall improvement of the urban system, generally limited economic resources are needed [9]. The inventory of buildings is an essential procedure for the acquisition of data for the evaluation of large-scale seismic vulnerability [10]. The available strategies usually take into consideration survey forms to collect information on urban buildings based mainly on the parameters necessary for the seismic vulnerability assessment, i.e., the system type resistant to lateral loading, the regularity, the maintenance conditions and the presence of existing damages. The application of the survey forms allows us to fully contextualise, especially in the case of heterogeneous urban centres, the various structural types located there.

Therefore, the vulnerability can be defined as the potential of buildings to suffer a certain level of damage when subjected to a seismic event. Furthermore, the concept of vulnerability pertains to a system of basic concepts involved in risk analysis. This kind of definition, which is definitely vague, requires considerable refinements in order to become an operational tool for various purposes, as the estimate of seismic risk, development of earthquake scenarios or development of strategies for risk mitigation [11–13].

In general, the predisposition of a building to be damaged by a seismic impact depends on many aspects that are mainly concerned with the construction type, the quality of materials, the construction methods and the maintenance. Many studies [14–16] have shown how the lack of these issues makes structures inefficient so as to cause extensive damage during earthquakes. Indeed, the masonry buildings sample within the historical centres very often is characterised by a static inadequacy mainly due to an obsolete construction technique that does not guarantee an adequate safety standard. The lack of connections among orthogonal walls combined with the poor mechanical characteristics of masonry blocks and mortars are usually the main causes of incipient collapses caused by seismic phenomena.

However, the dynamic response of buildings is also influenced by site effects, which play an important role in the evaluation of the expected global damage. Site effects are responsible for the amplification of seismic waves in surface geological layers. In fact, surface motion can be strongly amplified if geological conditions are unfavourable [17].

Generally, specific geological, geomorphologic and geo-structural settings of restricted areas can induce a high level of shaking on the ground surface even in occasions of low-intensity/magnitude earthquakes. This effect is called site or local amplification. An appropriate geotechnical model based on the effects induced by local amplification phenomena certainly aims to favour a better prediction of the damage increase caused by the different soil conditions. It is principally due to the seismic motion transfer from hard deep soils to soft superficial ones and to effects of seismic energy focalization owing to the typical geometrical setting of the deposits [18,19].

A reliable and easy method for large scale analysis developed in Reference [20] allows us to estimate the macroseismic intensity increment derived from specific soil category so as to properly define, taking into account the local amplification factors, the global vulnerability of building stocks. In this framework, the macroseismic method allows for a direct estimation of the problem as it allows directly linking the cause (expected damage) to the effect (local seismic amplification) according to the EMS-98 scale.

Based on these considerations, a sub-urban sector of the historical centre of Senerchia, in the district of Avellino, was selected as a case study. The proposed work aims to evaluate the effects of local amplification varying the topographic class (from T1 to T4) and the type (from A to E) of soils foreseen in the new Italian Technical standard NTC18 [21]. The influence of soil conditions was considered by implementing a macroelement model of a typical masonry aggregate of the investigated study area with the goal to plot damage scenarios expected under different earthquake moment magnitudes and site-source distances.

2. The Historical Centre of Senerchia

2.1. Geographical-Historical Background

The municipal territory (Figure 1) is mostly mountainous and hilly, being characterised on the South-East side by the Picentini Mountains.

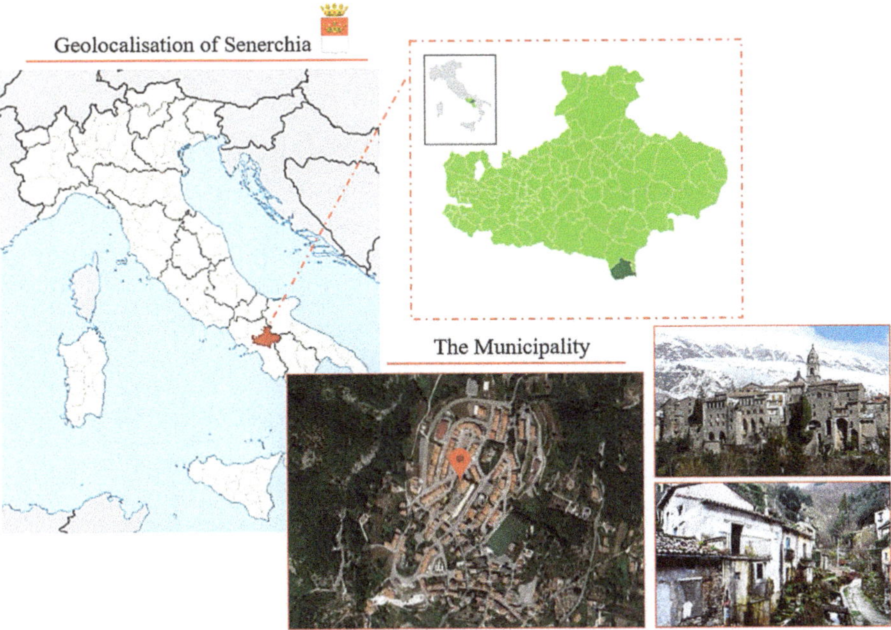

Figure 1. The geographical localisation of the municipality of Senerchia.

The village is located 600 meters above sea level in the High Valley of the Sele, a hilly area on the eastern side of the Picentini mountains group, stretching up to the slopes of Boscotiello mount. Besides the Sele, which laps the municipal territory, the other watercourses are Vallone Rovivo, Vallone Forma, Piceglia, Fiumicello, Rovivo, Pozzo San Nicola, Acquabianca and Vallone Badoleia (better known as Vallone Varleia).

Senerchia has ancient origins dating back to the IX century of the vulgar era. The Picentine population, who lived in the steep, expensive rocks of the valley, settled after the Second Punic War, came to settle in the Roman *oppidum* with the name of SENA HERCLEA.

The birth of Senerchia, therefore, is due to the Longobards, who were the protagonists of the birth of most of the castles and villages of the provinces of Avellino and Salerno.

The first settlement nucleus, located on the foothills upstream of the church of San Michele Arcangelo, was made of partial winding surrounded by parts of walls leading, also due to the orographic feature of the site, towards the castle-church binomial typical of fortified centres.

The type of housing is called the "house on a slope". As in the second war, the population did not want to abate the use of building on the rock, even the second housing expansion along the Vallone river saw the buildings erected on the rocky coast [22].

2.2. The 1980 Irpinia Earthquake

The Irpinia earthquake occurred on 23 November 1980 at 7:34 pm. A strong shock (X degree of the Mercalli scale) lasting about 90 s, with a hypocenter of about 12 km depth, striking an area of 17,000 km^2 that extended from Irpinia to Vulture, straddling the provinces of Avellino, Salerno and Potenza [23].

The municipalities most severely affected by the earthquake were Castelnuovo di Conza, Conza della Campania, Laviano, Lioni, Sant'Angelo dei Lombardi, Senerchia, Calabritto and Santomenna (Figure 2).

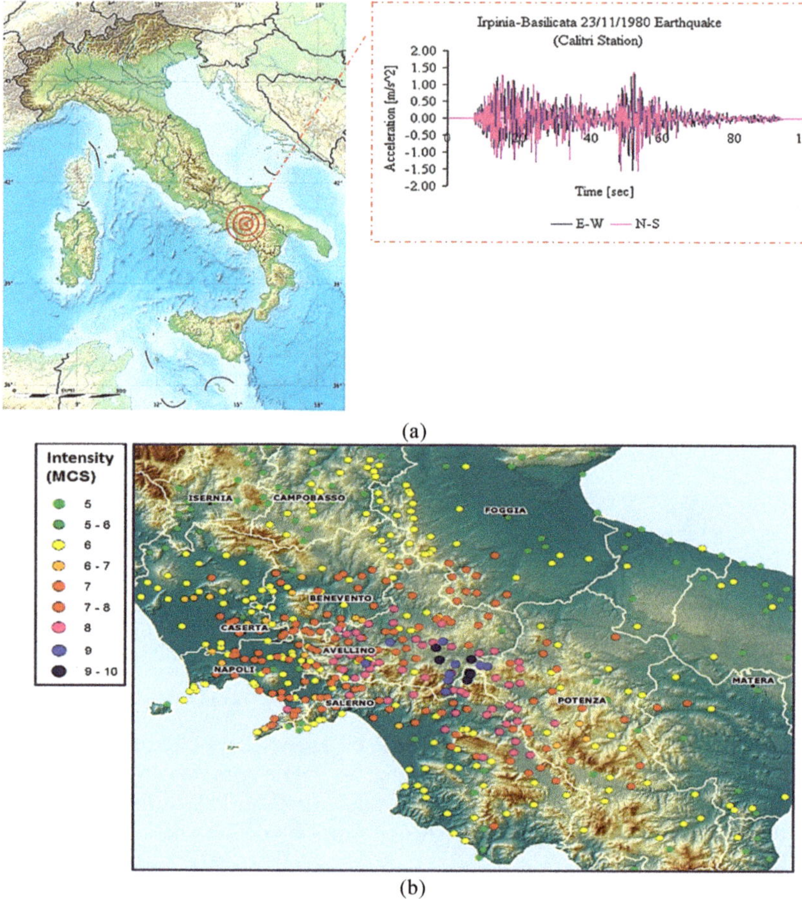

Figure 2. The Irpinia earthquake effects [23]: (**a**) time history main shock; (**b**) distribution of seismic intensities.

The seismic effects, however, spread to a much wider area affecting practically the entire central part of the southern area of the Italian peninsula [24,25]. Many injuries and collapses also took place in Naples involving many dilapidated or damaged buildings and old dwellings in tuff stones.

In particular, in the Poggioreale district of Naples, a building along Stadera Street collapsed, probably due to construction defects, causing 52 deaths. The three most damaged provinces were those of Avellino (103 municipalities), Salerno (66 municipalities) and Potenza (45 municipalities). Thirty-six municipalities in the epicentre area had about 20,000 destroyed or unusable homes. In 244 (far from the epicentre) municipalities of the provinces of Avellino, Benevento, Caserta, Matera, Foggia, Naples, Potenza and Salerno, another 50,000 houses suffered serious and medium-large damages. Additional 30,000 homes were only slightly affected by the earthquake [26].

The 1980 Irpinia earthquake was characterised by three distinct phenomena along different fault segments that occurred in about 40 s. The rupture spread from the hypocenter involving fault segments along the Marzano mountains, Carpineta and Cervialto. After about 20 s, the break propagated towards the southeast in the direction of the Piana di San Gregorio. The last fault segment affected by the breaking process, after 40 s, was located in the northeast of the first segment.

The earthquake replicas were distributed along the entire length of the fault and involved an extended focal volume between the four faults implicated. The fracture reached the earth's surface, generating a clearly visible fault slope for about 35 km, as depicted in Figure 3 [27].

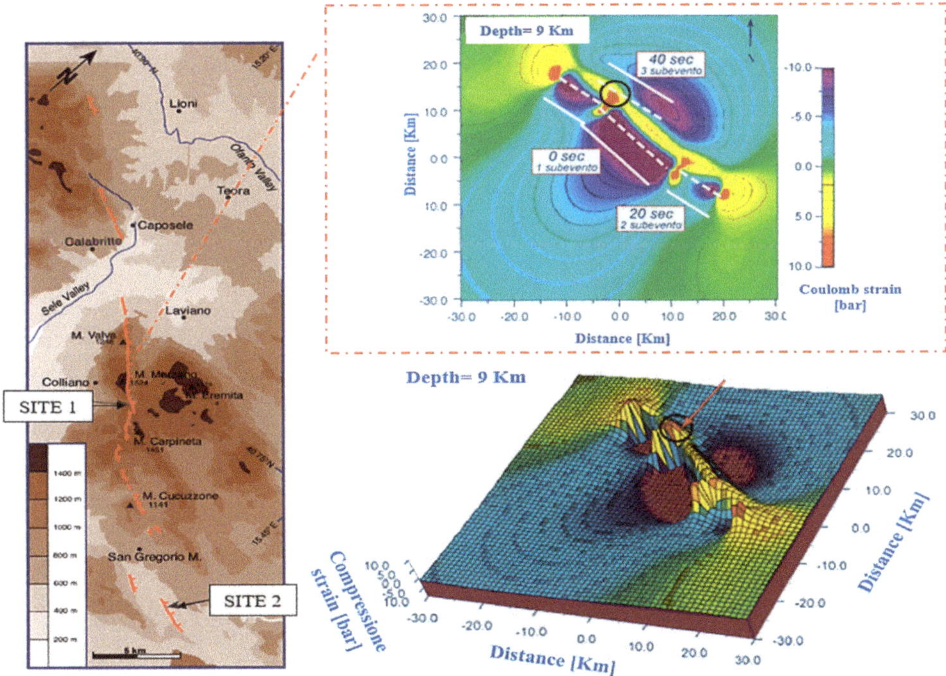

Figure 3. The fault mechanisms that occurred in the 1980 Irpinia earthquake.

3. Typological and Structural Characterisation of the Buildings Sample

3.1. The CARTIS Form

The classification of the construction typologies is a very important task for large-scale seismic risk assessments. In this framework, an effective methodology [28] allowed a new grouping technique based on an analysis of covariance on probabilistic seismic demand models. This grouping method, mainly calibrated on bridges, reduces sub-classes from all possible sub-class combinations. The proposed method is relevant and could also be applied to buildings.

In the urban area under investigation, however, the CARTIS form was adopted as the analysis tool to identify the most relevant structural typologies.

The CARTIS form [29] was developed by the PLINIVS research centre of the University of Naples "Federico II" during the ReLUIS 2014–2016 project, the "Development of a systematic methodology for the assessment of exposure on a territorial scale based on the typological characteristics/structural of buildings", in collaboration with the Department of Italian Civil Protection (DPC).

The CARTIS form aimed to detect the prevalent ordinary building typologies in municipal or sub-municipal areas, called urban sectors, characterized by typological and structural homogeneity.

The form is divided into four sections: Section 0 for the identification of the municipality and the sectors identified therein; Section 1 for the identification of each of the predominant typologies characterizing the generic sub-sector of the assigned municipality; Section 2 for the identification of general characteristics of each typology of the constructions; Section 3 for the characterisation of the structural elements of all individuated construction typologies.

Thus, focusing on the case study, the historical centre of Senerchia is composed of two building compartments (C01 and C02). In both sectors, the buildings, made of local limestone, are grouped in aggregates and are characterized by the lack of effective connections among orthogonal walls, which could prevent out-of-plane mechanisms. The small houses characterising the historical centre buildings, composed of walls with an average thickness of 0.8 m, are illuminated by a few small windows. The storey average height is about 3.50 m. The horizontal structures, as well as roofs, are generally made of timber beams (Figure 4).

Figure 4. The street views of some building typologies inside the inspected historical centre of Senerchia.

The study herein illustrated is conducted on the C01 urban sector composed of 43 structural units (S.U.), as reported in Figure 5.

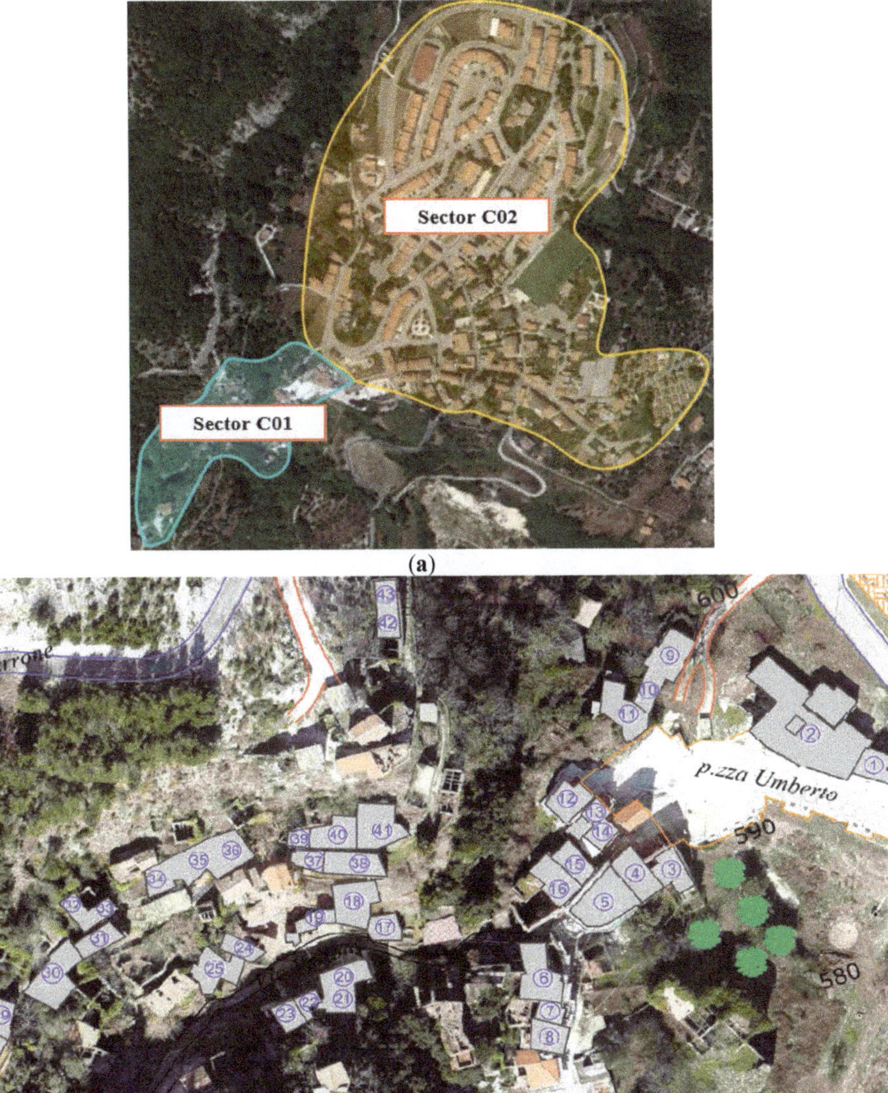

Figure 5. The study area: (**a**) urban sectors; (**b**) building inventory within the C01 sector.

3.2. Seismic Vulnerability Assessment

In order to implement a rapid seismic assessment procedure for examined building aggregates, the new form of vulnerability depicted in Table 1 is used [30,31].

Table 1. The vulnerability form for historical building aggregates.

Parameters	Class Score, S_i				Weight, W_i
	A	B	C	D	
1. Organization of vertical structures	0	5	20	45	1.00
2. Nature of vertical structures	0	5	25	45	0.25
3. Location of the building and type of foundation	0	5	25	45	0.75
4. Distribution of plan resisting elements	0	5	25	45	1.50
5. In-plane regularity	0	5	25	45	0.50
6. Vertical regularity	0	5	25	45	0.50
7. Type of floor	0	5	15	45	0.80
8. Roofing	0	15	25	45	0.75
9. Details	0	0	25	45	0.25
10. Physical conditions	0	5	25	45	1.00
11. Presence of adjacent building with different height	−20	0	15	45	1.00
12. Position of the building in the aggregate	−45	−25	−15	0	1.50
13. Number of staggered floors	0	15	25	45	0.50
14. Structural or typological heterogeneity among adjacent S.U.	−15	−10	0	45	1.20
15. Percentage difference of opening areas among adjacent facades	−20	0	25	45	1.00

This new form is based on the vulnerability index method proposed some decades ago by Benedetti and Petrini [32]. The new investigation form appropriately conceived for building aggregates is achieved by adding five new parameters, which take into account the effects of mutual interaction under earthquakes among aggregated structural units, to the ten basic parameters of the original form developed by the above-mentioned researchers.

The additional parameters, deriving from previous studies found in the literature [30], are summarized as follows:

– *Parameter 11: Presence of adjacent buildings with different heights*

The in-elevation interaction among adjacent buildings has a significant effect on the seismic response of S.U. The optimal configuration occurs for adjacent buildings with the same height (class A) due to the mutual confinement action. Additionally, a building adjacent to taller buildings (on one or both sides, class B) can suffer less damage than one adjacent to buildings with less height (classes C and D).

– *Parameter 12: Position of the building in the aggregate*

This parameter provides indications about the in-plane interaction among S.U. In particular, the case of an isolated building corresponds to class D (most unfavourable condition), while the intermediate, corner and heading conditions are related to classes A, B and C, respectively. It is worth noting that the inclusion in aggregate, regardless of the position of the structural unit, always gives rise to the reduction of seismic vulnerability.

– *Parameter 13: Number of staggered floors*

In the case of an earthquake, staggered floors are responsible for pounding effects among adjacent S.U., which could activate out-of-plane collapse mechanisms. The best condition is the absence of staggered floors (class A), while one (class B), two (class C) or more than two (class D) staggered floors increase the vulnerability.

– *Parameter 14: Structural or typological heterogeneity among adjacent S.U.*

This parameter takes into account that adjacent buildings can be built with different construction technologies or have structural heterogeneity. In the case of "homogeneous" buildings, i.e., made of the same masonry type, the vulnerability remains unchanged compared to that of the isolated building (class C). On the other hand, the case of structural heterogeneity (i.e., a masonry S.U. near to an RC structure) is the most favourable seismic condition. Finally, S.U. can be placed next to another masonry unit with worst (class B) or better mechanical properties (class D).

– *Parameter 15: Percentage difference of opening areas among adjacent facades*

This parameter affects the seismic response of S.U. because it is responsible for the distribution of horizontal actions among façades of adjacent buildings. Neglecting the case of no difference of opening areas (class A), the S.U. can be placed between buildings with lower (classes B and C) or higher (class D) percentages of openings.

The methodology is based on the evaluation of a vulnerability index, I_V, for each S.U. as the weighted sum of the 15 parameters listed in Table 1. The estimated parameters are distributed in 4 vulnerability classes (A, B, C and D, from the best to the worst), characterised by a given score (also with negative sign in case of vulnerability reduction), which a correspondent weight, W_i, is assigned to, changing it from a minimum of 0.25 for the less important parameters up to a maximum of 1.20 for the most important ones. Further information on how the scores and classes were determined are found in Reference [30].

Based on these considerations, the vulnerability index, I_V, is calculated according to the following equation:

$$I_V = \sum_{i=1}^{15} S_i * W_i \tag{1}$$

In detail, the survey form adopted is applied to a generic building with specific constructive and technological characteristics. Substantially, each parameter is associated with vulnerability classes, with a specific score, S_i, and specific weight, W_i.

Subsequently, the vulnerability index value is normalized in the range [0−1] by means of Equation (2), assuming, from this moment, the notation V_I.

$$V_I = \left[\frac{I_V - (\sum_{i=1}^{15} S_{\min} * W_i)}{\left| \sum_{i=1}^{15} [(S_{\max} * W_i) - (S_{\min} * W_i)] \right|} \right] \tag{2}$$

where I_V is the vulnerability index deriving from the previously properly defined form; $(S_{min} \cdot W_i)$, equal to −125.50, represents the sum of scores associated to the vulnerability class *A* of each parameter multiplied by respective weights; $(S_{max} \cdot W_i)$, equal to 495.00, represents the sum of scores associated to the vulnerability class *D* of each parameter multiplied by the respective weights; and the denominator, equal to 620.50, represents the total vulnerability.

Appropriate vulnerability curves [33] are obtained to estimate the propensity of the damage of the analysed building stock (Figure 6). More in detail, these curves express the probability

P[LS | $I_{EMS\text{-}98}$] that a building reaches a certain limit state "LS" at a given seismic intensity "$I_{EMS\text{-}98}$", which is defined according to the European macroseismic scale EMS-98.

In particular, as mathematically expressed by Equation (3), the vulnerability curves depend on three variables: the vulnerability index (V_I), the seismic hazard, expressed in terms of macroseismic intensity ($I_{EMS\text{-}98}$) and the ductility factor *Q*, ranging from 1.0 to 4.0, which describes the ductility of a certain typological class. In the case under study, a ductility factor *Q* of 2.3 is considered [34].

$$\mu_D = 2.5 * \left[1 + \tan h \left(\frac{I_{EMS-98} + 6.25 * V_I - 13.1}{Q} \right) \right] \tag{3}$$

As reported in the previous figure, the vulnerability curves are derived for a sample of buildings representative of the construction types found in the inspected area. However, for a more complete representation of the expected damage, the mean typological vulnerability curves are herein represented together with other curves taking into account the variability of damage in the vulnerability range ($V_m - \sigma$; $V_m + \sigma$; $V_m + 2\sigma$; $V_m - 2\sigma$) according to Reference [35].

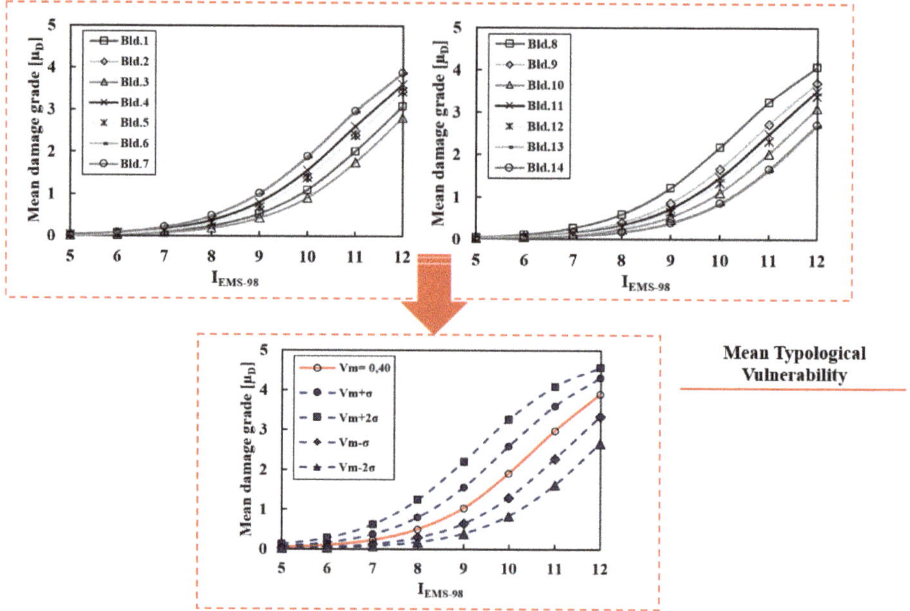

Figure 6. The mean typological vulnerability curves for examined buildings.

3.3. Parametric Estimation of Seismic Impact Scenarios

The earthquake is associated with the accumulation of stresses in particular points of the lithosphere, between the surfaces in contact with ancient faults or in other areas. When these stresses exceed the shear strength, there are shifts and breakages with the release of energy in the form of seismic volume waves (*P*-waves and *S*-waves). These waves radiate at different speeds in all directions with roughly spherical wavefronts. There is, therefore, a progressive attenuation of the energy contained by the seismic waves from the epicentre to the different sites where earthquake effects are felt.

The seismic waves also undergo other modifications, which are linked to reflection and refraction phenomena at the interface between layers of different characteristics (attenuation by scattering) and to the internal damping of the soils. They have a further attenuation of the energy content with the distance and a "verticalization" of the propagation direction of seismic waves.

However, the severity of seismic damage to buildings of similar structural characteristics would be a regular and decreasing function with the distance from the epicentre, which a progressive reduction of the expected acceleration, a progressive increase in terms of the duration and an increase of frequencies are associated to Reference [36].

The seismicity prediction of a given site can be achieved by adopting appropriate seismic attenuation laws, which are empirical formulations calibrated on the statistical analysis of a sample of instrumental or macroseismic data. The main feature of these formulations is to estimate the value of a seismic parameter (i.e., accelerations, velocity, seismic intensity, etc.) as a function of other synthetic factors, such as magnitude (M_w), epicentre (R) or hypocentre (h) distances.

Generally, the calibration of attenuation laws is based on the analysis of a catalogue of seismic events occurred in a given site of interest [37].

According to References [38,39], attenuation laws in terms of spectral accelerations (S_a) and peak ground accelerations (PGA) were derived in North America using a stochastic approach for the prediction of ground motions.

Other studies [40] derived attenuation equations in terms of seismic intensity (MMI) from instrumental recordings of events that occurred in the United States.

Similarly, research performed in Europe [41] provided important insights into the adoption of seismic attenuation laws based on the maximum expected acceleration.

In the current investigation, the analysis of the damage scenarios of the inspected urban compartment of Senerchia is carried out by means of the following seismic attenuation law [42]:

$$I_{EMS-98} = 6.39 + 1.756 * M_w - 2.747 * (R + 7) \quad (4)$$

where M_w is the moment magnitude and R (measured in Km) is the site-source distance. To this purpose, the historical earthquakes in the examined area were taken from the Italian Macroseismic Database DBMI-15 (National Institute of Geophysics and Volcanology) [43]. In particular, the seismic events of Area Nolana (1805), Irpinia (1982) and Potentino (1990), which gave rise to moment magnitudes of 4, 5 and 6, respectively, were selected (Figure 7). The selection of these magnitude sets has allowed us to plot the expected damage scenarios.

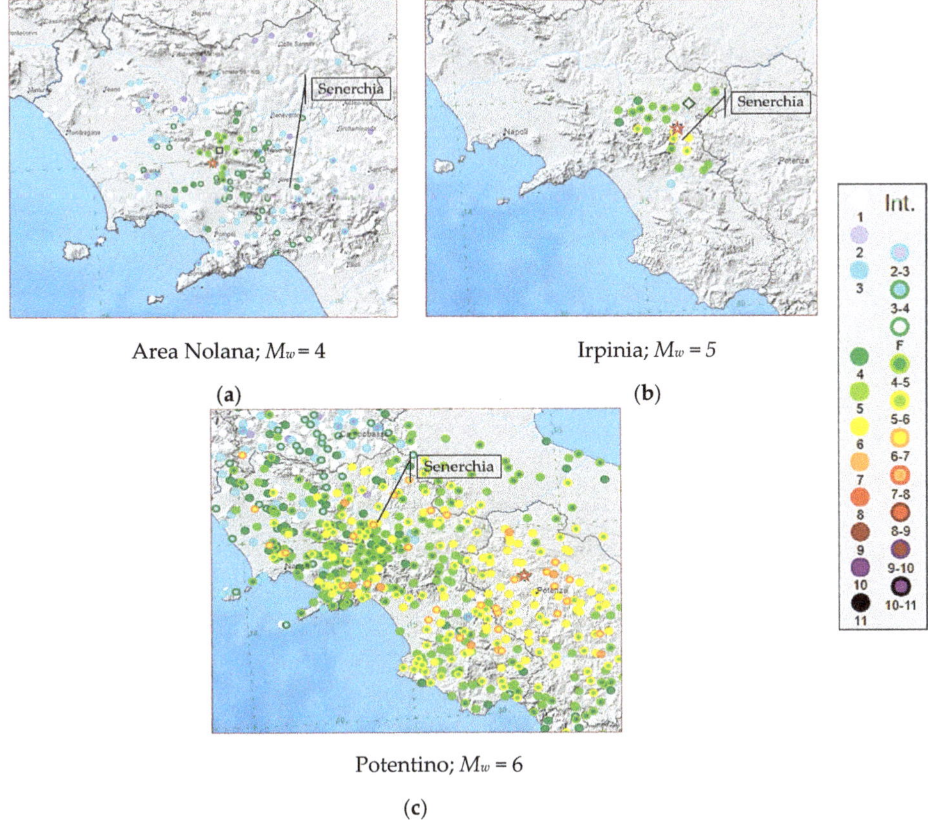

Figure 7. The historical earthquakes selected for the case study area: (**a**) Area Nolana; (**b**) Irpinia; (**c**) Potentino [43].

In order to predict the seismicity of the study area, a set of magnitudes (in the range (4–6)) and epicentre distances (in the range (5–20 Km)) are considered. The macroseismic intensities correlated to the earthquake magnitudes on the basis of Equation (4) are reported in Table 2.

Table 2. The correlation between the moment magnitude, M_w, and macroseismic intensity, $I_{EMS\text{-}98}$.

Magnitude M_w	Macroseismic Intensity $I_{EMS\text{-}98}$			
	R = 5 Km	R = 10 Km	R = 15 Km	R = 20 Km
4	VII	VI	V	IV
5	VIII	VII	VII	VI
6	X	IX	VIII	VII

Subsequently, 12 damage scenarios at different magnitudes and site-source distances are evaluated, as shown in Figure 8.

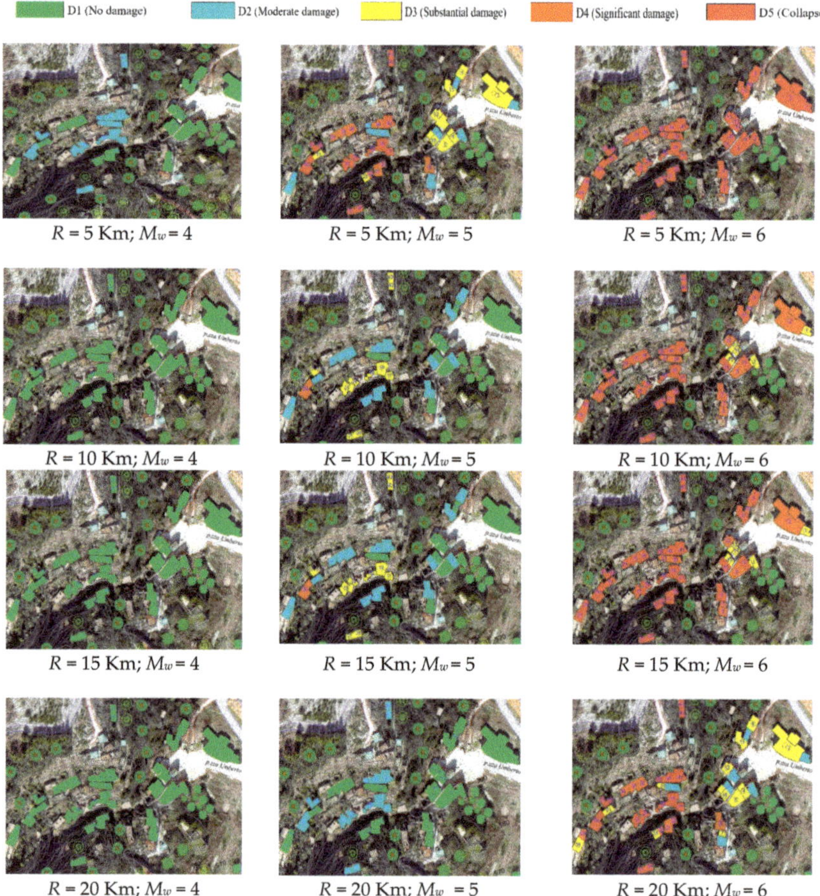

Figure 8. The seismic impact damage scenarios in the investigated urban area.

The correlation between the mean damage grade, μ_D, and the damage thresholds, D_K, were developed according to the EMS-98 scale [44].

From the analysis results, it appears that for M_w = 4 about 40% of the building sample suffers damage D2, while for M_w = 6, a damage threshold D5 (collapse) is attained for the entire urban-sector. Instead, when M_w = 5, a more variable damage distribution is achieved.

4. Geo-Hazard Effects

4.1. Local Site Conditions

The analysis of effects induced by the local seismic hazard represents an important issue for a more effective prediction of the damage scenarios of the investigated area and, therefore, for the implementation of risk mitigation plans. In fact, seismic site effects are related to the amplification of seismic waves in superficial geological layers. The surface ground motion may be strongly amplified if the geological conditions are unfavourable (e.g., sediments). Consequently, the study of local site effects is an important aspect to take into account for the assessment of seismic hazard and physical vulnerability of buildings, since damage due to earthquakes may thus be aggravated.

According to Reference [20], site effects are estimated by a macroseismic approach, which allows us to determine the expected damage through the calibration of a suitable local amplification coefficient.

Even though it is considered a discrete quantity, the macroseismic intensity is the main parameter for directly correlating the seismic input to damage. Furthermore, as previously analysed, a scenario analysis aims to estimate the damage level at the territorial scale instead of predicting the response of a certain structure at a specific site [45].

4.2. Geological and Geotechnical Classification of the Study Area

Focusing on the case study, the area of the historic centre of Senerchia is located to the southwest of the town, geographically in the foothills of the Magnone limestone-dolomite mountain. This area belongs to the Western area of the Southern Apennine Range (Picentini mountains), on the right of the High Valley of Sele.

The territory is characterised by the presence of heterogeneous deposits composed of more or less regular alternations of lithotypes. These deposits have a propensity for failure induced by pre-existing discontinuity surfaces with various ages and dimensions. In particular, as shown in Figure 9, the municipal territory mainly consists of two distinct areas, i.e., limestone and dolomite limestone (*CLU*) and deposits of ground debris (a_3).

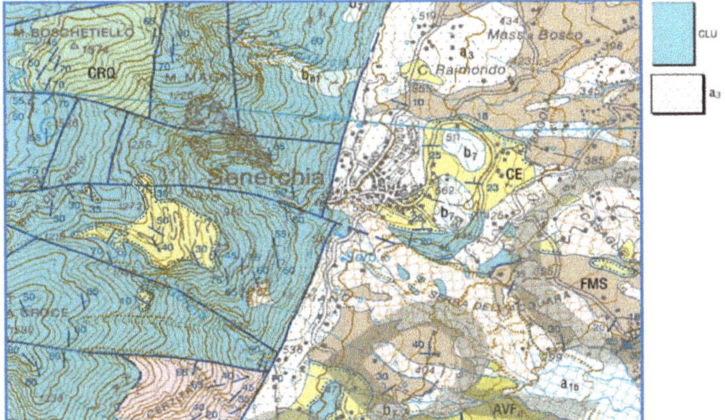

Figure 9. The geology of the study area.

For the seismic microzonation of this historic centre, data from on-site surveys made available by the Municipal Administration are used. Furthermore, on-site tests are carried out to identify the stratigraphy of the ground and the propagation velocity of the seismic waves measured by the geophone.

Figure 10 shows the areas where the geological and geotechnical characterisation tests were carried out.

First, the Down-Hole test is performed, aiming at estimating the propagation velocity of the seismic waves. The seismic wave velocity values obtained from the survey show that the soil under examination is of category B [21], i.e., rocks and deposits made of very thick coarse-grained soils or very fine-grained soils.

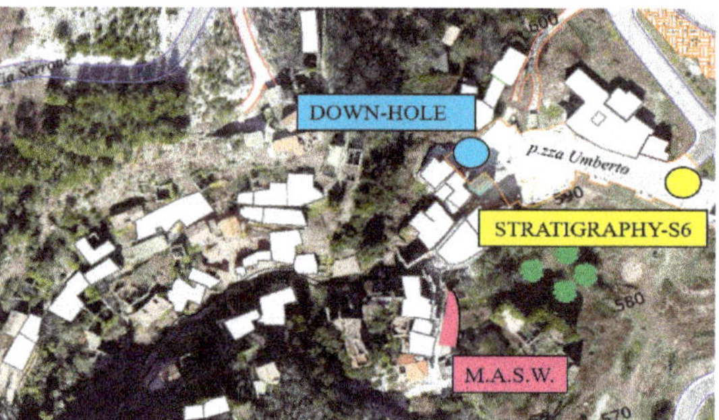

Figure 10. The identification of the in-situ tests.

Subsequently, the M.A.S.W test is carried out. This test consists of placing a series of equidistant geophones on the ground surface, which record the effects generated by external perturbations produced by a source (hammer strike on an instrumented plate). The data acquired and processed by a software show an average propagation velocity of 812 m/s, which corresponds to category A soil (rigid soil) [21].

Finally, the third test is executed by the Municipal Administration of Senerchia. This test, indicated with "S6" in Figure 10, provides a stratigraphy of the ground consisting of deposits of pyroclastic material and calcareous debris. Therefore, the subsoil category associated with this ground type is the C one [21].

The seismic microzonation study gives rise to the identification of 3 geomorphologic areas, as required by the municipal plan (Figure 11).

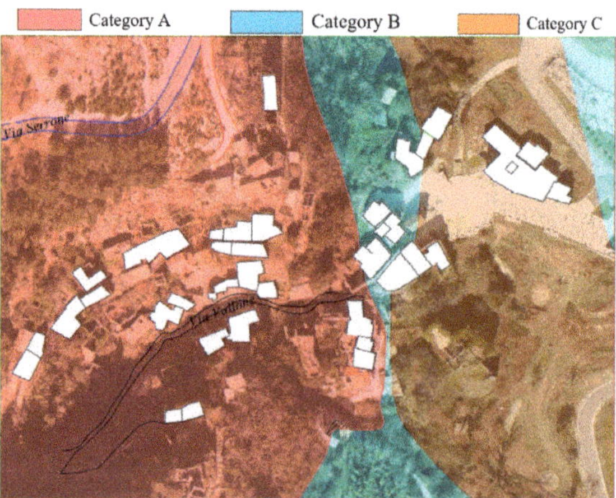

Figure 11. The geomorphologic characterisation of the area of Senerchia.

4.3. *Macroelement Numerical Analysis of an Aggregate Case Study*

In the current work a building aggregate typical of the urban tissue of Senerchia is selected as a case study representative of the typological class of buildings within the investigated area (Figure 12).

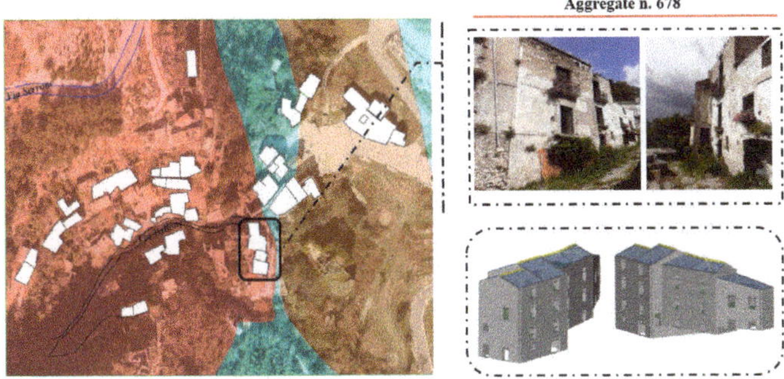

Figure 12. The identification and structural macroelement model of a typical urban aggregate.

Parametric non-linear static analyses are conducted with the 3Muri macro-elements software [46]. This software is based on the equivalent frame model, which schematises masonry walls as a set of single-dimension macro-elements represented by columns, beams and nodes [47]. The columns are vertical elements, which support vertical loads transmitted by the beams. The rigid nodes consist of masonry parts, confined between beams and columns, which result in them being undamaged due to the earthquake. In the performed analysis, the evaluation of the seismic risk factor can be defined as the ratio between the seismic demand acceleration, *D*, and the structure capacity one, *C* [48].

Subsequently, on the aggregate model described above, the evaluation of an appropriate local seismic amplification coefficient, f_{PGA}, is carried out in order to consider the seismic intensification effects due to different soil categories. This coefficient is defined as the ratio between the factor α_A, when the rigid soil (type A) is taken into account, and the factor α_{Soil}, based on the different geological conditions (topographic class and soil category) taken one by one.

The general formula to calculate f_{PGA} is defined as follows:

$$f_{PGA} = \frac{\alpha_A}{\alpha_{soil}} \tag{5}$$

Moreover, for an accurate assessment of the local seismic response, parametric analyses are performed in the two main directions (X and Y) of the building aggregate by varying the type soil and topography class according to the Italian Code [21].

Therefore, based on the achieved numerical results, the values of f_{PGA} are determined for each soil category (A, B, C, D, E) and for each topographic class, T_i (i = 1, 2, 3, 4). The results are summarized in Tables 3 and 4, respectively.

Table 3. The effect of the soil category variation on the local seismic amplification factor.

Soil Category	α_{soil}		$f_{PGA,X}$	$f_{PGA,Y}$
	X Direction	Y Direction		
A	0.614	0.420	-	-
B	0.487	0.322	1.26	1.30
C	0.410	0.269	1.50	1.56
D	0.361	0.214	1.70	1.96
E	0.388	0.252	1.58	1.66

Table 4. The effect of the topographic class variation on the local seismic amplification factor.

Top. Category	α_{soil}		$f_{PGA,X}$	$f_{PGA,Y}$
	X Direction	Y Direction		
T_1	0.614	0.420	-	-
T_2	0.512	0.349	1.20	1.20
T_3	0.512	0.349	1.20	1.20
T_4	0.439	0.299	1.40	1.40

The results show that the most unfavourable condition is obtained for the soil class D, which is associated with the local amplification coefficients equal to 1.70 (X direction) and 1.96 (Y direction). Similarly, considering the variation of the topographic category, it can be seen that the worst condition is given by T_4 (hillside with slope >30°). Based on the obtained results and depending on the geologic conditions detected for the case study, the mean damage degree μ_D of the investigated structural units determined in Section 3.3 can be amplified by the maximum f_{PGA} factor between those of the two analysis directions depicted in Tables 3 and 4 leading to the more correct expected damage grade $\mu_{D,s}$:

$$\mu_{D,s} = \mu_D \times f_{PGA} \tag{6}$$

Consequently, the new damage scenarios for each set of moment magnitudes and distances from the epicentre are plotted (Figure 13).

Figure 13. The new seismic damage scenarios considering the local amplification factor due to the geologic conditions.

For a synthetic representation of the damage occurred, Figure 14 shows the damage distribution obtained for the above-described combinations of magnitude and site-source distance.

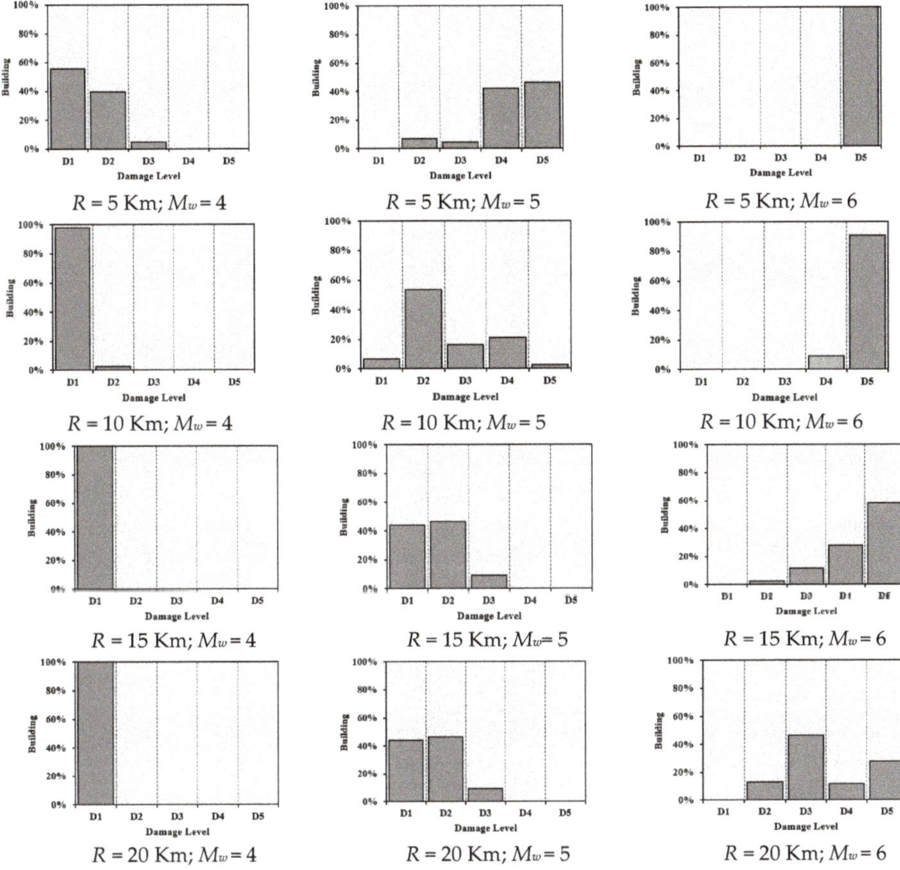

Figure 14. The damage distribution considering local seismic amplification effects.

Comparing the new damage scenario (Figure 14) to that reported in Section 3 (see Figure 8), it is possible to estimate the damage increase due to site effects, which is variable from 2% to 50%, as shown in Table 5.

Table 5. The global damage increases when considering the local site effects.

Magnitude M_w	Global Damage Increase (%)			
	R = 5 Km	R = 10 Km	R = 15 Km	R = 20 Km
4	5	2	-	-
5	50	42	28	14
6	-	20	49	47

Finally, the expected damages with and without local effects are shown in Figure 15 in the cases of magnitude 5 and epicentre distances of 5 Km and 20 Km.

It is noted that for an epicentre distance value of 5 km and magnitude M_w = 5, when considering the local site effects, damage thresholds D2 and D3 are attained with an occurrence probability less than that of the basic case, when the soil influence is only marginally evaluated. Contrarily, the D4 and D5 damage levels significantly increase due to the site effects. On the other hand, considering the same magnitude and an epicentre distance of 20 km, the site effects reduce the D1 damage level, but increase the D2 threshold and also introduce the D3 one.

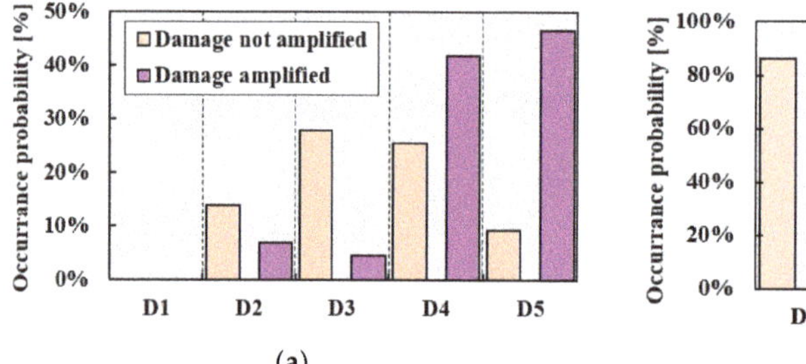

(a)

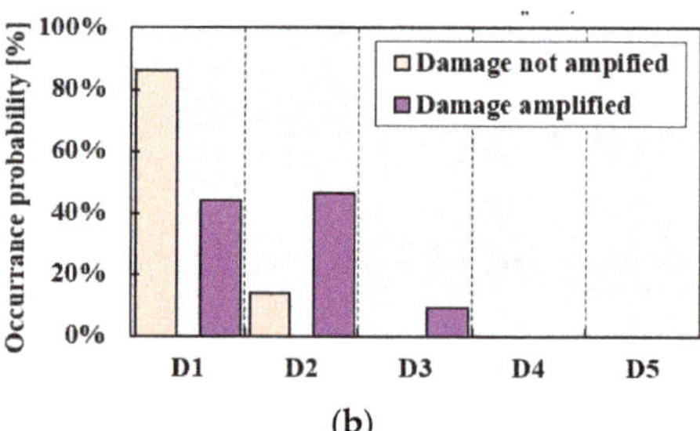

(b)

Figure 15. The damage scenarios with and without local seismic effects: comparison of results for M_w = 5 and R = 5 Km (**a**) and M_w = 5 and R = 20 Km (**b**).

5. Final Remarks

The study proposes the first results of a novel methodology to analyse the seismic vulnerability of masonry building aggregates located in historic centres considering the influence of geologic conditions. An urban sector of the historical build-up of Senerchia, in the district of Avellino (Italy), was identified as the study area for the application of the proposed analysis method.

Firstly, the characterisation of the typological classes of the urban sector examined was done by means of the CARTIS form, which has allowed us to classify also the building compounds from the structural point of view. From on-site recognition, the masonry buildings with timber floors were the most recurring typological class.

Subsequently, the seismic vulnerability of the inspected urban sector was estimated through a vulnerability index method conceived for building aggregates. It is worth noting that, on average, the global vulnerability index of the sector is equal to 0.40. Moreover, the mean vulnerability curves of inspected typological classes were obtained in order to evaluate their susceptibility at damage by varying the seismic intensity according to the EMS-98 scale.

From the analysis results, it was detected that, for moderate values of seismic intensity ($I_{EMS\text{-}98}$ < X), the expected damage is not relevant, but for higher values (X < $I_{EMS\text{-}98}$ < XII), the expected damage would cause an incipient collapse of the analysed building sample.

In the next analysis step, the damage scenarios of the investigated urban sector, based on a specific seismic attenuation law, were estimated for different moment magnitudes and site-source distances. The results obtained showed that when the site-source distance increases, the expected damage tends to diminish, attaining a stable condition for large distances.

Contrary, when the magnitude increases, the most unfavourable condition occurred for M_w = 6, when, independently from the epicentre distance, most of the buildings should collapse.

Finally, the parametric numerical analyses on a case study building aggregate were carried out for the definition of local seismic amplification factors, taking into account the different topographic classes and categories of soils defined in the new Italian standard NTC18. These factors have allowed us to increase the expected damages of buildings due to local amplification phenomena linked to geologic conditions. From the new damage scenarios, it was detected that the site effects lead to a damage increment variable from 2% to 50%, which was much more marked at the smallest considered distance. In addition, local seismic effects were considerable for larger magnitudes. In fact, for the magnitude M_w = 4, the global vulnerability averagely increases by 5%, while, for M_w = 5, the site effects produce a mean damage increment of 50%. Moreover, seismic amplification factors due to the soil condition increase the occurrence probability of attaining the largest damage thresholds.

As a conclusion, the seismic vulnerability and risk analysis of the investigated area represent the starting point to extend the method herein implemented—based on the influence of geologic conditions—to the whole territory of Senerchia, as well as to that of other municipalities, for the development of more correct analysis strategies for seismic risk mitigation.

Author Contributions: Methodology, A.F.; Investigation, N.C.; Data Curation, N.C. and A.F.; Writing—Original Draft Preparation, N.C.; Writing—Review & Editing, A.F.

Funding: This research did not receive any external funding.

Acknowledgments: The Authors would like to acknowledge Prof. Antonio Santo from University of Naples "Federico II", who provided the results of the geological-geotechnical study.

Conflicts of Interest: The authors declare no conflict of interest.

References

1. Galli, P.; Camassi, R. Rapporto Sugli Effetti del Terremoto Aquilano del 6 Aprile 2009. Available online: http://www.protezionecivile.gov.it (accessed on 2 January 2019). (In Italian)
2. Galli, P.; Castenetto, S.; Peronace, E. Rilievo macrosismico MCS Speditivo 1, Rapporto Finale. Available online: http://www.protezionecivile.gov.it (accessed on 3 January 2019). (In Italian)
3. National Institute of Geophysics and Volcanology (INGV). Rapporto Di Sintesi Sul Terremoto in Centro Italia Mw 6.5 del 30 Ottobre 2016. INGV Working Group on the Earthquake in Central Italy, 2016; pp. 1–49. Available online: https://ingvterremoti.files.wordpress.com (accessed on 9 January 2109). (In Italian)
4. Barbat, A.H.; Carreño, M.L.; Pujades, L.G.; Lantada, N.; Cardona, O.D.; Marulanda, M.C. Seismic vulnerability and risk evaluation methods for urban areas. A review with application to a pilot area. *Struct. Infrastruct. Eng.* **2010**, *6*, 17–38. [CrossRef]
5. Porter, K.A.; Beck, J.L.; Shaikhutdinov, R. Simplified estimation of economic seismic risk for buildings. *Earthq. Spectra* **2004**, *20*, 1239–1263. [CrossRef]
6. Zuccolo, E.; Vaccari, F.; Peresan, A.; Panza, G.F. Neo-Deterministic and Probabilistic Seismic Hazard Assessments: A Comparison over the Italian Territory. *Pure Appl. Geophys.* **2011**, *168*, 69–83. [CrossRef]
7. Basaglia, A.; Aprile, A.; Spacone, E.; Pilla, F. Performance-based Seismic Risk Assessment of Urban Systems. *Int. J. Archit. Herit.* **2018**, *12*, 1131–1149. [CrossRef]
8. Weißhuhn, P.; Müller, F.; Wiggering, H. Ecosystem Vulnerability Review: Proposal of an Interdisciplinary Ecosystem Assessment Approach. *Environ. Manag.* **2018**, *61*, 904–915. [CrossRef] [PubMed]
9. Cara, S.; Aprile, A.; Pelà, L.; Roca, P. Seismic Risk Assessment and Mitigation at Emergency Limit Condition of Historical Buildings along Strategic Urban Roadways. Application to the "Antiga Esquerra de L'Eixample" Neighborhood of Barcelona. *Int. J. Archit. Herit.* **2018**, *12*, 1055–1075. [CrossRef]
10. Jiménez, B.; Pelà, L.; Hurtado, M. Building survey forms for heterogeneous urban areas in seismically hazardous zones. Application to the historical center of Valparaíso, Chile. *Int. J. Archit. Herit.* **2018**, *12*, 1076–1111. [CrossRef]
11. Formisano, A.; Florio, G.; Landolfo, R.; Mazzolani, F.M. Numerical calibration of a simplified procedure for the seismic behaviour assessment of masonry building aggregates. In Proceedings of the 13th International Conference on Civil, Structural and Environmental Engineering Computing, Crete, Greece, 6–9 September 2011; p. 28.
12. Formisano, A.; Chieffo, N.; Fabbrocino, F.; Landolfo, R. Seismic vulnerability and damage of Italian historical centre: A case study in the Campania region. *AIP Conf. Proc.* **2017**, *1863*, 450007.
13. Rapone, D.; Brando, G.; Spacone, E.; De Matteis, G. Seismic vulnerability assessment of historic centers: Description of a predictive method and application to the case study of scanno (Abruzzi, Italy). *Int. J. Archit. Herit.* **2018**, *12*, 1171–1195. [CrossRef]
14. Clementi, F.; Gazzani, V.; Poiani, M.; Lenci, S. Assessment of seismic behaviour of heritage masonry buildings using numerical modelling. *J. Build. Eng.* **2016**, *8*, 29–47. [CrossRef]
15. Angelillo, M.; Lourenço, P.B.; Milani, G. Masonry behaviour and modelling. In *Mechanics of Masonry Structures*; CISM International Centre for Mechanical Sciences, Courses and Lectures; Angelillo, M., Udine, Eds.; Springer: Vienna, Austria, 2014; Volume 551, pp. 1–26.

16. Formisano, A.; Marzo, A. Simplified and refined methods for seismic vulnerability assessment and retrofitting of an Italian cultural heritage masonry building. *Comput. Struct.* **2017**, *180*, 13–26. [CrossRef]
17. Sá, L.; Morales-Esteban, A.; Durand Neyra, P. The 1531 earthquake revisited: Loss estimation in a historical perspective. *Bull. Earthq. Eng.* **2018**, *16*, 1–27. [CrossRef]
18. D'Ayala, D.; Ansal, A. Non linear push over assessment of heritage buildings in Istanbul to define seismic risk. *Bull. Earthq. Eng.* **2012**, *10*, 285–306. [CrossRef]
19. Evangelista, L.; del Gaudio, S.; Smerzini, C.; d'Onofrio, A.; Festa, G.; Iervolino, I.; Landolfi, L.; Paolucci, R.; Santo, A.; Silvestri, F. Physics-based seismic input for engineering applications: A case study in the Aterno river valley, Central Italy. *Bull. Earthq. Eng.* **2017**, *15*, 2645–2671. [CrossRef]
20. Giovinazzi, S. Geotechnical hazard representation for seismic risk analysis. *Bull. N. Z. Earthq. Eng.* **2009**, *42*, 221–234.
21. Ministerial Decree, DM 20/02/2018. "Updating of Technical Standards for Construction". Available online: http://www.gazzettaufficiale.it/eli/gu/2018/02/20/42/so/8/sg/pdf (accessed on 3 January 2018). (In Italian)
22. Calvanese, F.; Carchedi, F.; D'Amelio, R. Emigrazione e immigrazione in Campania: Il caso dell'Alto Sele. In *Materiali*; Ediesse: Roma, Italy, 2005; p. 336.
23. Scognamiglio, L.; Tinti, E.; Emolo, U.R.A.; Gallovic, F. Progetto S3—Scenari di Scuotimento in aree di Interesse Prioritario e/o Strategico. 2007. Available online: http://esse3.mi.ingv.it (accessed on 8 Januar 2019).
24. Pantosti, D.; Schwartz, D.P.; Valensise, G. Paleoseismology along the 1980 surface rupture of the irpinia fault: Implications for earthquake recurrence in the southern Apennines, Italy. *J. Geophys. Res. Solid Earth* **1993**, *98*, 6561–6577. [CrossRef]
25. Ameri, G.; Emolo, A.; Pacor, F.; Gallovič, F. Ground-Motion simulations for the 1980 M 6.9 Irpinia earthquake (Southern Italy) and scenario events. *Bull. Seismol. Soc. Am.* **2011**, *101*, 1136–1151. [CrossRef]
26. Pulinets, S.A.; Biagi, P.; Tramutoli, V.; Legen'ka, A.D.; Depuev, V.K. Irpinia earthquake 23 November 1980—Lesson from nature reviled by joint data analysis. *Ann. Geophys.* **2007**, *50*, 61–78. [CrossRef]
27. Pasquale, G.; Matteis, R.; Romeo, A.; Maresca, R. Earthquake focal mechanisms and stress inversion in the Irpinia Region (Southern Italy). *J. Seismol.* **2009**, *13*, 107–124. [CrossRef]
28. Mangalathu, S.; Jeon, J.; Padgett, J.E.; Desroches, R. ANCOVA-based grouping of bridge classes for seismic fragility assessment. *Eng. Struct.* **2016**, *123*, 379–394. [CrossRef]
29. Zuccaro, G.; Della Bella, M.; Papa, F. Caratterizzazione tipologico strutturali a scala nazionale. In Proceedings of the 9th National Conference ANIDIS, L'ingegneria Sismica, Italy, 20–23 September 1999.
30. Formisano, A.; Florio, G.; Landolfo, R.; Mazzolani, F.M. Numerical calibration of an easy method for seismic behaviour assessment on large scale of masonry building aggregates. *Adv. Eng. Softw.* **2015**, *80*, 116–138. [CrossRef]
31. Formisano, A.; Chieffo, N.; Mosoarca, M. Seismic Vulnerability and Damage Speedy Estimation of an Urban Sector within the Municipality of San Potito Sannitico (Caserta, Italy). *Open Civ. Eng. J.* **2017**, 1106–1121. [CrossRef]
32. Benedetti, D.; Petrini, V. Sulla vulnerabilita sismica di edifici in muratura: Proposta su un metodo di valutazione. *L'industria Delle Costr.* **1984**, *149*, 64–72.
33. Giovinazzi, S.; Lagomarsino, S. A Macroseismic Method for the Vulnerability Assessment of Buildings. In Proceedings of the 13th World Conference on Earthquake Engineering, Vancouver, BC, Canada, 1–6 August 2004.
34. Lagomarsino, S. On the vulnerability assessment of monumental buildings. *Bull. Earthq. Eng.* **2006**, *4*, 445–463. [CrossRef]
35. Maio, R.; Ferreira, T.M.; Vicente, R.; Estêvão, J. Seismic vulnerability assessment of historical urban centres: Case study of the old city centre of Faro, Portugal. *J. Risk Res.* **2016**, *19*, 551–580. [CrossRef]
36. Elnashai, A.S.; Di Sarno, L. *Fundamentals of Earthquake Engineering*, 1st ed.; Elnashai, A.S., Di Sarno, L., Eds.; p. 374. Available online: https://fac.ksu.edu.sa/sites/default/files/fundamentals_of_earthquake_engineering.pdf (accessed on 2 January 2019).
37. Vilanova, S.P.; Fonseca, J.F.B.D. Probabilistic seismic-hazard assessment for Portugal. *Bull. Seismol. Soc. Am.* **2007**, *97*, 1702–1717. [CrossRef]
38. Toro, G.R.; Abrahamson, N.A.; Schneider, J.F. Model of Strong Ground Motions from Earthquakes in Central and Eastern North America: Best Estimates and Uncertainties. *Seismol. Res. Lett.* **1997**, *68*, 41–57. [CrossRef]

39. Atkinson, G.M.; Boore, D.M. Stochastic point-source modeling of ground motions in the Cascadia region. *Seismol. Res. Lett.* **1997**, 74–85. [CrossRef]
40. Atkinson, G.M.; Kaka, S.L.I. Relationships between felt intensity and instrumental ground motion in the Central United States and California. *Bull. Seismol. Soc. Am.* **2007**, *97*, 497–510. [CrossRef]
41. Ambraseys, N.N.; Simpson, K.A.; Bommer, J.J. Prediction of Horizontal Response Spectra in Europe. *Earthq. Eng. Struct. Dyn* **1996**, *25*, 371–400. [CrossRef]
42. Crespellani, T.; Garzonio, C.A. Seismic risk assessment for the preservation of historical buildings in the city of Gubbio. In Geotechnical engineering for the preservation of monuments and historic sites. In Proceedings of the International Symposium on Geotechnical Engineering for the Preservation of Monuments and Historic Sites, Napoli, Italy, 3–4 October 1997.
43. Locati, M. DBMI15, the 2015 Version of the Italian Macroseismic Database, 2016 (In Italian). Available online: https://emidius.mi.ingv.it (accessed on 2 January 2019).
44. Grünthal, G. European Macroseismic Scale 1998. Available online: http://lib.riskreductionafrica.org/bitstream/handle/123456789/1193/1281.European%20Macroseismic%20Scale%201998.pdf?sequence=1 (accessed on 7 January 2019).
45. Formisano, A. Local- and global-scale seismic analyses of historical masonry compounds in San Pio delle Camere (L'Aquila, Italy). *Nat. Hazards* **2017**, *86*, 465–487. [CrossRef]
46. S.T.A data srl, "3Muri 10.9.0—User Manual". Available online: www.3muri.com/documenti/brochure/it/3Muri10.9.0_IT.pdf (accessed on 11 January 2019). (In Italian)
47. Maio, R.; Vicente, R.; Formisano, A.; Varum, H. Seismic vulnerability assessment of an old stone masonry building aggregate in San Pio delle Camere, Italy. In Proceedings of the 2nd European Conference on Earthquake Engineering and Seismology, Istanbul, Turkey, 24–29 August 2014.
48. Formisano, A. Theoretical and Numerical Seismic Analysis of Masonry Building Aggregates: Case Studies in San Pio Delle Camere (L'Aquila, Italy). *J. Earthq. Eng.* **2017**, *21*, 227–245. [CrossRef]

Article

Impacts of Earthquakes on Energy Security in the Eurasian Economic Union: Resilience of the Electricity Transmission Networks in Russia, Kazakhstan, and Kyrgyzstan

Dmitrii Iakubovskii [1,2,*], Nadejda Komendantova [3,4], Elena Rovenskaya [3,5], Dmitry Krupenev [1,2] and Denis Boyarkin [1,2]

1 Melentiev Energy Systems Institute of the Siberian Branch of the Russian Academy of Science (ESI SB RAS), 664033 Irkutsk, Russia; krupenev@isem.irk.ru (D.K.); boyarkin_denis@mail.ru (D.B.)
2 Institute of Power Engineering, Irkutsk National Research Technical University, 664074 Irkutsk, Russia
3 International Institute for Applied Systems Analysis (IIASA), Schlossplatz 1, A-2361 Laxenburg, Austria; komendan@iiasa.ac.at (N.K.); rovenska@iiasa.ac.at (E.R.)
4 Department of Environmental Systems Science, Institute for Environmental Decisions (ETH), 8092 Zurich, Switzerland
5 Faculty of Computational Mathematics and Cybernetics, Lomonosov Moscow State University, 119991 Moscow, Russia
* Correspondence: yakubovskii.dmit@mail.ru; Tel.: +7-924-605-0539

Received: 14 December 2018; Accepted: 17 January 2019; Published: 21 January 2019

Abstract: In our research, we focus on the reliability of the interconnected electricity supply system of three countries of the Eurasian Economic Union (EAEU)—Russia, Kazakhstan, and Kyrgyzstan. We apply a mathematical model to evaluate the reliability of the electricity supply system under the threat of earthquakes. Earthquakes can damage elements of electricity grids and, considering the interconnectivity of electricity supply systems in the EAEU, effects in the aftermath of earthquakes can be far-reaching and even transboundary. This necessitates the development of coordinated policies and risk management strategies to deal with electricity outage risks in the EAEU. In our study, the earthquake probability is derived from seismic zone maps, while damage events are computed using maps of energy power systems. In addition, we determine which elements of the system are susceptible to failure due to an earthquake of a given magnitude. We conduct a scenario analysis of earthquakes and their impacts on the reliability of the power supply system, considering potential energy losses and threats to energy security. An analysis of the resilience of electricity transmission grids allows us to determine the critical interconnection lines in terms of exposure to earthquake risk, as well as exposure to total systemic loss. We also identify the most critical interconnection lines where power outages can lead to the destabilization of the entire power supply system. Some examples of such lines are at the border of Kazakhstan and Kyrgyzstan, where power outages can lead to serious economic costs and electricity outages.

Keywords: risk assessment; natural hazards; earthquake risk; energy security; reliability of power supply; Eurasian Economic Union (EAEU); integration process; common electricity market

1. Introduction

Regional integration processes, like the Eurasian Economic Integration, are significantly influencing national energy security, due to the growing interconnectivity of critical infrastructures such as electricity transmission lines. The Eurasian Economic Union (EAEU), which currently consists of Armenia, Belarus, Kazakhstan, Kyrgyzstan, and Russia, creates a basis for such integration. The EAEU is currently facing a process of integration of its energy markets due to the planned creation of a common electricity market by 2019. Member states are already preparing strategies and plans for the integration of their electricity markets. The common electricity market will be based on existing trading platforms, while the pricing mechanism, as well as various other important aspects such as common quality standards, still need to be established [1], and existing barriers (described in detail in the "obstacles register" in the energy field [2]) will need to be overcome.

Electricity transmission infrastructure is a critical challenge for the EAEU countries as current electricity transmission grids date from the time of the Soviet Union and are in need of maintenance and modernization [3]. The liberalized trade of electricity within the EAEU, which is foreseen to result from the creation of a common electricity market, will change the load distribution on the elements of the grids and may, thus, lead to unexpected system stresses. These may, in turn, result in power outages and cascading effects of these outages on the entire electricity supply system of the Union.

The main goal of the EAEU's planned common electricity market is to provide a reliable and affordable electricity supply to households, as well as to the private and public sector. Thus, energy security is essential for the economic prosperity of all EAEU member states. At minimum, the reliability of electric power systems and interconnections should be kept at an acceptable level, while efforts should be made to increase it where possible. In order to achieve this, it is necessary to ensure the timely planning of the energy power systems (EPSs) that have to be developed, while ensuring the reliability of the power supply. Apart from having an economic function in that it guarantees the satisfaction of current and future energy demands in the EAEU region, a reliable common electricity system also has a role to play in the larger context of the development of integration processes, both within the EAEU and between the EAEU and its neighboring countries.

However, furthering integration processes on the electricity markets of EAEU countries and the interconnectivity of electricity supply system brings not only benefits, but also risks. Such risks include cascading power outages that originate in one part of the grid and then spread to its other parts and, potentially also, to the territories of other countries. A major challenge in the context of energy security in the EAEU is to guarantee a reliable, continuous, and sufficient energy supply considering all relevant existing risks in any part of the interconnected electricity transmission system and their systemic effect on the entire system.

The question around the reliability of electricity supply in the wake of natural hazards gained the attention of the EAEU states in the framework of energy strategies with plans specified up to 2030 [4–6] and, also, at the level of the EAEU Board [7]. Even though such interruptions are not frequent, each power outage or blackout affects several millions of people [8,9] and causes vast economic losses and damages [10].

Integrated risk assessment and risk governance are essential to ensure the reliability of a system as complex as the electricity transmission system within the EAEU. The risk assessment should also address the issue of an optimal balance between the generation, transmission, and distribution of electricity in each node, and should be based on plausible scenarios of future electricity demand and electricity generation. Such a risk and reliability assessment should be carried out periodically, taking into account newly introduced equipment, EPS elements, and various external factors affecting the operation of the system [4–7]. Thus, the concepts of the EAEU countries provide a direction for the development of the common energy system of this Union that takes the required high reliability of EPSs into account [11].

In this paper, we focus on three out of the five EAEU countries: Russia, Kazakhstan, and Kyrgyzstan, and on earthquakes as one of the major natural hazards that can cause shortages of electricity supply in these three countries. Armenia and Belarus were excluded, as Armenia is not physically linked to the three countries that were included, and Belarus is located in the western part of the Eastern European platform [12], where the probability of earthquakes strong enough to cause a disconnection of elements of the electricity infrastructure is very small [13].

Several kinds of natural hazards, including earthquakes, landslides, cyclones, heat waves, and manmade risks such as cyber threats, are able to affect the physical reliability of the electricity grids infrastructure, as well as to influence its transmission capacity in these three countries. We focus on earthquakes as the most frequent risk in the Central Asian region [14].

Other authors, for example [11], have assessed the impacts of earthquakes on the electricity transmission infrastructure for each EAEU member country separately. Yet, to the best of our knowledge, an integrated assessment taking into account interconnections between the electricity transmission systems of the three countries is not available. This paper addresses this gap in the literature.

In this work, we use a modification of the simplified reliability assessment approach, based on the so-called "$N - i$ criterion" [15]. The "$N - i$ criterion" is the rule according to which the elements remaining in operation within a Transmission System Operator's control area after the occurrence of a contingency are capable of accommodating the new operational situation without violating operational security limits. The proposed methodology includes three stages. At the first stage, we generate system states resulting from switching off one, two, or three system elements (lines and/or nodes), imitating their collapse due to an earthquake. We sort through all possible combinations of the disabled system elements. At the second stage, we optimize the allocation of flows across lines such that the total power shortages are minimized. At the third and last stage, we assess the significance of each line depending on its reliability and the probability of failure from the effects of the earthquake. The main modification we make here, compared to [14], is that, at the first stage, we use the probabilities of strong earthquakes that are capable of destroying lines, instead of the probabilities of line failures. Using this approach, we investigate the impacts of earthquakes on the electricity transmission system, taking into consideration the existing interconnections within the EAEU region and the ongoing integration processes.

2. Background

2.1. The Eurasian Economic Union and the Common Electricity Market

The EAEU was established in January 2015 and currently includes Russia, Belarus, Kazakhstan, Armenia, and Kyrgyzstan. The goal of the EAEU is to strengthen political, social, economic and environmental cooperation among its member states. This cooperation also includes the creation of common energy markets [1,16]. The electricity market is an essential part of the common energy market relying on existing electricity interconnections within EAEU countries and between them. Its drivers and barriers, as well as the roadmap and strategy up to the year 2025, are analyzed in [17]. The creation of common energy and electricity markets within the EAEU region has the potential for generating new opportunities for cooperation and larger spheres of integration [18].

The Russian EPS is one of the largest energy systems in the world. It includes many complex processes, such as production, transmission, distribution of electricity, and centralized operational and technological management. The system covers a territory of about 7000 km from west to east, and 3000 km from north to south. This system consists of seven United Power Systems (UPSs), six of which—UPS Center, UPS South, UPS North-West, UPS Middle Volga, UPS Ural, and UPS of Siberia—operate in a synchronous (parallel) mode, while the seventh, UPS East, operates in isolation from the other Russian UPSs [19]. The Russian EPS includes over 700 power stations whose capacity is greater than 5 MW, and many more smaller ones. As of 1 January 2017, the total installed capacity amounts to 236,343.63 MW, with each UPS operating with a surplus of electricity generation. The system is huge, and various procedures and types of protective equipment have been put in place to ensure a safe and stable operation. Automatic devices that disconnect broken and unstable parts of the system from the rest of the network prevent local collapses in the system from spreading further.

Kazakhstan's EPS is also complex and peculiar in terms of topology and energy balance. It is divided into three large energy regions, each of which consists of a number of regional formations. These are the UPS West, UPS North, and UPS South. UPS West is located in the western part of the country and occupies a relatively small area. It is isolated from the rest of the energy regions of Kazakhstan. UPS West usually has an electricity deficit and exports electricity from Russia through its interconnections with this country. UPS North and UPS South are more advanced. They have several connections among themselves and, also, interconnections with Russia and Kyrgyzstan. UPS North has various connections with different UPSs of Russia, which works for both export and import. A special broadband line between UPS West and UPS North is currently under construction. In Kazakhstan, electric power generation is carried out by 118 power plants using various forms of production. As of 1 January 2017, the total installed capacity of Kazakhstan's power plants is 22,055.5 MW, and the available capacity is 18,789.1 MW [20].

The EPS of Kyrgyzstan is mainly based on a multitude of hydro power plants (HPP) and the plan for the development of the country's electric power industry includes the construction of many additional hydroelectric power stations with varying generation ability. At the same time, the system is a single whole without any division. The current electricity demand and generation in the country are in a balanced equilibrium. The total amount of generation is 3797.90 MW. The EPS of Kyrgyzstan has a small number of interconnections with Kazakhstan.

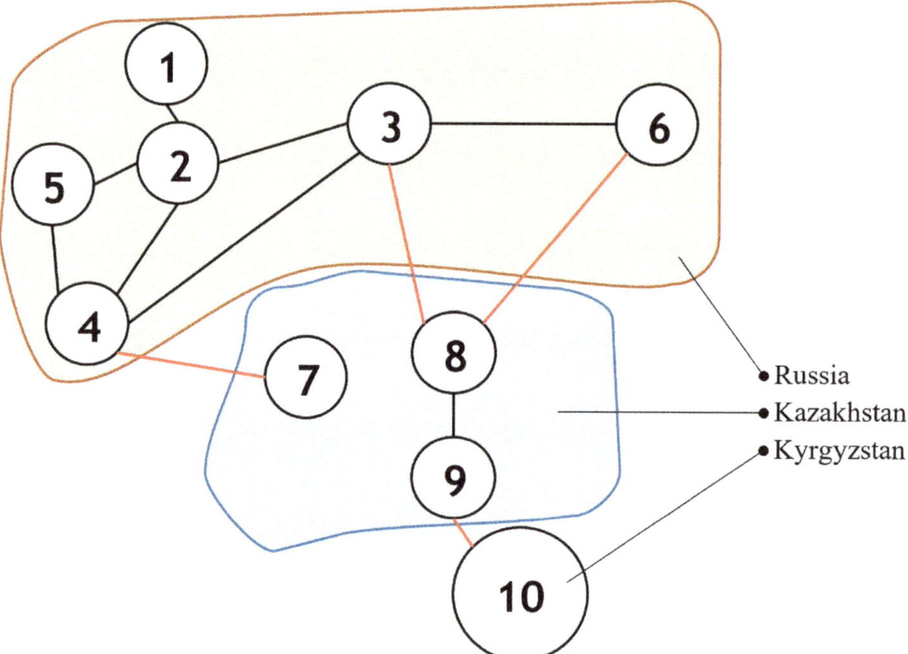

Figure 1. The diagram depicts the national energy power systems (EPSs) and united power systems (UPSs) described above schematically. For convenience of modeling, the authors of this paper have enumerated the UPSs.

Based on various sources of public data, including reports from the national Ministries of Energy, national energy companies, magazines, and other reference materials, we collected a dataset containing information on seismic zone maps. These included probabilities of earthquakes, the EPS schemes of each country, and information on each power plant, as well as data on resistance parameters for the seismic evaluation of each type of EPS element used in the calibration of our model.

Figure 1 schematically depicts the national EPSs and UPSs described above. In total, the system under consideration in this paper contains 10 nodes (UPSs) with 12 connections between them. We did not include Russia's UPS East here because it has no connection with the other considered UPSs.

2.2. The Need to Address the Risks of Power Outages and Blackouts

During the last decade, the number of power outages—in other words short- or long-term states of electric power loss, in a given area or section of a power grid, caused by equipment failure resulting from a failure of the power supply—was getting more frequent. The number of large-scale blackouts, that is to say, when electric power is cut off for a general region especially due to shortages, mechanical failure, or overuse by consumers, also increased significantly. However, even small outages can have disastrous effects on unprepared businesses and the common market. Today, the task of protecting electricity transmission systems has become a greater challenge than ever before, as several blackouts have occurred over the last five years. These were caused by different factors, one of which is the growing interconnectivity of electricity transmission systems. Previous blackouts in Europe had severe consequences. There are three historical cases of blackouts in Europe, namely, the 2003 blackout that affected Italy and Switzerland, the blackout in Sweden and Denmark that happened during the same year, and the 2006 blackout in Germany. The 2003 blackout, that started in Switzerland and then also affected Italy, left 56 million people without electricity. The second blackout of 2003 affected 1.6 million people in Sweden and 2.4 million people in Denmark. It also resulted in 4700 MW of load being lost in Sweden and 1850 MW in Denmark. The blackout of 2006 in Germany lasted for up to two hours. This was a major blackout which affected more than 15 million people. The blackout had cascading effects on people in Poland, Benelux countries, France, Portugal, Spain, Greece, the Balkans, and even Morocco [21,22].

The EAEU region is experiencing both of the above—small local electricity failures and huge blackouts. Recent examples include the unexpected power outage in Siberia [10] and the far east of Russia, and periodic power outages in the Almaty region of Kazakhstan. UPS East also experienced a blackout on the evening of August 2017. The press service of the Russian Ministry of Energy commented that the collapse was due to a short circuit in one of the transmission lines with a voltage of 220 KV. This caused disruptions in the United Energy System of the East (UES East) and, because of this blackout, there was a mass shutdown of consumers in the Amur Region, the Khabarovsk Territory, and Primorye [9]. One of the most influential incidents occurred on 15 April 2009, when, due to an accident in the Central Asian energy system, the consumers of Almaty and the Almaty region were disconnected from electricity. The reason was the emergency shutdown of the high-voltage lines 500 kV Toktogul-Lachin and Toktogul-Frunze in Kyrgyzstan, which led to the shutdown of the Toktogul HPP line (Kyrgyzstan) and the transmission of capacity for the transit of 500kV to the North–South Kazakhstan line [8].

Power outages and blackouts have a huge impact on billions of customers. They can also cause problems in the economy, disturb the functioning of markets, and slow down economic development and growth. All of these factors can lead to an increase in social pressures in a society. The reliability of the electricity supply system is also affected by multiple risks or multi-risks, including cascading and systemic risks. These include many different natural hazards, such as earthquakes, floods, heat waves, and storms that can destroy infrastructure and power generation stations, or cut off power lines. All of these hazards are among the main causes of power outages. Although risk assessment for single risks exist, problems often lie in the area of risk governance, as risk mitigation strategies are frequently implemented to address individual risks, rather than their cascading or systemic effects.

The interconnected electricity system of the EAEU region requires special attention in light of its vulnerability in terms of the huge territory it covers and the several thousand kilometers of electricity transmission lines that run across the region. The probability of earthquakes occurring is high in Kazakhstan and Kyrgyzstan when one takes into consideration that these countries are located in the Mediterranean–Trans-Asian seismic belt. The vulnerability of electricity transmission lines in these countries is increased due to the state of the electricity infrastructure, most of which is overloaded and needs renovation, as well as the huge deserted areas where infrastructure is deployed. According to the Central Asia Earthquake Risk Reduction Forum (2015), the Central Asian region is one of the regions in the world that is most vulnerable to many natural hazards, among which, earthquakes are the most catastrophic.

The global seismic hazard map [23] shows that the Central Asian region is in a zone with high seismic activity where earthquakes are frequent, and there is a probability for high intensity earthquakes, which are really damaging to electricity transmission infrastructure.

Even though the probability of earthquake risks in Russia is significantly lower, the electricity transmission infrastructure in the country might also be impacted due to interconnectivity with the electricity transmission grids in Central Asian countries. In addition, considering the large territory of Russia, the regions of Siberia, the Caucasus and the Southern Federal District are under the influence of this seismic belt and, accordingly, have a high probability of earthquakes occurring.

Thus, the EAEU is a large union of countries, each of which is vulnerable to one or another type of natural hazard risk. As the first steps towards a common electricity market have already been taken through the preparation of common rules and management mechanisms, in order to maintain the reliability of supply within this system, it is, however, necessary to assess the impacts of natural hazard risks.

3. Methodology

Our methodology included three steps, each including several stages. The first step was the process of collecting data. During the second, the focus was on the mathematical model, scenarios, and simulations. The third step involved the analysis of the simulation results.

3.1. Data

The data collection included five stages. First, we collected seismic zone maps for each of the three countries: the Russian Federation, the Republic of Kazakhstan, and the Kyrgyz Republic. Maps were taken from public sources [24–26].

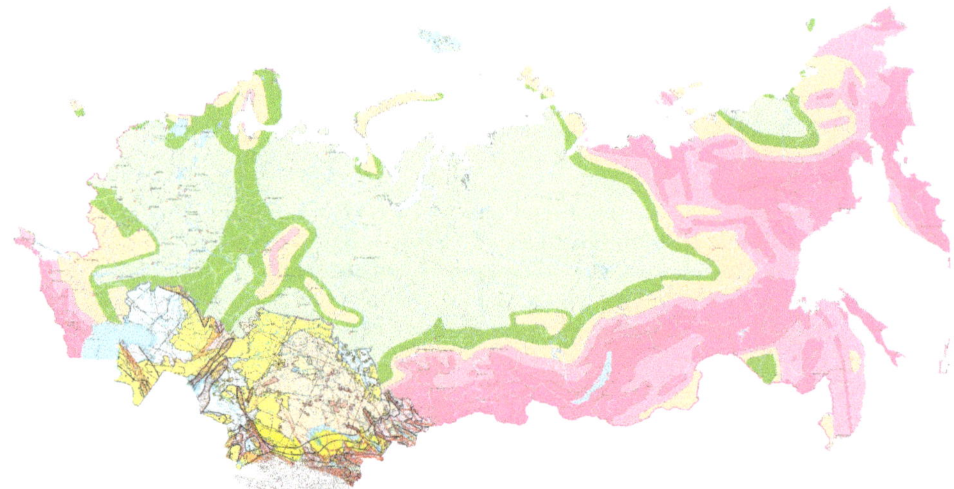

Figure 2. Combined seismic zone maps of (Russia, Kazakhstan, and Kyrgyzstan). Source: (1) http://seismos-u.ifz.ru; (2) http://www.seismo.kg; (3) http://geolog.at.ua.

As represented in Figure 2, we used three different types of maps. Different seismic zone maps were developed for each country included in this study. The maps were developed using various methodologies but contained comparable datasets. The Russian Federation map consists of four maps. Each of these contains zones delineated by various colors that indicate the intensity of the earthquake and the probability of system failure. The Kazakhstan map comprises polylines and signs of earthquake intensity scores and probabilities. The seismic zone map of Kyrgyzstan also contains polylines and signs of the frequency of earthquakes with various intensity scores. These scores were developed using data from the Institute of Seismology NAS of Kyrgyzstan on the probability of an earthquake with its frequency of occurrence in years.

Second, we assembled EPS schemes of these countries, which we also received from public sources, such as seismological institutes or other geophysical sources [27–29].

Figure 3 depicts a combined map with public information of EPS in this region. We used a standard method from classical methodology and created a simplified scheme (Figure 1).

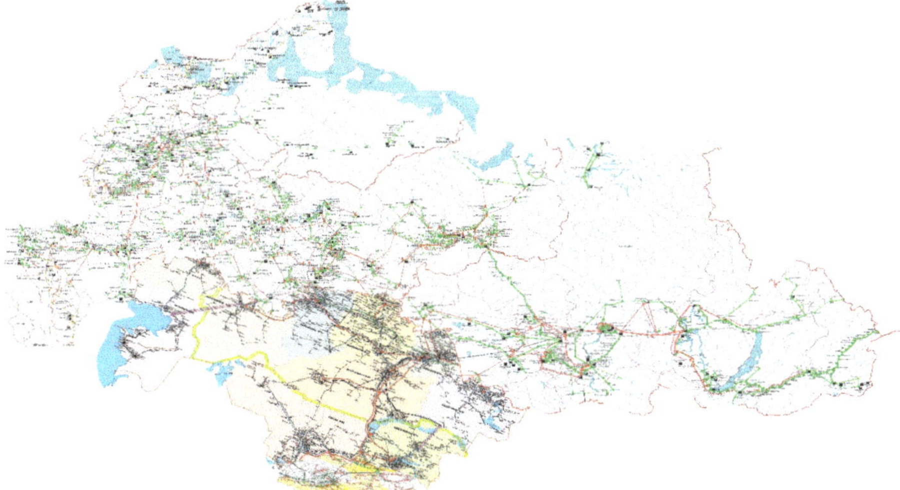

Figure 3. Combined EPS map of Russia, Kazakhstan, and Kyrgyzstan. Source: (1) http://energybase.ru/map; (2) http://www.kegoc.kz; (3) http://nesk.kg/.

Thirdly, we collected information for each power plant in the three abovementioned countries, as well as for each interconnection between them [30,31]. In order to have accurate information about each station in these systems, we conducted an in-depth data search. As a result, we compiled a catalog of stations and interconnections. During the fourth stage, we collected data on resistance parameters for the seismic evaluation of each type of EPS element. Then, we combined all the data, overlapping the maps of the EPS schemes and the seismic zones.

Finally, we compared the resistance parameters of each EPS element on the schemes and the seismic score of areas on the seismic zone maps where it dislocates, and identified the probability of an earthquake occurring for each EPS element in our study area (Figure 4). Thus, each station and line were given a probability of failure in accordance with the probability of an earthquake occurring that is severe enough to destroy or damage elements of the system.

Figure 4. Combined and overlapped EPS and seismic zone map of Russia, Kazakhstan, and Kyrgyzstan. Source: (1) http://seismos-u.ifz.ru; (2) http://www.seismo.kg; (3) http://geolog.at.ua; (4) http://energybase.ru/map; (5) http://www.kegoc.kz; (6) http://nesk.kg/.

3.2. Methods

The methodology of this research is based on the classical theory of reliability. The standard methodology includes several stages. First, the scheme of the system is created. Then, the probability of failure is determined for each EPS element. This probability of failure for each EPS element, Q_k, is determined using statistical data and is based on real events.

$$Q_k = \frac{(C_k * R_k)}{T},\ k = 1, \ldots, K \tag{1}$$

For lines, the probability is calculated according to Formula (1), where C_k is the total number of earthquakes for the period investigated (count events), R_k is the repair time for an EPS element (hours), T is the time when the events under investigation occurred, and K is the number of EPS elements. Next, a model with a set of nodes and lines is created, where the node contains data on demand and generation, and the line denotes its capacity. Following this, various situations are simulated, and the power deficits in the system are calculated. Ultimately, according to the received data, we determine the reliability of certain objects in the system [32,33]. We modified the methodology that was used to assess the reliability of the interconnected electricity systems of the three countries. We also included the methodology for assessing the reliability of power systems developed at the Melentiev Energy Systems Institute of the Siberian Branch of the Russian Academy of Sciences, to obtain distributions with mathematical expectations of the risk value for working and non-working lines. Below is a description of this method.

As already noted, there are several methodologies of reliability assessment. We use the main methodology of the classical theory of reliability to calculate the probability of EPS element failure. It is important that input data contain a set of different events that influence EPS elements. The set contains only events that cause damage that can provoke the disabling of an element and/or part of the system. According to the classical theory, the probability of EPS element failure can be calculated using Formula (1).

However, in the proposed methodology, we used the probability of an earthquake occurring as the probability of failure for an element located in the affected area, when the intensity of the earthquake exceeded the threshold of the inherent stability of the EPS elements. This approach is very effective because, already at this stage, it allows us to analytically determine which of the system elements can be corrupted and the associated probability.

During the next step, where the simulation of the system operation took place, we used the basic approaches of the classical reliability theory and simulated a large set of scenarios. To do this, we specifically created a distribution flow model based on the EPS scheme. Next, we applied a deterministic approach and used the method of scenario formation according to the "$N - i$ criterion", where i is the number of system elements that were switched off at once, and that did not have a zero probability of failure. In the third step, each scenario was optimized by minimizing the power shortage. This simulation gave us information about deficits in the system.

During this step, we worked with all the information obtained in the first and the second steps. We applied the collected information to create two distributions for each line (interconnection) between the EPSs of investigated countries. The first distribution considered the sum of multiple deficits and probabilities when the line worked, and the second distribution considered the sum of multiple deficits and probabilities when the line did not work. We calculated the risk value of losses within a set containing a working line in focus:

$$M = \sum_{n=1}^{N} d_n * p_n, \tag{2a}$$

and the risk value of losses in a complementary set:

$$\overline{M} = \sum_{n=1}^{\overline{N}} d_n * p_n, \tag{2b}$$

where d_n is the deficit in node n, p_n is the full probability of a node n to collapse due to an earthquake, N is the set of nodes an investigated line forms part of, and $\overline{N}$ is the set of nodes that the line being investigated is not a part of. The last step is aimed at defining the value of importance of each line (interconnection).

3.3. Simulations

In this part of the research work, we reviewed the model [34] for the estimation of power shortages in electric power systems with quadratic (in the value of power transmitted) power losses in transmission lines. The current model is rather common, and realistically represents the flow distribution. The problem under investigation is one of minimization of power shortages with quadratic power losses. A problem description is necessary in order to determine the optimal distribution flow in full EPS considering the defined values for each node of working generation and required consumer demand, as well as the capacity and coefficient of power losses for each line.

The mathematical formulation employs the minimization of the sum of power deficits:

$$\min_{y} \sum_{i=1}^{n} (\overline{y}_i - y_i). \tag{3}$$

We must also consider the balance constrains with quadratic power loses in nodes:

$$x_i - y_i + \sum_{j=1}^{n} (1 - a_{ji} z_{ji}) z_{ji} - \sum_{j=1}^{n} z_{ij} = 0,\ i = 1, \ldots, n,\ i \neq j. \tag{4}$$

It is very important to note that flow can only be in one direction. We added flow direction constrains to account for this:

$$z_{ji} * z_{ij} = 0,\ i = 1,\ldots,n,\ j = 1,\ldots,n,\ i \neq j. \tag{5}$$

The constrains are as follows:

$$0 \leq y_i \leq \overline{y}_i,\ i = 1,\ldots,n, \tag{6}$$

$$0 \leq x_i \leq \overline{x}_i,\ i = 1,\ldots,n, \tag{7}$$

$$0 \leq z_{ij} \leq \overline{z}_{ij},\ i = 1,\ldots,n,\ j = 1,\ldots,n,\ i \neq j, \tag{8}$$

where x_i is the power generation in node i, $\overline{x}_i$ is the current maximal generation for node i, y_i is the power demand that must be covered for node i, and $\overline{y}_i$ is the current maximal demand for node i. z_{ij} is the value of the flow from node i to node j, z_{ji} is the value of the flow, but for the opposite direction, $\overline{z}_{ij}$ is the capacity of the flow between node i and j, and $\overline{z}_{ji}$ is the capacity of the flow for the opposite direction. By running the model, we get incoming flow with power loses where a_{ji} is specified as positive coefficients of unit power losses during the transmission from node j to node i, $i \neq j, i = 1,\ldots,n, j = 1,\ldots,n$.

An important step in this work was creating various states of the system. We defined a large set of states by "$N - i$" criteria, where $= \{1,2,3\}$. The $N - 1$ criterion is a minimum system security measure that the system operator should use to model the transmission network to address redundancy while avoiding potential power interruptions and/or system failures. The $N - 1$ criterion set means that there are various nonrepeating states with a single EPS element that is not working (failed, switched off, broke, etc.). [34] The other criteria have the same meaning, but the set contains two and three not-working elements in one set. For example, we generated a start set with numbers of each EPS element, and created three sets by applying these criteria (other sets contain a number of not-working elements).

We solved our optimization problem (3–8) by applying the steepest descent algorithm [35]. In addition, the problem was transformed from one with constraints to an unconstrained nonlinear programming problem. This became possible with the use of the penalty method, which let us use simple optimization methods, such as gradient and steepest descent, for solving some complex problems. The main changes were in the objective function, where we added all constraints in the penalty form. This implies that small changes or divergences will be influential in the fully optimized function because of quadratic penalties. The main objective function is to get the minimal power shortage, using optimized and starting values of loading. We can calculate the deficit in each node $d_n = \overline{y}_n - y_n$. After the simulations, we got a set of deficits for each node in each stage, and used them to calculate the risk value of losses (2a, 2b).

3.4. Probabilities

We used the following calculation algorithm to generate a forecast for the emergency of the power system elements for 2018:

1. We used the formula for the exponential change in the probability of a failure-free operation of an element of the system:
$$Q = 1 - e^{-\lambda \widetilde{T}}, \tag{9}$$
where Q is the probability of failure, λ is the intensity of equipment failures, and $\widetilde{T}$ is the period of accumulation of the probability of failure;
2. Since data on the breakdown of each line in the next 50 years had already been found, we used the probability of failure for the lines;

3. We substituted the available values in Formula (9) and calculated the intensity of equipment failures:

$$\lambda = \frac{ln(1-Q)}{\tilde{T}}. \tag{10}$$

4. If the intensity of failures and the last strong earthquake in the surrounding areas were defined, we calculated the probability of equipment failure, for example, the power line being interconnected.

4. Results

Our research allowed us to develop two sets of results. The first are analytical results obtained during the search for and processing of input data. These results concern the probability of failure in each line. The second set of results were obtained during the modeling and simulation of the operation of systems in different states. These results are about the importance of each line for the interconnected electricity system. Therefore, these two sets of results allow us to discuss the likelihood of risk and the seriousness of concerns around it.

4.1. Analytical Analyses of Earthquake Influence on Interconnections

The first set of results focused on the lines of intersystem connections. These lines have different voltage levels and varying capacity. In addition, there is a very high probability of failure in each of them. To simulate the operation of the system, information about all power plants and intersystem links in seismically hazardous areas was collected. Our research allowed us to identify 264 power generation stations, of which 204 were located in the seismic zones with a high intensity of earthquakes and a high risk for the destruction of power generation stations. Among all identified power plants, 77.2% are in seismically active zones, and are subject to partial or total destruction by earthquakes with all negative economic consequences.

Our research also allowed us to identify 95 interconnection lines, from which we excluded 28 with a voltage of 110 KV and small capacity, as our research focused mainly on lines with a voltage of more than 110 KV. We excluded the low voltage power lines because, usually, such lines are used as distribution networks for the transmission of energy to private consumers and, even if these lines were to be used for transmitting energy between state systems, they still have little influence on the interconnected electricity transmission system.

We focused on the remaining 67 lines, aiming to determine the risk of failure of the electricity supply. We determined that only 14 lines are really at risk of being destroyed by an earthquake. Thus, 20.8% of the lines can be destroyed as a result of the earthquake hazard. This risk can influence the operation of the system and can cause a reduction in the necessary electric power supply to neighboring regions.

To calculate the current probability of failure of ties (for the moment of 2018) we use the algorithm from "3.4 Probabilities" and then use Formula 9 to define the probability of failure of the line for the current year. We also applied the data from two huge earthquakes that occurred between Kazakhstan and Kyrgyzstan in 1992 and 2003 [36,37], and a 2001 earthquake that occurred in an area with a connection between two energy regions of Kazakhstan. To compute the probabilities of interconnections between Russian power plants (the Siberian node) and Kazakhstan, we used a 2003 earthquake while, for the interconnections between the Southern Russian node and Kazakhstan, we used earthquakes from 2008 [38].

Table 1 contains full information about problem lines in all three countries.

Table 1. Parameters of trouble lines between nodes.

№	Country Code	Node Name	Point Name	Country Code	Node Name	Point Name	Voltage KV	Capacity MW	Probability of Earthquake (50 Years)	Probability of Failure of Line (2018)
1	KG	-	Bystrovka	KZ	South	Zapadnaja	220	135	0.05	0.01832
2	KG	-	Glavnaja	KZ	South	Almaty	220	135	0.05	0.01061
3	KG	-	Glavnaja	KZ	South	Shu	220	135	0.31	0.125187
4	KG	-	Bishkek	KZ	South	Shu	500	900	0.1	0.021671
5	KG	-	Bishkek	KZ	South	Zhambyl	500	900	0.1	0.037264
6	KG	-	Bishkek	KZ	South	Zhambylskaja GRES	220	135	0.1	0.037264
7	KZ	South	GPP-2	KZ	North	Kumkol	220	135	0.005	0.001181
8	KZ	North	Ust'-Kamenogorsk	RU	IES Siberia	Rubcovsk	500	900	0.05	0.010609
9	KZ	North	Aktjubinskaja	RU	IES Ural	Orsk	220	135	0.05	0.007086
10	KZ	North	Kempirsaj	RU	IES Ural	Orsk	220	135	0.05	0.007086
11	KZ	North	Ul'ke	RU	IES Ural	Novotroick	500	900	0.05	0.007086
12	KZ	West	Ural'skaja	RU	IES m. Volga	Kinel'	220	135	0.05	0.007086
13	KZ	West	Stepnaja	RU	IES m. Volga	Juzhnaja	220	135	0.05	0.007086
14	KZ	West	Stepnaja	RU	IES m. Volga	Balakovskaja NPP	500	900	0.05	0.007086

Abbreviations: KG is Kyrgyzstan, KZ is Kazakhstan, and RU is Russia. Source: (1) http://seismos-u.ifz.ru; (2) http://www.seismo.kg; (3) http://geolog.at.ua; (4) http://energybase.ru/map; (5) http://www.kegoc.kz; (6) http://nesk.kg/; (7) http://energybase.ru; (8) http://www.segrp.ru/.

Our results show that the electricity supply through the lines on the Kazakhstan and Kyrgyzstan border has a high probability of being affected by an earthquake. The most vulnerable line is between the "Glavnaya" (Kyrgyzstan) and "Shu" points in the southern node of Kazakhstan. This line has the highest probability of failure, which is 0.31 (0.125187 for 2018) for a line with 135 MW of capacity. In addition, there are three lines that were continually being threatened by events with a probability of 0.1 (0.021671 and 0.037264 for 2018) and a high capacity level. These lines have a common start/finish point, which is "Bishkek" in Kyrgyzstan. Taking into consideration the small number of lines between Kyrgyzstan and Kazakhstan (10 lines total) we suppose that this situation requires particular attention because 60% of 10 existing lines with non-zero probability of failure is a lot.

The border between Kazakhstan and Russia included eight lines with a non-zero probability of failure. We would like to pay specific attention to the isolated "West" node. This node is in a working state but usually has a big deficit. The Russian Federation supports this region through electricity export to Kazakhstan. It is also worth mentioning that we found 11 lines between the "West" node of Kazakhstan and two nodes of Russia. These are mostly lines with small capacity, but all lines with a high capacity level have a non-zero probability of failure from earthquakes. The government of Kazakhstan recently started a program to construct a new line between the "West" and "North" nodes. This solution will probably help Kazakhstan to make their system more balanced and reliable. We also think that this region is important for further investigation, and could probably experience serious economic damage in the case of earthquakes occurring.

4.2. Assessment Values of Important Interconnections Investigated

The second set of results contained data about values of importance for each investigated line. We applied the "$N-1$" and "$N-2$" criteria and developed two distributions of mathematical expectations of the risk value of each line. The resulting distributions were ranked according to the mathematical expectation of the risk value of non-working lines. This approach makes it possible to determine the importance of lines, taking into account their influence on individual parts of the system (Figure 5).

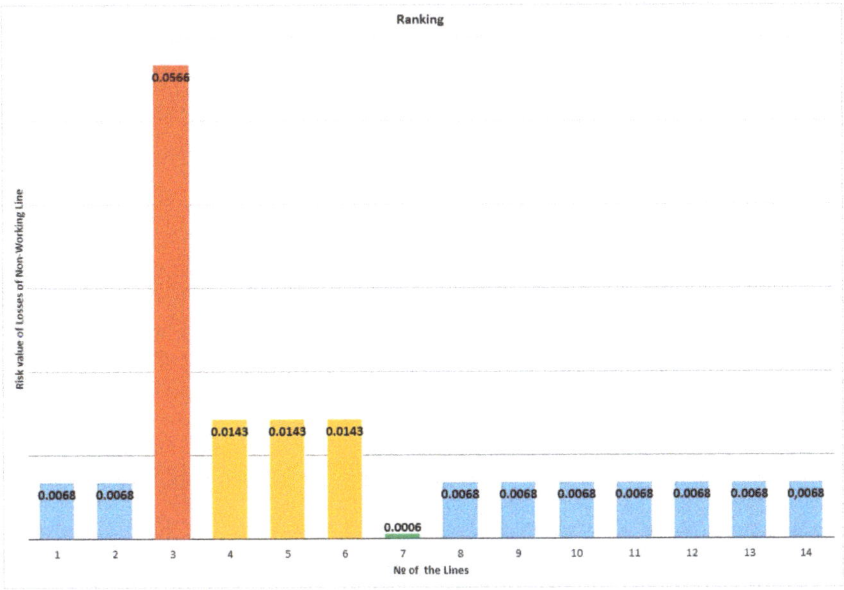

Figure 5. Histogram ranking the importance of lines by mathematical expectation of the risk value of non-working lines.

This chart is described by two axes, where one is called "Line №" (abscissa). All the numbers of the lines studied are located on this axis. The second is called "Probability of earthquake (50 years)" (ordinate), and is responsible for the height of the histogram column and allows one to visually compare the likelihood of the output of the investigated lines going out of order.

We developed four categories for all lines according to their impact on the electricity transmission system, and on the probability of failure of electricity supply:

- The first category, marked in red, contains lines with a high level of influence on the system. This means that, in the event of a power failure and disconnection of this line (failure), the region will be in a deficit state.
- The second category, which is marked in orange, characterizes the lines as an average necessity, which means that, in the event of certain circumstances, the disconnection of these lines can adversely affect the system and there will be a deficit.
- The third category, marked in yellow, are lines of which the influence is noticeable but not significant.
- The fourth category, marked in green, are lines without any visible influence on the system, as a whole or on its separate regions.

Table 2 contains the mathematical expectation of the risk value of working and non-working lines in the region investigated.

The current results show us that all lines between Kazakhstan and Russia have the same mathematical expectation of risk value. This means that there is no difference between interconnections between Russia and the isolated and deficit "West" node, or the connected and surplus "North" and "South" nodes.

As the high probability of failure has a big influence on the mathematical expectation of the risk value, we can see that the difference in the mathematical expectation of the risk value of the non-working line is bigger if the probability of failure is higher than usual. We found this observation for lines between Kazakhstan and Kyrgyzstan.

Our current model and data are a first-dimension approach, and consider only fully working systems with failed elements after earthquakes. This means that the real situation can be different because the full system will not work with full power transmitting and demand. Any element should be in a state of scheduled maintenance or failed. It is also worth considering the graphic of demand for each month, day of the week, and hour of the day. Unfortunately, the majority of interesting information, such as graphics of demand, is not available in public access sources, and further research will require additional access to data.

Often, in the classical theory of reliability, one can find research where correction factors are applied to change the state of the system toward more real data. To reduce the level of generation to real thresholds, the generation of each node is multiplied by 0.6–0.8. Such a coefficient simulates the operation of a real system, because the initial data contains a list of all generating stations. However, in reality, many of them may not work at full capacity, for example, HPPs due to low water levels or different combined heat and power (CHP) plants due to possible changes or fuel shortages. This factor reduces the amount of generation, simulating the location of generating stations. Correction coefficients were not applied in the studied system because such data is highly individual and, therefore, not readily available. Consequently, the studied EAEU EPS in its original position has no deficits and, when processing various scenarios, it has very modest deficits. However, applying the analytical approach, we can determine that if the coefficients are used, the amount of generation in Kyrgyzstan will be sharply reduced and the balance will be violated, and the South node of Kazakhstan will be in a state of small deficit close to the equilibrium of the load and generation. Thus, the disconnection of interconnections between Kazakhstan and Kyrgyzstan can lead to a deficit in Kyrgyzstan, which can have a negative economic effect.

Table 2. Values of mathematical expectations of trouble lines (working and non-working).

№	Rank (by Groups)	Country Code	Node Name	Point Name	Country Code	Node Name	Point Name	Risk Value of Losses of Working Line	Risk Value of Losses of Non-Working Line
1	3	KG	-	Bystrovka	KZ	South	Zapadnaja	0.8693	0.0068
2	3	KG	-	Glavnaja	KZ	South	Almaty	0.8693	0.0068
3	1	KG	-	Glavnaja	KZ	South	Shu	0.8195	0.0566
4	2	KG	-	Bishkek	KZ	South	Shu	0.8618	0.0143
5	2	KG	-	Bishkek	KZ	South	Zhambyl	0.8618	0.0143
6	2	KG	-	Bishkek	KZ	South	Zhambylskaja GRES	0.8618	0.0143
7	4	KZ	South	GPP-2	KZ	North	Kumkol	0.8755	0.0006
8	3	KZ	North	Ust'-Kamenogorsk	RU	IES Siberia	Rubcovsk	0.8693	0.0068
9	3	KZ	North	Aktjubinskaja	RU	IES Ural	Orsk	0.8693	0.0068
10	3	KZ	North	Kempirsaj	RU	IES Ural	Orsk	0.8693	0.0068
11	3	KZ	North	Ul'ke	RU	IES Ural	Novotroick	0.8693	0.0068
12	3	KZ	West	Ural'skaja	RU	IES m. Volga	Kinel'	0.8693	0.0068
13	3	KZ	West	Stepnaja	RU	IES m. Volga	Juzhnaja	0.8693	0.0068
14	3	KZ	West	Stepnaja	RU	IES m. Volga	Balakovskaja NPP	0.8693	0.0068

Abbreviations: KG is Kyrgyzstan, KZ is Kazakhstan, and RU is Russia.

5. Conclusions and Discussion

Our results show that the energy security of the EAEU region is affected by the existence of interconnections that are vulnerable to the threat of earthquakes, and where disruptions of the electricity supply will have high impacts on the reliability of electricity supply within the EAEU region. These interconnections are situated between Kazakhstan and Kyrgyzstan, as well as between the isolated "West" node of Kazakhstan and Russia. Power supply interruptions at these lines can seriously influence the stability of the electricity transmission system and lead to huge economic losses in the affected regions.

The public data used for the study allowed us to establish that the hazardous seismic situation in the countries studied could be the reason for the decrease in the reliability of the EPSs. However, the first-dimension assessment allowed us to identify only a part of the troubling sections of the EPSs and their interconnections. For a better evaluation, additional detailed and expanded studies are required. As alternative approaches and methodologies, modifications in which the first stage of generating states is replaced by a scenario approach, or where the states are generated by an expert group, can be used. A more complex approach based on the Monte Carlo method [39] can also be applied to generate solutions to the presented problem. Recently, it became known that the Commonwealth of Independent States (CIS) countries are strengthening cooperation in the field of geodesy, cartography, and remote sensing of the Earth. The creation of a unified system of coordinates and heights in the CIS was discussed. In addition, an interstate technical committee could be created for the standardization. Such changes can, in many respects, positively influence further studies of this region, because one of the problems of researching energy systems for several countries is the lack of standardized maps and notations. These new studies could give us greater clarity, not only in the regions that have already been explored, but also in terms of expanding the geography of the research in general. We could also investigate the rest of the countries in Central Asia that have an interest in joining the common energy market. It is worth mentioning that the proposed approach could be used in any other region, or even for a group of countries, to conduct first-dimension research and identify problem areas in power systems.

Author Contributions: Conceptualization, D.I., N.K., E.R. and D.K.; software, D.I. and D.B., data curation, D.I. and D.B.; investigation, D.I., E.R., N.K. and D.K.; methodology, D.K., E.R. and D.I.; project administration, D.I., N.K.; supervision, N.K., E.R.; writing—original draft, D.I. and N.K.; writing—reviewing and editing, N.K., E.R., D.K.

Funding: This research was funded by Petr Aven, through the Young Scientists Summer Program (YSSP) at the International Institute for Applied Systems Analysis (IIASA), Laxenburg, Austria.

Acknowledgments: We are grateful to Alexei Gvishiani from the Russian Academy of Science, to Kamchybekov from the Institute of Seismology NAS of Kyrgyzstan as well as to Pavel Kabat, Brain Fath, Tanja Huber, and Aleksandra Cofala for their valuable support during this research.

Conflicts of Interest: The authors declare no conflict of interest.

References

1. Pastukhova, M.; Westphal, K. *A Common Energy Market in the Eurasian Economic Union: Implications for the European Union and Energy Relations with Russia*; SWP Comment 2016/C 09; SWP: Berlin, Germany, 2016; 8p.
2. Sarkisian, T.S. The development of common energy markets in the Eurasian economic union. *J. Izv. St.-Peterburg. Gos. Èkon. Univ.* **2017**, *1*, 65–69.
3. Gibadullin, A.A.; Bortalevich, S.I.; Kalenskaya, E.V. Problems of formation of the electric power system of the Eurasian Economic Union. *Russ. J. Resour. Conserv. Recycl.* **2016**, *3*. [CrossRef]
4. Energeticheskaya Strategiya Rossii na Period do 2030 goda. (Energy Strategy of Russia for the Period up to 2030). Available online: https://minenergo.gov.ru/node/1026 (accessed on 15 April 2018).
5. Koncepciya Razvitiya Ehnergetiki Kyrgyzskoj Respubliki na Period do 2030 goda. (The Concept of Energy Development of the Kyrgyz Republic for Period up to 2030). Available online: http://www.cawater-info.net/library/rus/concept-energo-kg.pdf (accessed on 15 April 2018).

6. Koncepciya Razvitiya Toplivno-Ehnergeticheskogo Kompleksa Respubliki Kazahstan do 2030 goda. (The Concept of Development of the Fuel and Energy Complex of the Republic of Kazakhstan until 2030). Available online: http://adilet.zan.kz/rus/docs/P1400000724 (accessed on 15 April 2018).
7. O Koncepcii Formirovaniya Obshchego Ehlektroehnergeticheskogo Rynka Evrazijskogo Ehkonomicheskogo Soyuza. (About the Concept of the General Electricity Market of the Eurasian Economic Union). Available online: http://www.garant.ru/products/ipo/prime/doc/70926140/ (accessed on 15 April 2018).
8. Jelektrosnabzhenie Almaty i Almatinskoj oblasti bylo priostanovleno v rezul'tate avarijnogo otkljuchenija vysokovol'tnyh linij v NJeS KR—AlmatyJenergoSbyt. (Electricity supply of Almaty and Almaty Region Was Suspended as a Result of Emergency Switching-off of High-Voltage Lines in the NES KR—AlmatyEnergoSbyt). Available online: http://www.zakon.kz/137890-jelektrosnabzhenie-almaty-i.html (accessed on 2 August 2017).
9. Prichinoj Blekauta na Dal'nem Vostoke stalo Korotkoe Zamykanie na LJeP: Hronologija «Konca Sveta». (The Reason of Blackout in the Far East Was a Short Circuit on the Power Line: The Chronology of the "End of Days"). Available online: https://ampravda.ru/2017/08/01/076328.html (accessed on 15 August 2017).
10. Avarija v Jenergosisteme Sibiri Privela k Otkljucheniju Jelektrosnabzhenija 3 Zavodov «Rusala». (The Power Outage in Energy System of Siberia Provoked the Disconnection of 3 Factories of "Rusal"). Available online: http://ifvremya.ru/avariya-v-energosisteme-sibiri-privela-k-otklyucheniyu-elektrosnabzheniya-3-zavodov-rusala/ (accessed on 8 August 2017).
11. Voropai, N.I.; Kovalev, G.F.; Kucherov, Y.N. *Kontseptsiya Obespecheniya Nadezhnosti v Elektroyenergetike. (The Concept of Ensuring Reliability in the Electric Power Industry)*; Publ. House «Energiya»: Moscow, Russia, 2013; 212p.
12. Eastern European Platform Map. Available online: https://www.bygeo.ru/materialy/karty/360-strukturnaya-karta-evropy.html (accessed on 16 April 2017).
13. V Belarusi Bylo 9 Zemletrjasenij. No bojat'sja Nechego. (There Were 9 Earthquakes in Belarus. However, There Is Nothing to Fear). Available online: http://www.energocentre.by/analytics/2015-v-belarusi-bylo-9-zemletryaseniy-no-boyatsya-nechego.html (accessed on 29 December 2017).
14. Thurman, M. Natural Disaster Risks in Central Asia: A Synthesis. 2011. Available online: http://www.undp.org/content/dam/rbec/docs/Natural-disaster-risks-in-Central-Asia-A-synthesis.pdf (accessed on 7 November 2018).
15. Bondarenko, A.F.; Gerikh, V.P. O traktovke kriterija nadejnosti n-1 (About interpretation of the reliability criterion N-1). *Elektricheskie stancii* **2005**, *6*, 40–43.
16. Eurasian Economic Commission [Official Internet Page]. Available online: http://www.eurasiancommission.org/ru/act/energetikaiinfr/Pages/default.aspx (accessed on 26 August 2017).
17. Baranchik, J. Na kakih Uslovijah Vozmozhno Sozdanie Jenergeticheskogo Sojuza EAJeS? (Requirement Conditions for EAEU Establishment). Available online: http://mirperemen.net/2017/05/na-kakix-usloviyax-vozmozhno-sozdanie-energeticheskogo-soyuza-eaes/ (accessed on 26 August 2017).
18. Nachal Svojo Formirovanie Jenergeticheskij Rynok Evrazii (Eurasian Common Energy Market Was Initiated to Creation). Available online: http://evrazia-povolzhye.ru/nachal-svoyo-formirovanie-energeticheskij-rynok-evrazii/ (accessed on 27 August 2017).
19. Ministry of Energy of the Russian Federation. Available online: https://minenergo.gov.ru/node/1161 (accessed on 15 July 2017).
20. Ministry of Energy of the Republic of Kazakhstan. Available online: http://energo.gov.kz/index.php?id=2957 (accessed on 17 July 2017).
21. Commission Staff Working Document Impact Assessment. Available online: https://ec.europa.eu/energy/sites/ener/files/documents/mdi_impact_assessment_main_report_for_publication.pdf (accessed on 2 January 2018).
22. OSCE. *Protecting Electricity Networks from Natural Hazards*; OSCE: Vienna, Austria, 2016; 122p.
23. Global Seismic Hazard Map. Available online: http://www.200stran.ru/maps_group5_item289.html (accessed on 6 July 2017).
24. Seismic Zones Map of Kazakhstan. Available online: http://www.seismo.kg/ru/ (accessed on 18 July 2017).
25. Seismic Zones Map of Kyrgyzstan. Available online: http://geolog.at.ua/ (accessed on 25 July 2017).
26. Seismic Zones Maps of Russia. Available online: http://seismos-u.ifz.ru/ (accessed on 20 July 2017).
27. EPS Schemes of Russia. Available online: http://energybase.ru/map (accessed on 26 July 2017).

28. EPS Schemes of Kazakhstan. Available online: http://www.kegoc.kz/ru (accessed on 29 July 2017).
29. EPS Schemes of Kyrgyzstan. Available online: http://nesk.kg/ (accessed on 12 August 2017).
30. Public Resource (1) of EPS Elements. Available online: http://energybase.ru/ (accessed on 1 August 2017).
31. Public Resource (2) of EPS Elements. Available online: http://www.segrp.ru/ (accessed on 4 August 2017).
32. Kovalev, G.F.; Lebedeva, L.M. *Nadezhnost' Sistem Elektroenergetiki (The Reliability of the Electric Power Systems)*; Nauka Publ.: Novosibirsk, Russia, 2015; 224p.
33. Li, W. *Probabilistic Transmission System Planning*; Wiley-IEEE Press: Hoboken, NJ, USA, 2011; 361p.
34. Zorkaltsev, V.I.; Lebedeva, L.M.; Perzhabinsky, S.M. Model for estimating power shortage in electric power systems with quadratic losses of power in transmission lines. In *Numerical Analysis and Applications*; Pleiades Publishing, Ltd.: New York, NY, USA, 2010; Volume 3, pp. 231–240.
35. Panteleev, A.V.; Letova, T.A. *Metody Optimizatsii v Primerakh i Zadachakh [Optimization Methods on Examples and Tasks]*; Vysshaya Shkola Publ.: Moscow, Russia, 2005; 544p.
36. Kazakhstan Earthquakes of 2003. Available online: http://www.zakon.kz/4446756-samye-silnye-zemletrjasenija-v.html (accessed on 26 December 2017).
37. Kirgizstan Earthquakes of 1992. Available online: http://knews.kg/2013/12/kyirgyizstan-seysmoopasnyiy-tri-tyisyachi-zemletryaseniy-v-god/ (accessed on 26 December 2017).
38. Kazakhstan Earthquakes of 2001, 2003, 2008. Available online: https://earthquake.usgs.gov/earthquakes/ (accessed on 4 January 2018).
39. Boyarkin, D.; Krupenev, D.; Iakubovskii, D.; Sidorov, D. Machine learning in electric power systems adequacy assessment using Monte-Carlo method. In Proceedings of the 2017 International Multi-Conference on Engineering, Computer and Information Sciences (SIBIRCON), Novosibirsk, Russia, 18–22 September 2017.

Article

Children's Psychological Representation of Earthquakes: Analysis of Written Definitions and Rasch Scaling

Daniela Raccanello *, Giada Vicentini and Roberto Burro

Department of Human Sciences, University of Verona, 37129 Verona, Italy; giada.vicentini@univr.it (G.V.); roberto.burro@univr.it (R.B.)

* Correspondence: daniela.raccanello@univr.it; Tel.: +39-045-802-8157

Received: 19 April 2019; Accepted: 7 May 2019; Published: 9 May 2019

Abstract: Natural disasters have a potential highly traumatic impact on psychological functioning. This is notably true for children, whose vulnerability depends on their level of cognitive and emotional development. Before formal schooling, children possess all the basic abilities to represent the phenomena of the world, including natural disasters. However, scarce attention has been paid to children's representation of earthquakes, notwithstanding its relevance for risk awareness and for the efficacy of prevention programs. We examined children's representation of earthquakes using different methodologies. One hundred and twenty-eight second- and fourth-graders completed a written definition task and an online recognition task, analyzed through the Rasch model. Findings from both tasks indicated that, in children's representation, natural elements such as geological ones were the most salient, followed by man-made elements, and then by person-related elements. Older children revealed a more complex representation of earthquakes, and this was detected through the online recognition task. The results are discussed taking into account their theoretical and applied relevance. Beyond advancing knowledge of the development of the representation of earthquakes, they also inform on strengths and limitations of different methodologies. Both aspects are key resources to develop prevention programs for fostering preparedness to natural disasters and emotional prevention.

Keywords: psychological representation of earthquakes; open-ended and closed-questions surveys; children; seismic hazard assessment; emotions; emotional prevention

---

1. Introduction

This study focuses on children's psychological representation of natural disasters such as earthquakes. Knowledge of individuals'—and in particular children's—representation of earthquakes is highly relevant from both a theoretical and an applied perspective. From a theoretical perspective, it provides information on how people develop complex representations of risk-related phenomena, revealing which are the most salient elements when people perceive them; this is also key information when studying risk awareness, e.g., [1]. From an applied perspective, awareness of children's knowledge of earthquakes is a preliminary step for prevention programs, aiming at promoting the mastery of adequate geological, behavioral, and psychological contents, also after the identification of lacking areas. However, scarce attention has been devoted to study how children develop concepts and representations of natural disasters.

Nowadays, these issues are increasingly important, in light of both the prevalence of seismic events and the higher exposure to them through all the mass media and social networks [2,3]. In other words, it is currently more probable for a child to become a direct or indirect victim of an earthquake than in the past.

From a psychological perspective, natural hazards, including earthquakes, have a potentially highly traumatic impact on individuals' functioning [4–6]. Documented traumatic consequences include impaired health (e.g., cardiovascular ailments, etc.), increased rates of psychopathology (e.g., posttraumatic stress disorder, distress, depression, etc.), and negative emotional impact (e.g., anxiety, fear, anger, feelings of threat, etc.), both for primary victims experiencing the events directly and for secondary victims indirectly affected through media exposure, e.g., [7–10]. This is notably true for children and adolescents, whose vulnerability depends on their level of cognitive and emotional development [11].

The aim of this study is to fill in a gap in the literature, investigating how children develop the representation of earthquakes using two different tasks, a written definition task and an online recognition task.

1.1. Children's Abilities to Represent the World

Before formal schooling, children spontaneously develop naïve physics, biology, and psychology theories on the world, enabling them to make differentiations between domains and also to draw connections between them [12–16]. Naïve theories include elements on both the nature of the phenomena, i.e., ontological elements, and their underlying causal mechanisms and interrelations [12–16].

With regard to the physical domain, the grasping of core beliefs about the nature of physical objects and physical causes begins in infancy, and a deeper comprehension develops gradually later on [14]. Such everyday understanding is central for the development of people's interaction with any middle-sized object present in the external world.

Concerning the biological domain, authors such as Carey [17] credit children with a naïve theory of biology from seven or eight years. Previously, they would accumulate encyclopedic knowledge of biological events, which would undergo a conceptual change only later on. However, other authors have suggested that preschool children also possess a biological theory distinct from a psychological theory [14,18,19]. This is indicated for example by their abilities to attribute the causes of phenomena such as illnesses to the contact with contaminated substances rather than to psychological reasons [19].

Finally, children begin to develop a psychological theory, or theory of mind, from an early age, interacting with the social environment [20,21]. Preschool children understand both ontological aspects and related causal processes [21–23]. Ontological elements of the psychological theory relate to the existence and nature of mental states which are different from the real world of physical objects, material states, and mechanic or behavioral processes. Psychological causality regards the understanding of the relation between internal states, such as affects or cognitions, and one's own or others' overt behaviors [21].

Therefore, primary school students do possess all the basic abilities to represent the world, included natural phenomena such as earthquakes. Nevertheless, possible wrong ideas included in their spontaneous representation of the world should be detected before formal school instruction, in order to optimize the outcomes of learning processes.

1.2. Knowledge of Children's Representation of Earthquakes

As regards specifically natural disasters such as earthquakes, only a few studies investigated children's representation of them, focusing more on factual knowledge rather than psychological knowledge [24–27].

For example, Ross and Shuell involved New Zealander kindergartners through to sixth-graders with different experience of earthquakes, and found that a scientific conceptualization of them was rare, even if they revealed a certain awareness of their causes and consequences [27]. Similar results emerged with a sample of Turkish first- to sixth-graders [25]. However, more recent findings with New Zealander nine- and ten-year-olds suggested that children are quite aware of the core characteristic of earthquakes [24]. Raccanello and colleagues [26] revealed that second- and fifth-graders demonstrate a more refined representation of earthquakes in case they had experienced them directly and at increasing ages, without gender differences [26]. Specifically, higher refinement was described in terms of a more complex knowledge, a more frequent use of emotional language, and the presence of more intense emotions relating to earthquakes.

Studying how people, and in particular children, perceive and represent natural disasters could help to improve their risk awareness in order to promote an adequate preparedness [1], in particular focusing on emotional prevention [28]. However, on the whole, scarce attention has been paid to examining children's representation of earthquakes.

1.3. Some Methodologies to Investigate Children's Representation of Earthquakes

When investigated, children's representation of earthquakes has been studied with exploratory methodologies such as interviews or focus groups, frequently including open-ended questions [24–27]. King and Tarrant [24] involved nine- and ten-year-olds in focus groups. Laçin Şimşek [25] used semi-structured interviews with open-ended questions with first- to sixth-graders. Raccanello and colleagues [26] investigated primary school children's representation of earthquakes by analyzing the content of oral definitions elicited individually. Self-report tasks such these have the great advantage of enabling direct access to people's inner states, and notwithstanding possible bias effects such as social desirability they are still among the privileged ways to explore people's representation of one phenomenon [29]. However, they are time consuming in terms of transcription and coding of the materials. Moreover, individual interviews make it possible to gather detailed data on single participants, but they cannot be used collectively, and on the contrary focus groups can be used collectively but do not give the possibility to have data enabling to examine individual differences. In addition, such tasks rely on production abilities that give the responders the possibility to use their own lexicon without constriction, e.g., [30–33]. Indeed, a great advantage of open-ended questions consists in the variety of nuances emerging from the answers, useful in exploratory studies and impossible to observe when using closed questions. Nevertheless, they are quite demanding in terms of cognitive resources, especially for children.

An alternative way of exploring children's definitions of earthquakes is to use written tasks. Such tasks can be administered collectively, saving time, a requisite which can be fully appreciated when considering the need for involving a large number of people within prevention programs. Moreover, they enable to investigate individual differences. It is worth noting that, particularly for primary school children, performance in written tasks strictly depends on their emerging literacy abilities. Therefore, it is interesting to explore whether earthquakes' representation of children who have just acquired basic writing abilities emerges as similar when studying oral versus written definitions. This issue has been neglected within the literature as yet.

In addition, both oral and written definition tasks still rely on production abilities, which are mastered later during development compared to recognition abilities. This is the case, for example, with what happens for language at early ages [34]. Frequently, data gathered with oral or written production tasks are operationalized in order to obtain a description of the phenomenon at issue in terms of frequency tables corresponding to coding categories, which are usually treated as linear measures [35–40]. Such a process risks to distort the results. Alternatively, a way to investigate children's representation of earthquakes could be through recognition tasks, in which for example children have to evaluate how much a set of presented stimuli conveys the concept of earthquake. Data gathered through these kinds of tasks can be analyzed through the Rasch model [41–43], which makes it possible to go beyond the cited distortions, producing proper linear measures. In particular, the Rasch model goes beyond the weak points of traditional approaches to measurement (i.e., the classic test theory) by giving priority to objective measurement of latent dimensions which are based on the principles of fundamental measurement [44]. In other words, the characteristics of the individuals are measured independently of the characteristics of the stimuli (or items) and the stimuli calibrations is independent from the characteristics of the individuals. Fundamental measurement is taken for granted in the physical sciences, whereas in the social sciences the raw scores and the sum or means of these scores are typically considered as measures of a dimension independently of whether or not they conform to the principles of fundamental measurement [45]. However, also within the psychological field this approach can be used to investigate how people represent characteristics of the world, quantifying them, e.g., [46–50]. In particular, a specific advantage of the Rasch model is the "conjoint measurement" [35]: It is possible to estimate the values for persons and stimuli that can be represented on the same interval-scale of latent units (logits). The logits of persons (i.e., person-location) are independent from the administered stimuli (test-free condition) and the logits of stimuli (i.e., stimulus-location) are independent from the persons (sample-free condition). The accuracy of Rasch scaling is quantifiable by means of precise fit statistics [51].

Finally, individuals' representation of earthquakes could be studied taking advantage of what is offered by information and communication technology (ICT). Nowadays, the use of technologies for assessing how individuals perceive, elaborate, and react in everyday life, also in relation to learning, is increasing [10,52–55]. Given the growth in technological developments, psychological constructs can therefore be assessed and also promoted through many different devices (personal computers, laptops, tablets, mobile phones, etc.). Such use has many advantages. One advantage is the assessment mode, i.e., the self-administration in absence of a data collector [56]. Another is the fact that the individuals may tend to give more socially desirable responses in interviewers' administration than in self-administration [57]. Moreover, with ICT instruments there is the possibility, differently from what happens in laboratory studies, to assess the construct of interest within the individual's environment, increasing ecological validity [58]. A further advantage in using technological devices such as personal computers, tablets, or mobile phones to collect data in psychological research studies seems to be connected with higher and easier participation [59]. Even if the use of ICT within the psychological field is becoming more and more frequent, the current state of knowledge about the dynamics of taking surveys or teaching through learning software or applications is not as advanced as necessary, and more scientific literature is needed to evaluate some methodological aspects. Specifically, scarce attention has been paid to examining, for example, the appropriateness of materials and procedures for users' characteristics such as age, developing applications (as an exception, see the Emotional Prevention and Earthquakes in primary school project, PrEmT project, in Italian Prevenzione Emotiva e Terremoti nella scuola primaria, in which a web application on earthquake-related emotions for children is developed) [28]; the use of reliable research design to investigate the efficacy of interventions using ICT (i.e., evidence-based interventions) [28,60,61]; or the perception on usability of people involved in online psychological surveys, with children, e.g., [62] or adults, e.g., [28]. Therefore, further research on such issues is needed, in order to develop online instruments which can accurately detect individuals' characteristics and representation of the world, and subsequent programs aiming to enhance them.

1.4. Aims and Objectives of the Present Study

The general aim was to investigate children's representation of earthquakes, generalizing previous findings using different tasks [26]. Previous work indicated that when second and fifth-graders were requested to define earthquakes orally [26], they demonstrated a more complex representation in cases where they had experienced earthquakes directly and at increasing ages, without gender differences. Specifically, the salience of content categories pertaining to the material domain, such as natural and man-made elements, was higher compared to the salience of content categories relating to the person-related domain, such as behavioral, biological, or affective elements. Children with direct experience of earthquakes also referred more frequently to natural compared to man-made elements. In addition, at increasing ages, children reported definitions with more differentiated content types.

Beyond its theoretical relevance, knowledge of how different instruments enable to gather children's representation of earthquakes, with their strength, differences, and limitations, is relevant at an applied level. For example, this is a key element when choosing the more suitable instruments to increase awareness on children's representation as a preliminary step for intervention programs, both in terms of prevention and support subsequent to natural disasters.

For the written definition task, we had two objectives.

(1a) The first objective was to examine the salience of different domains in children's representation of earthquakes. On the basis of previous findings [26], we expected salience to be higher for natural elements compared to man-made elements, and for man-made elements compared to human elements.

(1b) The second objective was to examine whether the salience of the different domains varied according to age and gender. Concerning age, we hypothesized children to refer more frequently to the different domains at increasing ages [26]. Given the absence of gender differences in the literature, we did not formulate any specific hypothesis for gender.

Also, for the online recognition task, we had two objectives.

(2a) The first objective was to develop some images representing earthquakes and to quantify how much each of them conveyed the concept of earthquake [47,48]. In particular, we aimed at identifying the characteristics of the most representative images in terms of domains. Scales including such images as stimuli could be used as measurement instruments to assess how an earthquake is perceived.

(2b) The second objective was to examine how much children are able to perceive an earthquake on the basis of given stimuli. Specifically, we investigated age and gender differences. We expected children to be better able to identify earthquake-related stimuli as representing an earthquake at increasing ages, in parallel with better abilities to define earthquakes [26]. As for the written definition task, we did not have any specific hypothesis for gender.

2. Materials and Methods

2.1. Participants

We involved a convenience sample of 128 Italian primary school children. Sixty-three children were attending the second grade (mean age: 7.53 years, $SD = 0.34$; 43% females) while 65 were attending the fourth grade (mean age: 9.63 years, $SD = 0.25$; 54% females). For each grade level, the students were divided into four classes. The participants came from a variety of socio-economic status levels. We collected the data during February 2019 in Northeastern Italy.

Most of the children had never experienced earthquakes directly ($n = 91$; 71%) and only a small percentage of children had experienced them ($n = 30$; 23%). For seven children (6%) this information was missing. Among children who experienced earthquakes at least once, 19 children (64%) experienced one earthquake; 10 children (33%) experienced earthquakes twice; and one child (3%) experienced earthquakes three times. None of the children who had experienced earthquakes at least once reported any damage.

2.2. Procedure

The study was carried out following American Psychology Association (APA) ethical guidelines, and it was approved by the Local Ethical Committee of the Department of Human Science, University of Verona (protocol n. 134535) as part of a larger project, the PrEmT project [28]. The PrEmT project has the objective to test an intervention to promote children's knowledge of earthquakes and earthquake-related emotions, through an evidence-based design. The sample included in this study was involved in the pre-intervention phase of the pilot phase of the PrEmT project. Therefore, the students had not been involved previously in activities related to this project. In addition, we had verified with the teachers that they had not studied specific knowledge of earthquakes within their school lessons.

We obtained the authorization for the participation of children first from the school head, and then we obtained the informed written consents from parents.

We administered two tasks in class during regular school hours, in two sessions maximum one week apart one from the other. The school teachers were present, but they did not intervene. For both tasks, we read all the instructions aloud, in order to help children's understanding and to diminish the weight of the requested cognitive resources. We told them that there were no right and wrong answers, and that it was important that each child respond sincerely. In a first session, we presented the first task, i.e., a written definition task. We gave to the students a printed questionnaire in which they had to define earthquakes. We allowed them all the time that they needed to complete the task; on the whole, they used a maximum of five minutes. The questionnaire also included other tasks not analyzed here. In a second session, we administered the second task, which was an online recognition task. We gave to each student a tablet with a wireless Internet connection. The completion of the second task took about 15 min. At the end, we thanked all the children for their participation, underlining that their answers could be useful to help children to cope with traumatic events such as earthquakes.

Before beginning the data collection, we had administered the tasks to two children of the same age range of our sample, to check their feasibility.

Concerning parents, when completing the informed consent form, they also completed a questionnaire on sociodemographic data and children's experience of earthquakes.

2.3. Materials and Coding

2.3.1. Written Definition Task

In the written definition task, we asked children to write what an earthquake is (e.g., *What is an earthquake? Write all the things that come to mind to explain to a child what an earthquake is*; in Italian: *Che cos'è un terremoto? Scrivi tutto quello che ti viene in mente per spiegare a un/a bambino/a che cos'è un terremoto*). We adapted the task from previous studies, e.g., [26,31].

We transcribed verbatim all the definitions and coded them for their complexity, in terms of content type, adapting previous coding schemes [26,30,31,63]. We coded the absence/presence (0/1) of contents regarding material (natural and man-made) and person-related domain, e.g., [26]. Contents were defined as phenomena in terms of entities but also related processes. As in previous research [26], the coding scheme was developed deductively integrating categories from the literature [24–27] and inductively on the bases of the contents within children's definitions [64]. Natural elements pertained to earthquakes' geological characteristics (e.g., *When the earth moves; It is a huge fissure; An earthquake is a shake which provokes landslides and fissures in the terrain*; in Italian: *Quando la terra si muove; è una crepa gigante; Un terremoto è una scossa che provoca frane e crepe nel terreno*). Man-made elements related to all things that were built by people (e.g., *It makes all the things at home swing; The earthquake is a thing that shakes the houses and destroys the houses; The earthquake is a thing that makes things move*; in Italian: *Fa dondolare tutte le cose a casa; Il terremoto è una cosa che fa tremare le case e distrugge le case; Il terremoto è una cosa che fa muovere le cose)*. Person-related elements included all aspects pertaining to three categories, namely people's behaviors, and excluded those related to building things (behavioral aspects; e.g., *To protect you from an earthquake you have to go under the table; When it comes, you have to get under something; Children go out and they write their names [...] so the teacher is sure that all the children are there and there aren't children who have fallen behind*; in Italian: *Per salvarsi dal terremoto andare sotto il tavolo; Quando viene devi andare sotto qualcosa; I bambini vanno fuori e scrivono i loro nomi [...] così la maestra è sicura che i bambini siano completi e non ci sono bambini che sono rimasti indietro*); people's body, in terms of biological domain, and including reference to deaths and injuries (biological aspects; e.g., *The earthquake can cause many injuries but also many dead people; Pieces of wall can fall on your head; Something that [...] kills people and animals*; in Italian: *Il terremoto può causare molti feriti ma anche molti morti; Ti possono cadere dei pezzi di muro in testa; Una roba che [...] uccide persone e animali*); and affect, referring to emotions, mood, feelings, and also mental health symptoms (affective aspects; e.g., *It is not a good thing; An earthquake frightens everyone; It causes much panic to people, but you have to stay calm because the panic provokes fear and distress and other things*; in Italian: *Non è una roba bella; Un terremoto fa venire paura a tutti; Fa molto panico alle persone ma bisogna stare calmi perché il panico provoca paura e disagio e altre cose*).

A first judge coded all the definitions, while a second judge coded 30% of them for reliability. The mean percentage agreement was 95% for natural domain, 98% for man-made domain, and 100% for person-related domain. Disagreements were solved through discussion between judges.

2.3.2. Online Recognition Task

In the online recognition task, we asked the children to evaluate how much different graphic stimuli are representative of an earthquake. The task had previously been used with adult participants [65]. For completing the task, each child used a tablet in which we had set the link to an online survey, functioning through a common Internet browser. The survey was developed and administered using the Cognitive Metrix Survey Software, CMSS [54,66], a customization of the LimeSurvey open-source project [67].

The graphic stimuli were drawings developed ad-hoc to refer to the different elements of earthquakes that had resulted in be salient in children's representation [26]. We had an illustrator create a total set of 16 drawings (Table 1). All the drawings for the online recognition task had to include at least one element directly pertaining to the earthquake, and we chose to represent elements indicating shaking of the environment (see the red signs in Table 1). The environment was formed only by natural elements for half of them (i.e., flat ground with a fissure; a mountain with a fissure and rocks falling) and also by man-made elements for the other half (i.e., a house with a fissure and tiles falling; a bridge with a fissure). Each set of images could vary for the different presence of person-related elements, given that we were interested in studying experimentally how the different presence of person-related elements could differently influence children's representation, isolating the effect of the different characteristics of human beings. We therefore set four experimental conditions for person-related elements. Again, we chose categories of person-related elements spontaneously reported by children defining earthquakes [26]. Person-related elements could be absent, they could refer to a behavioral reaction (i.e., a stylized person who is escaping), to biological damage (i.e., a stylized person with bandages), or to an emotional reaction (i.e., a scared face). It is worth noting that previous research confirmed that for both children and adults fear is the most salient emotion associated with earthquakes [26,68]. We did not include combinations of the three person-related elements (e.g., biological damage and emotional reaction in the same drawing) because the number of stimuli would have been too high, in relation to children's attention abilities.

Table 1. Graphic stimuli developed for the study, varying for natural vs. man-made elements, and for person-related elements (absent, behavioral, biological, affective).

Person-Related Elements	Absent	Behavioral	Biological	Affective
Natural elements			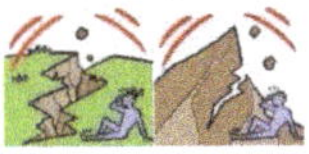	
Man-made elements				

The respondents were presented with the 16 drawings, and for each one we asked them to evaluate the degree to which it was representative of an earthquake (e.g., *How much does this drawing make you think of an earthquake*?; in Italian: *Quanto questa figura ti fa pensare a un terremoto?*) on a 5-point Likert type scale (1 = *not at all*, 5 = *very much*). We randomized the order of presentation of the images. We read aloud all the questions in order to favor children's understanding of the task.

2.3.3. Sociodemographic Data and Experience of Earthquakes

In a written questionnaire for parents, we asked children's date of birth, gender, and family socioeconomic status (in terms of information on parents' instruction and job).

We also investigated children's experience of earthquakes, asking to the parents: (a) whether their children had ever experienced an earthquake (0 = *no*, 1 = *yes*); (b) in cases where it had happened, how many times it had happened; (c) in cases where it had happened, what kind of damage they had experienced (1 = *no damage*; 2 = *damage to properties*; 3 = *psychological damage to relatives*; 4 = *physical damage to relatives*; 5 = *death of relatives*; 6 = *psychological damage to oneself*; 7 = *physical damage to oneself*).

2.4. Data Analysis

First, we applied Generalized Linear Mixed Models (GLMM), as implemented in the glmer functions in the lme4 package [69] within the R-software (Version 3.6.0, R Core Team, Wien, Austria) [70], to explore the salience of different content types, and the influence of class level (second, fourth-graders) and gender (males, females) on responses (objectives 1a, 1b, and 2b). We performed Mixed Model ANOVA Tables (Type 3 tests) via likelihood ratio tests implemented in the afex package of the R-software [71–74]; for applications see, for example, [32,75–81]. For the various analyses, class level and gender were the between-subject factors; content types were the within-subject factors. For binary dependent variables, the GLMM used the binomial family and logit link-function; in the case of count dependent variables, the GLMM used the Poisson family and logarithmic link-function (log link-function). For binary dependent variables, descriptive statistics reported in the text refer to logit-scale, which is a scale in which the unit of measurement is based on the log of the probability of occurrence of an event divided by the probability of non-occurrence of the same event. For example, when both probabilities are 50%, the logit value is equal to zero; when the probability of occurrence is higher than the probability of non-occurrence, the logit value is positive; and when the probability of occurrence is lower than the probability of non-occurrence, the logit value is negative. We conducted post hoc tests on the significant results using the Bonferroni correction implemented in the emmeans package [82]; this adjustment method makes it possible to deal with Type I errors by multiplying the *p*-values by the number of comparisons. We also reported the estimates (*EST*) as a measure of effect size [83]. The estimates corresponded to log odds ratios for binary dependent variables, and to log rate ratios for count dependent variables [82]. Given that estimates are log values, for positive estimates, the effect size is small when $0.00 < EST \leq 0.50$, medium when $0.50 < EST \leq 1.00$, and large when $EST > 1.00$; for negative estimates, effect size is small when $0.00 > EST \geq -0.50$, medium when $-0.50 > EST \geq -1.00$, and large when $EST < -1.00$ [84]. The level of significance was $p < 0.05$.

Second, we applied the Rasch model [35,41–43] to quantify how much each drawing conveyed the concept of earthquake (objective 2a). The Rasch analysis was carried out using the RUMM2030 software (Version 5.1, RUMM Laboratory, Perth, Australia) [85,86]. First, we checked the Rasch model assumptions: monotonicity [87,88]; local independence [87,89]; unidimensionality [87,88,90]; absence of differential item (i.e., stimulus for our case) functioning (DIF) or stimulus bias [87,88,91,92]. Then, we considered the reliability in terms of the Person Separation Index (PSI) and the Cronbach's alpha [88,93–95]. Both indexes make it possible to examine the consistency of the responses. Nevertheless, the two indexes differ for different aspects: (a) the Cronbach's alpha can be calculated with complete data only, while the PSI can also be calculated when there are random missing data; (b) the Cronbach's alpha is calculated on the raw scores, while the PSI is based on the estimated locations of the persons, which are non-linear transformation of the raw scores. We also checked the model fit. Finally, targeting—which indicates how well the measurement range of the scale matches the distribution of the calibrating sample [88,93,96]—is expressed here as floor and ceiling effects and targeting index [96].

There were no missing answers.

3. Results

3.1. Written Definition Task (Objectives 1a and 1b)

Definitions' Content Types: Salience, Age, and Gender Differences

First, we carried out a preliminary analysis to compare the three person-related content types. We included person-related elements (behavioral, biological, and affective elements) as the fixed effect;

participants as the random effect; and person-related elements scores as binary dependent variables. No significant effects emerged.

Second, we ran a GLMM with content type (natural, man-made, and person-related), class level, and gender as fixed effects; participants as the random effect; and content type scores as binary dependent variables. We found a significant effect of content type, $\chi^2(2) = 58.15$, $p < 0.001$. Post hoc tests indicated that the scores were significantly higher, $z = 5.34$, $p < 0.001$, $EST = 1.82$, for natural elements ($M = 0.78$, $SD = 0.26$, 95% CI: 0.73 to 0.82) compared to man-made elements ($M = 0.37$, $SD = 0.29$, 95% CI: 0.32 to 0.42), and in turn significantly higher, $z = 4.38$, $p < 0.001$, $EST = 1.54$, for man-made elements compared to person-related elements ($M = 0.11$, $SD = 0.20$, 95% CI: 0.08 to 0.15). No class level or gender differences emerged. See Figure 1 for response rates relating to content types (natural, man-made, and person-related) expressed in logit-scale.

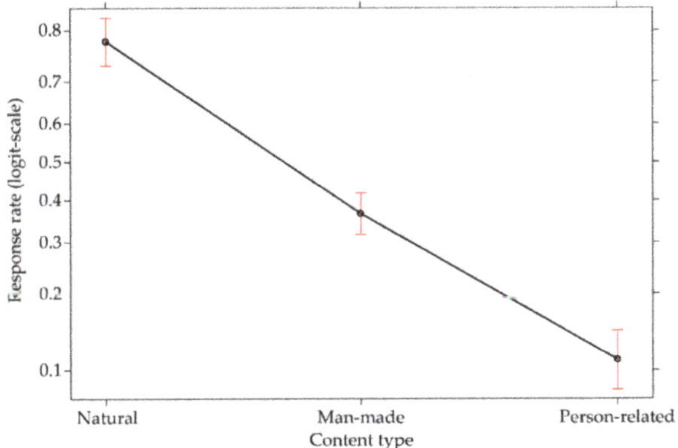

Figure 1. Response rates relating to content types (natural, man-made, and person-related) expressed in logit-scale.

3.2. *Online Recognition Task (Objectives 2a and 2b)*

3.2.1. Rasch Model: Scaling of Drawings

Concerning the Rasch model assumptions, the monotonicity request was satisfied, i.e., the responses for all the stimuli were used consistently (the difficulty thresholds were ordered) [97]. There was no local dependence: For all the stimuli, the residual correlations were lower than 0.2 [89]. The unidimensionality was confirmed by a post hoc paired *t*-test on separate estimates for each respondent (derived from subsets of stimuli identified by a Principal Component Analysis of the residuals): The percentage of significant tests was less than 5% of all the tests [90]. Lastly, no DIF was found [91], tested by a two-way analysis of variance for each stimulus, comparing scores across each level of person factor (gender and class level).

With regard to the model fit for the individual stimulus and the person level, each residual was in the range between −/+ 2.5. The summary fit residual statistics were expected to approach a mean of zero and a standard deviation of 1 [91]. Precisely, the summary residual statistics calculated values for stimuli were 0.185 (mean) and 0.713 (standard deviation). For persons, they were −0.264 (mean) and 1.587 (standard deviation).

The analysis of reliability for the total sample showed a PSI value equal to 0.895 (above 0.85 as the minimum requirement for individual person measurement) [98] and a Cronbach's alpha of 0.918, suggesting consistency in the children's responses.

The targeting of the scale to the sample shows graphically how well individual stimulus difficulties and individual person abilities can be matched on common scale [41]. The average person ability and standard deviation indicate how well the scale is targeted to the sample [88,92]. Analysis of targeting also entails the assessment of floor and ceiling effects. Figure 2 shows that the overall targeting was very good. The difference between average person ability and average stimulus difficulty was equal to 0.726, namely the persons were a little more skilled than stimulus difficulty.

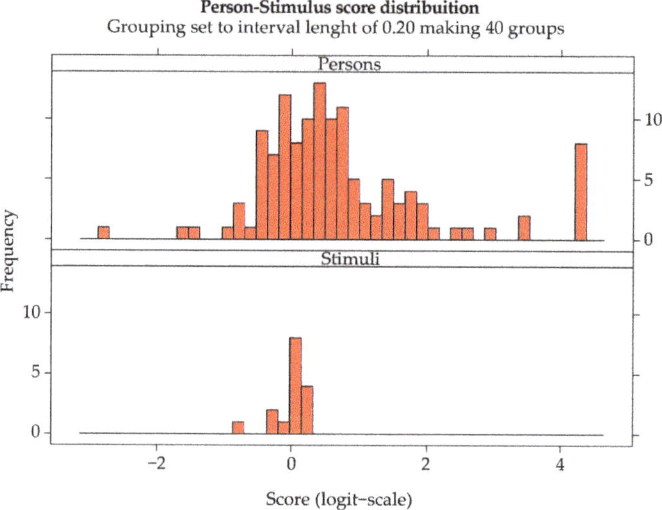

Figure 2. The upper bars represent groups of persons with the same total score; the lower bars represent groups of stimuli with the same total score. Between the two graphs, the logit scale is represented. In this kind of graph, the average stimulus score is always set at 0.0 logits. Overall, the targeting is very good, as there is a good matching between person-scores and stimulus-scores.

See Figure 3 for the representation of the scaling of the stimuli along one dimension, according to how much they convey the concept of earthquake, grouped by content type.

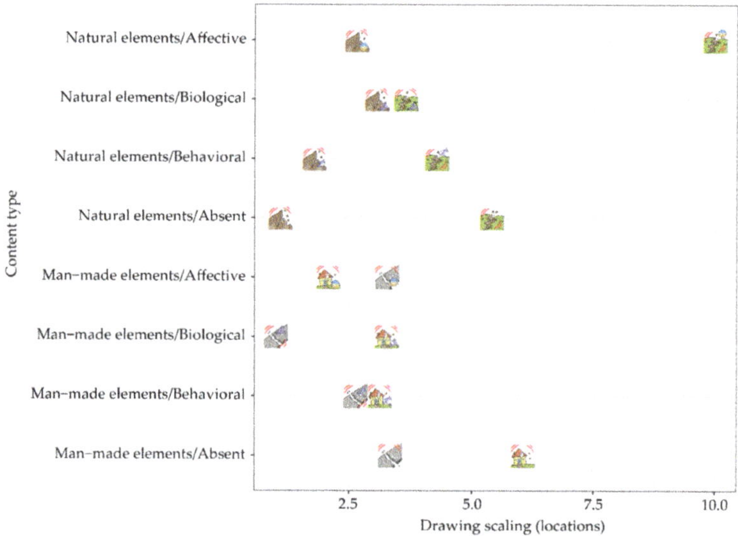

Figure 3. Scaling of the stimuli along one dimension, according to how much they convey the concept of earthquake, grouped by content type.

It is worth noting that the drawing perceived as the most representative of an earthquake was the one depicting a fissure in the plain terrain, in which we had added the scared face. This finding confirmed the salience of natural versus built elements in children's representation of earthquakes. In addition, it suggests that children differentiate secondary effects of earthquakes, i.e., landslides, from primary effects, and consider the first more peripheral in their representation of earthquakes. Finally, we underline that fear is the person-related element which mostly characterizes children's representation.

3.2.2. Drawings: Age and Gender Differences

Finally, we conducted a GLMM with class level (second- and fourth-graders) and gender as the fixed effects; participants as the random effect; and the person-locations as calculated through the Rasch model for each participant as the count dependent variable. The analysis yielded a significant effect of class level, $F(1) = 5.37$, $p = 0.022$. Scores were higher for fourth ($M = 64.055$, $SD = 16.142$, 95% CI [60.547, 67.562]) compared to second-graders ($M = 58.266$, $SD = 11.691$, 95% CI [54.787, 61.745]). No gender differences emerged.

4. Discussion and Conclusions

Acknowledging the relevance of prevention programs for fostering preparedness to natural disasters [1], we investigated children's representation of earthquakes with different tasks.

The literature on children's psychological development demonstrated that children already possess quite a large "toolbox" for understanding the nature of the physical, biological, and psychological phenomena and their interrelations [12–16]. Such knowledge can be considered as a prerequisite in order to acquire and refine knowledge of earthquakes. Supporting and extending previous studies [24–27], our findings indicated that on the whole knowledge and representation of earthquakes involve both natural and geological characteristics, as well as people-related characteristics. Such categories include all the elements pertaining to the things that are built by humans, but also all those elements relating to their functioning, in terms of behavioral, biological, and affective domain.

From the analyses of the definitions, it emerged that, in children's representation of earthquakes, natural elements such as geological ones are the most salient, followed by man-made elements, and in turn by person-related elements, extending and generalizing previous findings with oral definition tasks [26]. Therefore, Italian primary school children possess a basic knowledge of earthquakes, differentiating elements pertaining to different domains, and giving particular relevance to the core geological issues characterizing earthquakes. At the same time, they have an initial awareness of the different elements to which damage can be associated, reporting more frequently macroscopic and external consequences for the structures which are built by the individuals, but without neglecting the consequences for humans. Specifically, behaviors such as escaping, biological consequences such as being hurt, or affective reactions assume all a similar relevance.

Data from the recognition task supported these findings. First, we examined how much each drawing conveyed the idea of earthquake. To achieve this goal, we analyzed the scores given to the different exemplars of our set of drawings, which included natural or built elements combined with the four experimental conditions relating to person-related elements, that could be absent, behavioral, biological, or affective. Quantifying, through scientific approaches which respect the properties of the fundamental measurement (i.e., the Rasch model) [42,44,45,51], how much the different stimuli conveyed the concept of earthquake could be a first step for the development of specific scales for assessing people's perception of natural disasters such as earthquakes. Second, we found that the drawing that conveyed most the idea of earthquake was the one that included natural but not built elements. Among the two possible images, the most representative was specifically the one representing the fissure in the ground, and not the landslide, suggesting children's ability to identify the core characteristic of earthquakes as distinguished by their secondary effects. In addition, we highlight that the most representative image also included emotional elements, indicating that children already grasp the somehow overwhelming emotional burden associated with disasters. This element could be interpreted as a sign of the relevance of planning prevention programs focused on emotional prevention specifically devoted to developing children's understanding of earthquake-related emotions and regulation strategies, an issue currently still neglected [28]. For an exception, see the PrEmT project [28]. It is worth noting that providing pre-drawn representations and using ICT allows for a larger sample and the use of a range of statistical tools, but it does not make it possible to give value to the rich abilities already possessed by primary school children to represent the world. We could state that this particular activity investigated how well children connected with adult representations rather than only revealing how they represented the events themselves.

As regards age differences in children's representation of earthquakes, the analyses reported in our study only partially revealed them. Concerning written definitions, our results did not reveal age differences, differently from Raccanello and colleagues' study [26], who found that, in oral definitions of earthquakes, older children reported more information compared to younger children. It could be that the distance in terms of competence (i.e., for producing written definitions) between the younger and the older children is presumably higher for oral tasks than for written tasks. However, in Raccanello and colleagues' study [26], second-graders were compared to fifth-graders, and so the different age range for our sample (i.e., two years instead of three years) could have been related to the absence of this effect. Nevertheless, examining the means for the sample of second and fourth-graders included in our study, we can note that the difference among them is in the expected direction, with older students reporting a higher variety of contents (even if the difference between them was not statistically significant). Moreover, it is interesting to note that the recognition task was somehow more sensitive to age differences compared to the production task, given that through such task we detected age differences in the expected direction. At increasing ages, children were better able to interpret a given stimulus as conveying the concept of earthquake. In other words, older children demonstrated a more refined ability to identify images relating to earthquakes (all of them included a fissure and shaking of the ground) as conveying the intended meaning. Empirical evidence relating to the development of this ability at increasing ages can result very useful. For example, knowledge of which elements of earthquakes' representation are salient at different ages can be considered in order to develop instruments to be used as alert signals for emergency situations.

From an applied perspective, it is very important to develop instruments for identifying the naïve ideas of children on the entities and functioning of the physical, biological, and psychological world for maximizing the possibility of success of interventions focused on them. In school contexts, it is central to detect possible wrong ideas included in children's naïve theories as soon as possible, in order to discard them and increase the possibility that formal instruction is associated with learning outcomes which last over time [99]. To guarantee the efficacy of science education, it is highly relevant that students acknowledge their intuitive ideas, that teachers give them scientific facts and explanations in accordance with their level of cognitive development taking into account their naïve understandings, and that this process unfolds along with the promotion of critical thinking and reasoning abilities [99]. Similarly, within prevention programs aiming at increasing children's knowledge of earthquakes, possible wrong ideas included in their spontaneous representations of earthquakes should be detected early, in order to increase the success of such programs in ameliorating children's learning both at short-term and at long-term. Therefore, identifying instruments helping to describe accurately children's representation is a key step for the development of prevention programs aiming to enhance preparedness to disasters.

Schools play a central role also in the aftermath of natural disasters such as earthquakes, as revealed in studies conducted in a country that has experienced multiple earthquakes in more recent years, New Zealand [100]. For example, Bateman and Dandy [101] described a case study with four preschoolers who talked with their teachers about their experiences of the 2011 Christchurch earthquake; they analyzed turn-taking utterances and revealed that such practice can help to recover from the emotional stress associated with earthquakes and prevent from further stress. Mutch and Gawith [102] described three different ways through which schools helped children to come to terms with their experiences of several earthquakes (always in Christchurch) within a UNESCO-funded project: In one school children, parents, and teachers wrote an illustrated book together; in a second school they created a mosaic story on antecedents, core, and aftermath of the earthquakes; and in a third school they conducted interviews about the earthquakes. Lessons learnt from how children narrate and reconstruct the meaning of what they experienced, developing shared memories within a community, can give useful suggestions also from a prevention perspective. Based on the findings of these studies [100–102], we could argue that story-telling, both individual and collective, can be a useful methodology to help children and significant adults to build the meaning of experiences, also before a disaster happens. Such a tool enables both to build factual appropriate knowledge, and to help to reflect on the emotional nuances of how natural disasters can impact children's life, in an attempt to give them the resources to "gain perspective and distance" within their coping processes (p. 54) [102].

When reflecting on the methodological gain of our findings, we could argue that both production and recognition tasks, operationalized though oral, written, or online instruments (see Raccanello and colleagues [26] for oral production tasks, i.e., definition task; see this paper for written production tasks, i.e., definition task, and online recognition tasks) are reliable methodologies enabling to gather knowledge of children's representations on earthquakes. We already anticipated that, generally, production and recognition tasks differ in terms of the cognitive resources requested to master them. In addition, production tasks including open-ended questions and recognition tasks in terms of closed-ended questions differently enable to shed light on the different nuances with which the individuals perceive, elaborate, and represent reality, more differentiated and richer for the first compared to the latter, e.g., [30–33]. However, data gathered through tasks such as the online recognition task that we used have the possibility to be treated with analyses models, i.e., the Rasch models [35–40], which go beyond the typical limitations of the traditional measurement approaches, with advantages in terms of the quality of the assessed representation. Finally, the three methodologies have different constraints in terms of managing time and number of people which can be involved simultaneously. Oral production tasks are the most time-consuming, followed by written production tasks, and finally by online recognition tasks. To study individual differences, oral tasks should be administered individually, while written and online tasks can be administered also collectively.

Beyond these general strengths and limitations, the different tasks can be examined specifically focusing on how they helped to detect and describe children's representation of earthquakes. On the one hand, the salience of the different domains was basically confirmed across the three tasks. Our analyses of the contents of oral definitions supported previous findings with oral definition tasks [26] and extended them. Therefore, the three different tasks can be considered reliable in reaching their basic goals. On the other hand, age differences were not so well detected by a task requiring the use of writing abilities, compared to the recognition task. Particularly the latter, beyond requiring a lower amount of cognitive resources, could also take advantage of many of the benefits of using ICT [28,58], included the higher attractiveness particularly motivating for children.

On the whole, our findings suggest that when planning the specific methodology to be used in a prevention program, but also in a support intervention with children, scientists and practitioners should be particularly cautious in their choices, reflecting on and balancing both benefits and disadvantages in utilizing different instruments.

Our study suffers from limitations related to the nature of the instruments used, as already described. Moreover, there are also other useful methodologies to examine children's representation of earthquakes, such as analyzing drawings, that we did not use. Such instruments could be utilized in future research. There are specific limitations relating to the tasks that we developed. For example, for the online recognition task we did not consider all the possible combinations between the elements at issue, e.g., representing biological and affective damage in the same drawing. Future studies could examine more deeply the salience of many combinations of elements pertaining to different domains for children's representation of earthquakes. In addition, it is worth noting that we involved only a small sample of primary school students, and only in relation to earthquakes. Future research could verify whether our findings extend to younger children, but also to adolescents, adults, and elderly people, in order to examine more thoroughly how the representation of earthquakes changes with development. In addition, children in our sample rarely experienced directly earthquakes, and when they had experienced them, they luckily had not suffered from personal damage. Further studies could take into account more deeply the role of personal exposure, comparing samples who for example experienced directly, indirectly, or did not experience at all earthquakes. Finally, it would be highly relevant to extend the study of individuals' representations of phenomena such as earthquakes to a larger range of disasters, examining both natural and technological disasters, as a step to support awareness on citizens' resources for developing disaster-resilient societies.

Author Contributions: Conceptualization, D.R. and R.B.; methodology, D.R., G.V. and R.B.; software, G.V. and R.B.; formal analysis, R.B.; investigation, D.R., G.V. and R.B.; data curation, D.R., G.V. and R.B.; writing—original draft preparation, D.R., G.V., and R.B.; writing—review and editing, D.R.; visualization, R.B.; supervision, D.R.; project administration, D.R.; funding acquisition, D.R. and R.B.

Funding: This work was supported by the Fondazione Cariverona, Bando Ricerca Scientifica 2017, Italia.

Acknowledgments: The authors would like to thank all the participants and the teachers of the involved schools. They also would like to thank Veronica Barnaba, Erminia Dal Corso, Emmanuela Rocca, Giulia Rosolen, and Lorenza Marchiori for their help in the project. They finally thank Elisa Ferrari for drawing the images.

Conflicts of Interest: The authors declare no conflict of interest. The funders had no role in the design of the study; in the collection, analyses, or interpretation of data; in the writing of the manuscript, or in the decision to publish the results.

References

1. Scolobig, A.; De Marchi, B.; Borga, M. The missing link between flood risk awareness and preparedness: Findings from case studies in an Alpine Region. *Nat. Hazards* **2012**, *63*, 499–520. [CrossRef]
2. INGV, Istituto Nazionale di Geofisica e Vulcanologia [National Institute of Geophysics and Volcanology]. Available online: www.ingvterremoti.wordpress.com (accessed on 24 February 2019).
3. Shoshani, A.; Slone, M. The drama of media coverage of terrorism: Emotional and attitudinal impact of the audience. *Stud. Confl. Terror.* **2008**, *31*, 627–640. [CrossRef]
4. Fergusson, D.M.; Boden, J.M. The psychological impacts of major disasters. *Aust. N. Z. J. Psychiatry* **2014**, *48*, 597–599. [CrossRef]
5. Galambos, C.M. Natural disasters: Health and mental health considerations. *Health. Soc. Work* **2005**, *30*, 83–86. [CrossRef]
6. Neria, Y.; Nandi, A.; Galea, S. Post-traumatic stress disorder following disasters: A systematic review. *Psychol. Med.* **2008**, *38*, 467–480. [CrossRef]
7. Ben-Zur, H.; Gil, S.; Shamshins, Y. The relationship between exposure to terror through the media, coping strategies and resources, and distress and secondary traumatization. *Int. J. Stress Manag.* **2012**, *19*, 132–150. [CrossRef]
8. Furr, J.M.; Comer, J.S.; Edmunds, J.M.; Kendall, P.C. Disasters and youth: A meta-analytic examination of posttraumatic stress. *J. Consult. Clin. Psychol.* **2010**, *78*, 765–780. [CrossRef] [PubMed]
9. Masten, A.S.; Osofsky, J.D. Disasters and their impact on child development: Introduction to the special section. *Child. Dev.* **2010**, *81*, 1029–1039. [CrossRef]
10. Raccanello, D.; Burro, R.; Brondino, M.; Pasini, M. Relevance of terrorism for Italian students not directly exposed to it: The affective impact of the 2015 Paris and the 2016 Brussels attacks. *Stress Health* **2018**, *34*, 338–343. [CrossRef] [PubMed]
11. Kar, N. Psychological impact of disasters on children: Review of assessment and interventions. *World J. Pediatr.* **2009**, *5*, 5–11. [CrossRef] [PubMed]
12. Allen, M. Preschool children's taxonomic knowledge of animal species. *J. Res. Sci. Teach.* **2015**, *52*, 107–134. [CrossRef]
13. Schulz, L.E.; Gopnik, A. Causal learning across domains. *Dev. Psychol.* **2004**, *40*, 162–176. [CrossRef] [PubMed]

14. Wellman, H.M.; Gelman, S.A. Cognitive development: Foundational theories of core domains. *Annu. Rev. Psychol.* **1992**, *43*, 337–375. [CrossRef]
15. Wellman, H.M.; Inagaki, K.E. (Eds.) *The Emergence of Core Domains of Thought: Children's Reasoning about Physical, Psychological, and Biological Phenomena*; Jossey-Bass: San Francisco, CA, USA, 1997.
16. Zhu, L.; Liu, G. Preschool children's understanding of illness. *Acta Psychol. Sin.* **2007**, *39*, 96–103.
17. Carey, S. On the origin of causal understanding. In *Causal Cognition: A Multidisciplinary Debate*; Sperber, D., Premack, D., Premack, A.J., Eds.; Clarendon Press: Oxford, UK, 1995; pp. 268–302.
18. Inagaki, K.; Hatano, G. Children's understanding of mind-body relationships. In *Children's Understanding of Biology and Health*; Siegal, M., Peterson, C.C., Eds.; Cambridge University Press: Cambridge, UK, 1999; pp. 23–44.
19. Siegal, M.; Peterson, C.C. (Eds.) *Children's Understanding of Biology and Health*; Cambridge University Press: Cambridge, UK, 1999.
20. Hughes, C.; Leekam, S. What are the links between theory of mind and social relations? Review, reflections and new directions for studies of typical and atypical development. *Soc. Dev.* **2004**, *13*, 590–619. [CrossRef]
21. Wellman, H.M. *The Child's Theory of Mind*; Bradford Books/MIT Press: Cambridge, MA, USA, 1990.
22. Brown, J.; Dunn, J. Continuities in emotion understanding from three to six years. *Child Dev.* **1996**, *67*, 789–802. [CrossRef]
23. Wimmer, H.; Perner, J. Beliefs about beliefs: Representation and constraining function of wrong beliefs in young children's understanding of deception. *Cognition* **1983**, *13*, 103–128. [CrossRef]
24. King, T.A.; Tarrant, R.A.C. Children's knowledge, cognitions and emotions surrounding natural disasters: An investigation of Year 5 students, Wellington, New Zealand. *AJDTS* **2013**, *1*, 17–26.
25. Laçin Şimşek, C. Children's ideas about earthquakes. *Int. J. Environ. Sci. Educ.* **2007**, *2*, 14–19.
26. Raccanello, D.; Burro, R.; Hall, R. Children's emotional experience two years after an earthquake: An exploration of knowledge of earthquakes and associated emotions. *PLoS ONE* **2017**, *12*, e0189633. [CrossRef]
27. Ross, K.E.; Shuell, T.J. Children's beliefs about earthquakes. *Sci. Educ.* **1993**, *77*, 191–205. [CrossRef]
28. Raccanello, D.; Vicentini, G.; Brondino, M.; Burro, R. Technology-based trainings on emotions: A web application on earthquake-related emotional prevention with children. In *International Conference in Methodologies and Intelligent Systems for Technology Enhanced Learning*; Springer: Cham, Switzerland, in press.
29. Pekrun, R.; Bühner, M. Self-report measures of academic emotions. In *International Handbook of Emotions in Education*; Pekrun, R., Linnenbrink-Garcia, L., Eds.; Taylor & Francis: New York, NY, USA, 2014; pp. 561–579.
30. Gobbo, C.; Raccanello, D. Personal narratives about states of suffering and wellbeing: Children's conceptualization in terms of physical and psychological domain. *Appl. Cogn. Psychol.* **2011**, *25*, 386–394. [CrossRef]
31. Raccanello, D. L'adozione dal punto di vista dei bambini: Ruolo di età e coinvolgimento personale [Adoption from children's point of view: Role of age and personal involvement]. *Psic. Clin. Svil.* **2012**, *16*, 507–529.
32. Raccanello, D.; Hall, R.; Burro, R. Salience of primary and secondary school students' achievement emotions and perceived antecedents: Interviews on literacy and mathematics domains. *Learn. Individ. Differ.* **2018**, *65*, 65–79. [CrossRef]
33. Shaughnessy, J.J.; Zechmeister, E.B.; Zechmeister, J.S. *Research Methods in Psychology*; McGraw-Hill: New York, NY, USA, 2012.
34. Otto, B.W. *Language Development in Early Childhood Education*, 3rd ed.; Pearson: Boston, MA, USA, 2010.
35. Burro, R. To be objective in experimental phenomenology: A psychophysics application. *Springerplus* **2016**, *5*, 1720. [CrossRef] [PubMed]
36. Chimi, C.J.; Russell, D.L. The Likert scale: A proposal for improvement using quasi-continuous variables. In Proceedings of the ISECON, Washington, DC, USA, 5–8 November 2009.
37. Henson, R.K.; Hull, D.M.; Williams, C.S. Methodology in our education research culture. *Educ. Res.* **2010**, *39*, 229–240. [CrossRef]
38. Harwell, M.R.; Gatti, G.G. Rescaling ordinal data to interval data in educational research. *Rev. Educ. Res.* **2001**, *71*, 105–131. [CrossRef]
39. Jamieson, S. Likert scales: How to (ab)use them. *Med. Educ.* **2004**, *38*, 1212–1218. [CrossRef] [PubMed]
40. Knapp, T.R. Treating ordinal scales as interval scales: An attempt to resolve the controversy. *Nurs. Res.* **1990**, *39*, 121–123. [CrossRef]
41. Andrich, D. *Rasch Models for Measurement*; Sage Publications: Beverly Hills, LA, USA, 1988.

42. Bond, T.G.; Fox, C.M. *Applying the Rasch Model: Fundamental Measurement in the Human Sciences*, 2nd ed.; Includes Rasch software on CD-ROM; Lawrence Erlbaum Associates: Mahwah, NJ, USA, 2007.
43. Rasch, G. *Probabilistic Models for Some Intelligence and Attainment Tests;* Copenhagen, Danish Institute for Educational Research, expanded edition (1980) with foreword and afterword by B.D. Wright; The University of Chicago Press: Chicago, IL, USA, 1980.
44. Luce, R.D.; Krantz, D.H.; Suppes, P.; Tversky, A. *Foundations of Measurement;* Academic: San Diego, CA, USA, 1990; Volume 3.
45. Cavanagh, R.F.; Romanoski, J.T. Rating scale instruments and measurement. *Learn. Environ. Res.* **2007**, *9*, 273–289. [CrossRef]
46. Bianchi, I.; Savardi, U.; Burro, R. Perceptual ratings of opposite spatial properties: Do they lie on the same dimension? *Acta Psychol.* **2011**, *138*, 405–418. [CrossRef]
47. Burro, R.; Pasini, M.; Raccanello, D. Using Rasch Models for developing fast technology enhanced learning solutions: An example with emojis. In *International Conference in Methodologies and Intelligent Systems for Technology Enhanced Learning;* Springer: Cham, Switzerland, in press.
48. Burro, R.; Pasini, M.; Raccanello, D. Emojis' psychophysics: Measuring emotions in technology enhanced learning contexts. In *International Conference in Methodologies and Intelligent Systems for Techhnology Enhanced Learning;* Springer: Cham, Switzerland, 2019; Volume 804, pp. 70–78.
49. Burro, R.; Sartori, R.; Vidotto, G. The method of constant stimuli with three rating categories and the use of Rasch models. *Qual. Quant.* **2011**, *45*, 43–58. [CrossRef]
50. Burro, R.; Savardi, U.; Annunziata, M.A.; De Paoli, P.; Bianchi, I. The perceived severity of a disease and the impact of the vocabulary used to convey information: Using Rasch scaling in a simulated oncological scenario. *Patient Prefer Adherence* **2018**, *12*, 2553–2573. [CrossRef]
51. Wright, B.D. Fundamental measurement for psychology. In *The New Rules of Measurement: What Every Educator and Psychologist should Know;* Embretson, S.E., Hershberger, S.L., Eds.; Lawrence Erlbaum Associates: Hillsdale, NJ, USA, 1999; pp. 65–104.
52. Pasini, M.; Brondino, M.; Burro, R.; Raccanello, D.; Gallo, S. The use of different multiple devices for an ecological assessment. In *Methodologies and Intelligent Systems for Technology Enhanced Learning;* Springer: Cham, Switzerland, 2016; Volume 478, pp. 121–129.
53. Raccanello, D.; Brondino, M.; Crane, M.; Pasini, M. Antecedents of achievement emotions: Mixed-device assessment with Italian and Australian university students. In *Methodologies and Intelligent Systems for Technology Enhanced Learning;* Springer: Cham, Switzerland, 2016; Volume 478, pp. 183–191.
54. Raccanello, D.; Burro, R.; Brondino, M.; Pasini, M. Use of internet and wellbeing: A mixed-device survey. In *Methodologies and Intelligent Systems for Technology Enhanced Learning;* Springer: Cham, Switzerland, 2017; Volume 617, pp. 65–73.
55. Toepoel, V.; Lugtig, P. Online surveys are mixed-device surveys. Issues associated with the use of different (mobile) devices in web surveys. *MDA* **2015**, *9*, 155–162. [CrossRef]
56. Kreuter, F.; Presser, S.; Tourangeau, R. Social desirability bias in CATI, IVR, and web surveys: The effects of mode and question sensitivity. *Public Opin. Q.* **2008**, *72*, 847–865. [CrossRef]
57. Bowling, A. Mode of questionnaire administration can have serious effects on data quality. *J. Public Health* **2005**, *27*, 281–291. [CrossRef]
58. Shiffman, S.; Stone, A.A.; Hufford, M.R. Ecological momentary assessment. *Annu. Rev. Clin. Psychol.* **2008**, *4*, 1–32. [CrossRef]
59. Axinn, W.; Gatny, H.; Wagner, J. Maximizing data quality using mode switching in mixed-device survey design: Nonresponse bias and models of demo-graphic behaviour. *MDA* **2015**, *9*, 163–184. [CrossRef]
60. Flay, B.R.; Biglan, A.; Boruch, R.F.; Castro, F.G.; Gottfredson, D.; Kellam, S.; Moscicki, E.K.; Schinke, S.; Valentine, J.C.; Ji, P. Standards of evidence: Criteria for efficacy, effectiveness and dissemination. *Prev. Sci.* **2005**, *6*, 151–175. [CrossRef] [PubMed]
61. Gottfredson, D.C.; Cook, T.D.; Gardner, F.E.M.; Gorman-Smith, D.; Howe, G.W.; Sandler, I.S.; Zafft, K.M. Standards of evidence for efficacy, effectiveness, and scale-up research in prevention science: Next generation. *Prev. Sci.* **2018**, *16*, 893–926. [CrossRef]
62. Raccanello, D.; Brondino, M.; Pasini, M.; Landuzzi, M.G.; Scarpanti, D.; Vicentini, G.; Massaro, M.; Burro, R. The usability. In *International Conference in Methodologies and Intelligent Systems for Techhnology Enhanced Learning;* Springer: Cham, Switzerland, 2019; Volume 804, pp. 79–87.

63. Raccanello, D.; Gobbo, C. Understanding coexistence of internal states in children and adults. *Eur. J. Dev. Psychol.* **2015**, *12*, 158–176. [CrossRef]
64. Nolen, S.B. Young children's motivation to read and write: Development in social contexts. *Cogn. Instruct.* **2007**, *25*, 219–270. [CrossRef]
65. Raccanello, D.; Hall, R.; Burro, R. Formative evaluation of emojis to signal earthquake events. Unpublished work, 2019.
66. Burro, R.; Raccanello, D.; Pasini, M.; Brondino, M. An estimation of a nonlinear dynamic process using Latent Class extended Mixed Models. Affect profiles after terrorist attacks. *Nonlinear Dyn. Psychol. Life Sci.* **2018**, *22*, 35–52.
67. Schmitz, C. *LimeSurvey: An Open Source Survey Tool*; LimeSurvey Project: Hamburg, Germany, 2015; Available online: http://www.limesurvey.org (accessed on 19 April 2019).
68. Raccanello, D.; Barnaba, V.; Burro, R. Adults' representation of children's emotions and coping strategies related to earthquakes. In Proceedings of the 19th European Congress of Developmental Psychology, Athens, Greece, 29 August–1 September 2019.
69. Bates, D.; Maechler, M.; Bolker, B.; Walker, S. Fitting Linear Mixed-Effects Models Using lme4. *J. Stat. Softw.* **2015**, *67*, 1–48. [CrossRef]
70. R Core Team. *R: A Language and Environment for Statistical Computing*; R Foundation for Statistical Computing: Wien, Austria, 2019.
71. Barr, D.J. Random effects structure for testing interactions in linear mixedeffects models. *Front. Psychol.* **2013**, *4*. [CrossRef]
72. Barr, D.J.; Levy, R.; Scheepers, C.; Tily, H.J. Random effects structure for confirmatory hypothesis testing: Keep it maximal. *J. Mem. Lang.* **2013**, *68*, 255–278. [CrossRef] [PubMed]
73. Bastianello, T.; Majorano, M.; Burro, R. Children's attention in a word learning interactive context with educators and vocabulary growth. *Early Child. Dev. Care* **2018**. [CrossRef]
74. Bates, D.; Kliegl, R.; Vasishth, S.; Baayen, H. Parsimonious Mixed Models. 2015. Available online: https://arxiv.org/pdf/1506.04967.pdf (accessed on 19 April 2019).
75. Bianchi, I.; Branchini, E.; Burro, R.; Capitani, E.; Savardi, U. Overtly prompting people to "think in opposites" supports insight problem solving. *Think. Reason.* **2019**. [CrossRef]
76. Bianchi, I.; Canestrari, C.; Roncoroni, A.M.; Burro, R.; Branchini, E.; Savardi, U. The effects of modulating contrast in verbal irony as a cue for giftedness. *Humor* **2017**, *30*, 383–415. [CrossRef]
77. Branchini, E.; Bianchi, I.; Burro, R.; Capitani, E.; Savardi, U. Can contraries prompt intuition in insight problem solving? *Front. Psychol.* **2016**, *7*. [CrossRef]
78. Branchini, E.; Burro, R.; Bianchi, I.; Savardi, U. Contraries as an effective strategy in geometrical problem solving. *Think. Reason.* **2015**, *21*, 397–430. [CrossRef]
79. Canestrari, C.; Branchini, E.; Bianchi, I.; Savardi, U.; Burro, R. Pleasures of the mind: What makes jokes and insight problems enjoyable. *Front. Psychol.* **2018**, *8*. [CrossRef] [PubMed]
80. Raccanello, D.; Gobbo, C.; Corona, L.; De Bona, G.; Hall, R.; Burro, R. Long-term intergenerational transmission of memories of the Vajont disaster. *Memory* in press.
81. Rigon, J.; Burro, R.; Guariglia, C.; Maini, M.; Marin, D.; Ciurli, P.; Bivona, U.; Formisano, R. Self-awareness rehabilitation after traumatic brain injury: A pilot study to compare two group therapies. *Restor. Neurol. Neurosci.* **2017**, *35*, 115–127. [CrossRef] [PubMed]
82. Lenth, R. *Emmeans: Estimated Marginal Means, aka Least-Squares Means. R Package Version 1.3.3.*; The University of Iowa: Iowa City, IA, USA, 2019.
83. Olivier, J.; May, W.L.; Bell, M.L. Relative effect sizes for measures of risk. *Commun. Stat. Theory Methods* **2017**, *46*, 6774–6781. [CrossRef]
84. Chen, H.; Cohen, P.; Chen, S. How big is a big odds ratio? Interpreting the magnitudes of odds ratios in epidemiological studies. *Commun. Stat. Simul. Comput.* **2010**, *39*, 860–864. [CrossRef]
85. Andrich, D.; Sheridan, B.S.; Luo, G. *RUMM 2030: Rasch Unidimensional Models for Measurement*; Version 5.1; RUMM Laboratory: Perth, Australia, 2010.
86. RUMM Laboratory Pty Ltd. 1997–2017. Available online: www.rummlab.com (accessed on 19 April 2019).
87. Kreiner, S. The Rasch model for dichotomous items. In *Rasch Models in Health. Applied Mathematics Series*; Christensen, K.B., Kreiner, S., Mesbah, M., Eds.; ISTE Ltd.: London, UK; John Wiley & Sons, Inc.: Hoboken, NJ, USA, 2013; pp. 5–25.

88. Tennant, A.; Conaghan, P.G. The Rasch measurement model in rheumatology: What is it and why use it? When should it be applied, and what should one look for in a Rasch paper? *Arthritis Care Res.* **2007**, *57*, 1358–1362. [CrossRef]
89. Marais, I. Local dependence. In *Rasch Models in Health. Applied Mathematics Series*; Christensen, K.B., Kreiner, S., Mesbah, M., Eds.; ISTE Ltd.: London, UK; John Wiley & Sons, Inc.: Hoboken, NJ, USA, 2013; pp. 111–130.
90. Smith, E. Detecting and evaluating the impact of multidimensionality using item fit statistics and principal component analysis of residuals. *J. Appl. Meas.* **2002**, *3*, 205–231.
91. Pallant, J.F.; Tennant, A. An introduction to the Rasch measurement model: An example using the Hospital Anxiety and Depression Scale (HADS). *Br. J. Clin. Psychol.* **2007**, *46*, 1–18. [CrossRef]
92. Tennant, A.; Penta, M.; Tesio, L.; Grimby, G.; Thonnard, J.L.; Slade, A.; Lawton, G.; Simone, A.; Carter, J.; Lundgren-Nilsson, Å.; et al. Assessing and adjusting for cross-cultural validity of impairment and activity limitation scales through differential item functioning within the framework of the Rasch model: The PRO-ESOR project. *Med. Care* **2004**, *42*, I37–I48. [CrossRef]
93. Hobart, J.; Cano, S. Improving the evaluation of therapeutic interventions in multiple sclerosis: The role of new psychometric methods. *Health Technol. Assess.* **2009**, *13*, 1–177. [CrossRef] [PubMed]
94. Kreiner, S.; Christensen, K.B. Person parameter estimation and measurement in Rasch Models. In *Rasch Models in Health. Applied Mathematics Series*; Christensen, K.B., Kreiner, S., Mesbah, M., Eds.; ISTE Ltd.: London, UK; John Wiley & Sons, Inc.: Hoboken, NJ, USA, 2013; pp. 63–67.
95. Wright, B.D.; Masters, G.N. *Rating Scale Analysis*; MESA Press: Chicago, IL, USA, 1982.
96. Fisher, W.P.J. Rating scale instrument quality criteria. *RMT* **2007**, *21*, 1095.
97. Vanhoutte, E.K.; Faber, C.G.; van Nes, S.I.; Jacobs, B.C.; van Doorn, P.A.; van Koningsveld, R.; Cornblath, D.R.; van der Kooi, A.J.; Cats, E.A.; van den Berg, L.H.; et al. Modifying the Medical Research Council grading system through Rasch analyses. *Brain* **2012**, *135*, 1639–1649. [CrossRef] [PubMed]
98. Reeve, B.B.; Hays, R.D.; Bjorner, J.B.; Cook, K.F.; Crane, P.K.; Teresi, J.A.; Thissen, D.; Revicki, D.A.; Weiss, D.J.; Hambleton, R.K.; et al. Psychometric evaluation and calibration of health-related quality of life item banks: Plans for Patient-Report Outcomes Measurement Information System (PROMIS). *Med. Care* **2007**, *45*, 22–31. [CrossRef] [PubMed]
99. Vosniadou, S. The development of students' understanding of science. *Front. Educ.* **2019**, *4*. [CrossRef]
100. Mutch, C.; Gawith, E. The role of schools in disaster setting: Learning from the 2010-2011 New Zealand earthquakes. *Int. J. Educ. Dev.* **2015**, *41*, 283–291. [CrossRef]
101. Bateman, A.; Danby, S. Recovering from the earthquake: Early childhood teachers and children collaboratively telling stories about their experiences. *Disaster Prev. Manag.* **2013**, *22*, 467–479. [CrossRef]
102. Mutch, C.; Gawith, E. The New Zealand earthquakes and the role of schools in engaging children in emotional processing of disaster experiences. *Pastor. Care Educ.* **2014**, *32*, 54–67. [CrossRef]

Article

Dynamics of Socioeconomic Exposure, Vulnerability and Impacts of Recent Droughts in Argentina

Gustavo Naumann [1,*], Walter M. Vargas [2], Paulo Barbosa [1], Veit Blauhut [3], Jonathan Spinoni [1] and Jürgen V. Vogt [1]

1 European Commission, Joint Research Centre, 21027 Ispra, Italy; paulo.barbosa@ec.europa.eu (P.B.); jonathan.spinoni@ec.europa.eu (J.S.); juergen.vogt@ec.europa.eu (J.V.V.)

2 Consejo Nacional de Investigaciones Científicas y Técnicas—CONICET, C1428EGA Ciudad Autónoma de Buenos Aires, Argentina; vargas@at.fcen.uba.ar

3 Environmental Hydrological Systems, Albert-Ludwigs-University of Freiburg, 79098 Freiburg im Breisgau, Germany; veit.blauhut@hydrology.uni-freiburg.de

* Correspondence: gustavo.naumann@ec.europa.eu; Tel.: +39-0332-78-5535

Received: 11 December 2018; Accepted: 7 January 2019; Published: 12 January 2019

Abstract: During the last 20 years, Argentina experienced several extreme and widespread droughts in many different regions, including the core cropland areas. The most devastating recent events were recorded in the years 2006, 2009 and 2011. Reported impacts of the main events induced losses of more than 4 billion U.S. dollars and more than 1 million persons were reported to be directly or indirectly affected. In this paper, we analyse the drought risk in Argentina, taking into account recent information on drought hazard, exposure and vulnerability. Accordingly, we identified the most severe droughts in Argentina during the 2000–2015 period using a combination of drought hazard indicators and exposure layers. Three main events were identified: (1) during spring 2006 droughts peaked in the northeast of Argentina, (2) in 2009 precipitation deficits indicated a drought epicenter in the central Argentinian plains, and (3) in 2011 the northern Patagonia region experienced a combination of natural disasters due to severe drought conditions and a devastating volcanic eruption. Furthermore, we analysed the dynamics of drought exposure for the population and the main economic sectors affected by municipality, i.e., agriculture and livestock production. Assets exposed to droughts have been identified with several records of drought impacts and declarations of farming emergencies. We show that by combining exposure and vulnerability with drought intensity it is feasible to detect the likelihood of regional impacts in different sectors.

Keywords: drought; impacts; exposure; vulnerability; risk; policy

1. Introduction

Droughts are the result of a deficit or inadequate timing of precipitation over an extended period of time, often combined with high temperatures and increased water demands. Due to its insidious onset, and often long persistence, it is a complex phenomenon that propagates through the hydrological cycle, from meteorological drought to socioeconomic drought [1]. Accordingly, a drought disaster is related to an event that usually results in serious damage to socioeconomic and environmental systems [2]. The intensity of the disaster and its related damages depend on the combination of the drought severity and their spatial extension, the exposed assets in a certain location and their intrinsic vulnerability. Such drought impacts, e.g., on human societies, have been documented for selected case studies and applied to drought risk analyses from regional to global scales [3–7]. As a general result, the studies showed that certain demographic or socioeconomic factors drive the intrinsic vulnerability to drought-related impacts. Since drought risk is region- and sector-specific, however, each thematic investigation requires a specific set of hazard and vulnerability information to better predict drought impacts [8].

Today, a wealth of drought indices to measure drought intensity and duration are available, characterizing meteorological, soil moisture, hydrological or vegetation conditions [9]. Most commonly, simple indicators that are based on precipitation and standardized products (e.g., Standardized Precipitation Index; SPI) are applied for early warning systems [3]. Among them, drought indices that take into account an approximation of the water balance (like the Standardized Precipitation-Evapotranspiration Index; SPEI) have been shown to be good predictors of drought impacts [5,8,10].

The complex interactions between different economic sectors, cascading effects and indirect impacts hamper a quantification of the overall impacts of droughts. While a variety of different impact datasets exist (e.g., EM-DAT; DesInventar), none of them is complete and a direct linkage to the intensity of the natural hazard is often lacking [7]. Hence, efforts like the U.S. Drought Impact Reporter [11] or the European Drought Impact Report Inventory [7] to collect and standardize drought impact reports are valuable in order to display the multifaceted impacts and to analyse their link to the related drought characteristics. Reports of valid drought impacts can be retrieved from a variety of data sources, e.g., the media, official reports and/or scientific papers. In the few publicly available disaster databases, drought disasters are particularly underreported [11]. The general lack of structural damages, combined with an often prolonged drought duration and secondary and long-lasting effects, and the different further drivers that can increase or decrease the effects of drought (such as a heat wave in summer), make it extremely difficult to retrieve correct loss estimates spatially and monetarily. According to [12], droughts account for less than 7% of the total reported losses from natural hazards since 1960, while other sources document an increasing share for droughts [13]. In general, exists a significant gap between reported and real drought damages and losses that hinders a systematic analysis and quantification and that could help to inform the development of adequate drought management plans [12,14]. To be effective, such plans have to be adapted to local or national needs by taking into account the characteristics of the economic sectors and ecosystems potentially affected.

Despite its significant size, Argentina's economy relies greatly on large-scale rainfed agricultural production (7.6% GDP in 2016). Agriculture is a key sector that represents one fourth of Argentinian exports and one of the main sources of foreign currency income [15]. The country was adversely affected by recurrent droughts with impacts at the local and national level [16–18], but also their consequences are likely to affect food prices and availability at the global level [19]. In Argentina, only a fraction of the rural population is actually involved in farming, population relies mostly on markets for food supply. While the percentage of food-insecure persons in the country dropped from a peak of 10.6% after the economic crisis in 2002 to 1% of the total population in 2012, this issue is still a reality for particular vulnerable parts of the population [20]. Drought, however, can pose a significant problem to large-scale market-oriented farmers, who run the risk of cessation of activity in the most extreme case. Their coping capacity depends on the level of their financial resources for buffering production losses. The main mechanisms applied to cope with the drought risk, therefore, are insurance schemes, specific credits and benefits linked to the declaration of farming emergencies [21].

In this paper, we focus on systematically quantifying the drought risk in Argentina i) as a function of long-term hazard, exposure and vulnerability; and ii) dynamically as a combination of changes in drought conditions and the exposed assets. Information on drought impacts was collected from media news, official reports (national and provincial) and entries from the DesInventar disaster loss database. Our results represent a first step towards understanding how patterns of exposure emerge as a result of the interaction between changes in population structure and regional climate variability. Moreover, we discuss how these indices could be linked to reported impacts and public alleviation measures.

2. Materials and Methods

Following [4], we define drought risk in a contextual approach as a function of the natural hazard, the exposed assets and their vulnerability at a given moment. Drought risk can be understood as the probability of harmful consequences or likelihood of losses resulting from interactions between drought hazard, drought exposure and drought vulnerability. These interactions between the three determinants of drought risk can be represented in a mathematical form [22]; Risk = f(Hazard, Exposure, Vulnerability). The scores of local drought risk range on a scale of 0–1, where 0 represents the lowest risk and 1 is associated with the highest risk.

2.1. Drought Hazard

The hazard component is based on the SPEI [23], a standardized drought index that represents different features of the water balance and therefore is also sensitive to the variability and changes in climatic variables other than precipitation. Similar to the Palmer Drought Severity Index, it includes the effects of atmospheric water demand through the reference evapotranspiration, yet it has the advantage of aggregating variables over different time dimensions that allows identifying different drought types and impacts, similar to the SPI [24]. Recent studies evaluated linkages between the accumulation period of drought indices and impacts in various sectors based on empirical data [3,5,25–27]. These indicate that dependencies between drought indices and impacts are sector- and region-specific, with the best overall performance for the case of Europe obtained using the SPEI for a 12-month aggregation period [8].

For the analysis, we defined two classes of drought intensity (moderate and severe) and the thresholds inspired by the fixed thresholds classification as defined in [28]. Hence, moderate drought events are defined as SPEI $\in [-1.0; -1.99]$, which corresponds to a cumulative probability of 15.9%, while a severe drought (SPEI < -2.0) corresponds to a cumulative probability of 2.3%. Estimations of the related variables (precipitation and reference evapotranspiration) needed to compute the SPEI were obtained for the baseline period 1950–2015 using the CRU v.4.01 dataset [29].

2.2. Exposure

Exposure is the presence of people; livelihoods, environmental sectors and resources; infrastructure; or economic, social or cultural assets in places that could be adversely affected by a climate hazard [30]. In this work, we follow the model of drought exposure by [4], represented by agricultural lands, population and livestock distribution as well as water-stressed areas. These four spatially explicit layers are combined using a non-compensatory Data Envelope Analysis (DEA, [4]) to represent the drought exposure. Exposed assets are defined as the outcome of the intersection between climate and exposure information. A drought event was defined for the month and regions with SPEI values under -1 or below. Hence, the measure of exposure is the number of asset-months under drought, i.e., persons, livestock or crop land affected by SPEI below -1. In the following we briefly describe the various layers used.

Global agricultural lands in the year 2000: This data collection represents the proportion of land area used as cropland in the year 2000. Satellite data from MODIS and SPOT-VEGETATION were combined with agricultural inventory data to create this product [31]. The maps showing the extent and intensity of agricultural land use on Earth were compiled on a 5 min $\times$ 5 min latitude-longitude grid cell size.

Global Human Settlement Layer derived data (GHSL, 1 km resolution) was used to account for the population exposed to droughts. The GHSL population estimates (GHS-POP) correspond to the residential population for the year 2015 [32]. Population was consistently disaggregated from census or administrative units to grid cells, informed by the distribution and density of built-up areas as mapped in the GHSL global layer per corresponding period.

Gridded livestock of the world: This data collection provides modelled livestock densities of the world, adjusted to match official national estimates for the reference year of 2005, at a spatial resolution of 3 min of arc (5 × 5 km at the equator). The freely accessible maps are created through the spatial disaggregation of sub-national statistical data based on empirical relationships with environmental variables in similar agro-ecological zones [33].

2.3. Vulnerability

Vulnerability describes the characteristics and circumstances of a community, system or asset that make it susceptible to the damaging effects of a hazard [34]. Vulnerability to drought is estimated based on a multidimensional model incorporating social, economic and infrastructural factors. The underlying indicators are generic proxies that reflect the level of quality of different constituents of a civil society. Vulnerability can be conceptualized as determined by two dimensions, sensitivity (S) and adaptive capacity (AC). Sensitivity captures the characteristics of a community that influence its likelihood to experience harm while experiencing a drought event. Adaptive capacity is a function of both asset-based components of a community such as wealth and human capital that help to predict how flexible individuals may be in anticipating, responding to, coping with, and recovering from drought impacts. Here, each factor is characterized by a set of proxy indicators that are generalized at the national and sub-national scales. Fifteen indicators are selected in accordance with the work of [4,35] and substantiated by the vulnerability studies of [36,37]. Vulnerability to drought is computed as a 2-step composite model that derives from the aggregation of proxy indicators representing the economic, social and infrastructural factors of vulnerability at each geographic location, as similar as for the Drought Vulnerability Index [35]. In the first step, indicators for each factor are combined using a Data Envelopment Analysis (DEA) model [38]. In the second step, individual factors resulting from independent DEA analyses are arithmetically aggregated into a composite model of drought vulnerability. Details of the variables and dimensions adopted can be found in [4,35], while the relevant vulnerability proxies are depicted in the Appendix A.

2.4. Impact Information

The impact quantification was primarily based on a literature and media search in various web-based databases, including news outlets, Google Scholar and Scopus. The key terms used during the search were "Drought" OR "Impact" OR "Agriculture" OR "Argentina". Additionally, several kinds of word combinations were also used to find more references in Spanish and English. First-round screening was done based on a quick review of title, abstract, and keywords in the articles, official reports and news. After the first screening, a set of impact reports for the main drought events were retrieved from several sources: (1) A search in the national and international media was performed, (2) National and provincial official reports regularly published in the "Boletín oficial de la República Argentina" [39] were examined in order to quantify the location and amount of farming emergencies declared due to drought events as well as related measures taken; and (3) Entries from the DesInventar disaster loss database were explored. The DesInventar database contains a historical inventory of disasters for the period 1970–2009. For drought, it captures the temporal behaviour of e.g., affected people, losses expressed in US$, damages in crops or loss of cattle.

3. Results and Discussion

3.1. Regional Drought Risk

This section examines to what extent the concept of drought risk is relevant to understand how droughts adversely affect the population and the crop and livestock production. Drought affects agricultural production through its effects on yields and livestock, while the population is mainly affected through the reduction of freshwater availability and several indirect effects, like changes in food and energy prices, labour productivity, unemployment, etc. Static risk maps, like the one shown in Figure 1 might be useful to target districts or regions that are recurrently affected by droughts and historically are not able to cope with its impacts. An analysis of their particular vulnerability might help to tailor management and adaptation plans.

3.1.1. Drought Hazard

Figure 1 shows the overall hazard, exposure, vulnerability and risk for the country. The drought hazard is described as the frequency of events occurring at different intensities determined from the historical data (1950–2015). Even though droughts can occur everywhere, the most affected regions are the semi-arid Pampa regions, some areas in the humid Pampa, Patagonia and Cuyo. Dry periods in the region often can be linked to the cold phase (La Niña) of the El Niño-Southern Oscillation. The strongest signal is observed at the end of a La Niña event and in the year after [40]. During the recent past, there was a remarkable increase in precipitation over most of subtropical Argentina, especially since 1960. This has favoured agricultural yields and the extension of crop areas into marginal lands in semi-arid regions [41]. Apart from more favourable climate conditions, this extension was possible due to the application of new production technologies and genotypes, the enhanced global food demand, and the increase in grain prices.

3.1.2. Exposure

Exposure is a necessary, but not sufficient, determinant of risk. For similar drought events, similarly exposed areas might experience different impacts according to their levels of development, land-use planning and mitigation measures. Hence, neglecting vulnerability information, such as levels of development, land-use planning and mitigation measures, is insufficient. It is possible to be exposed but not vulnerable; however, it is also necessary to be exposed to be damaged by an extreme event [42]. Similar to the notion of vulnerability, exposure to drought is a multidimensional concept that varies across spatial and temporal scales. In this work, we considered persons, crops and livestock that are located in an area affected by drought hazard to be elements of exposure. According to [43], in Argentina the estimated total population in 2016 was around 44 million, where 3.5 million are considered as rural population. The population is unequally distributed, with about 60% of the population living in the Pampas region (21% of the total area), including 15 million people in Buenos Aires. Agriculture in Argentina provides around 7% of employment, and accounts for nearly 10% of the country's GDP. The total agricultural land in 2015 was 1.48 million km^2. Beef and other types of meat are some of the most important agricultural export products of the country, with a production of nearly 5 million tonnes of meat. According to FAOSTAT, in 2016 the livestock was composed of 52,636,778 cattle, 14,864,321 sheep, 5,119,438 pigs and 4,712,173 goats.

The regions most exposed to adverse drought impacts are mainly located in concordance with the core crop and livestock areas located in central Argentina (Buenos Aires, Córdoba, La Pampa, Santa Fe and Entre Ríos provinces). According to [44], during 2016, Argentina was the 3rd largest world producer of soybeans with 58,799,258 tonnes produced on a total area of 19,504,608 ha and the 4th producer of maize with 39,792,852 tonnes harvested on 5,346,593 ha of land. Regarding the production of cattle, during the same period, the country was the 6th largest producer with a stock of 52,636,778 head, and the 4th largest producer of meat (2,644,000 tonnes). The country is also a main exporter of fruits. Mainly in Patagonia and Cuyo (Mendoza, Rio Negro and Neuquén), the country produced almost 1 million tonnes of pears during 2016, making the country the second largest world producer and the first exporter of this type of fruit.

3.1.3. Vulnerability

Social vulnerability is linked to the level of well-being of individuals, communities and society, economic vulnerability depends on the economic status of individuals and communities, while infrastructural vulnerability is related to the basic infrastructure needed to maintain the production of goods. The social vulnerability to droughts is higher in the northern part of the country due to lacking infrastructure and slow social progress. González [45] stressed a similar pattern of social vulnerability in the country characterized with the historical backwardness in the Northeast and Northwest regions. In the Chaco region, changes in the land use from extensive grazing through mixed farming to industrial-scale soybean production were made possible by a trend of increasing rainfall and changes in public policy in the late 1970s [41,46]. This process reinforced the concentration of lands for industrial-scale production and excluded small farmers. With little access to credits, these traditional farmers have no access to capital for new technologies such as seeded, drought resistant pastures, or irrigation wells. However, during the early 2010s changes in public policies allowed smallholders to regulate land tenure [46]. Notwithstanding the prominent progress observed during recent years (for instance, increases in life expectancy, and a decrease in infant and maternal mortality rates), these regions remain highly vulnerable, mainly during the time of the recurrent economic crisis.

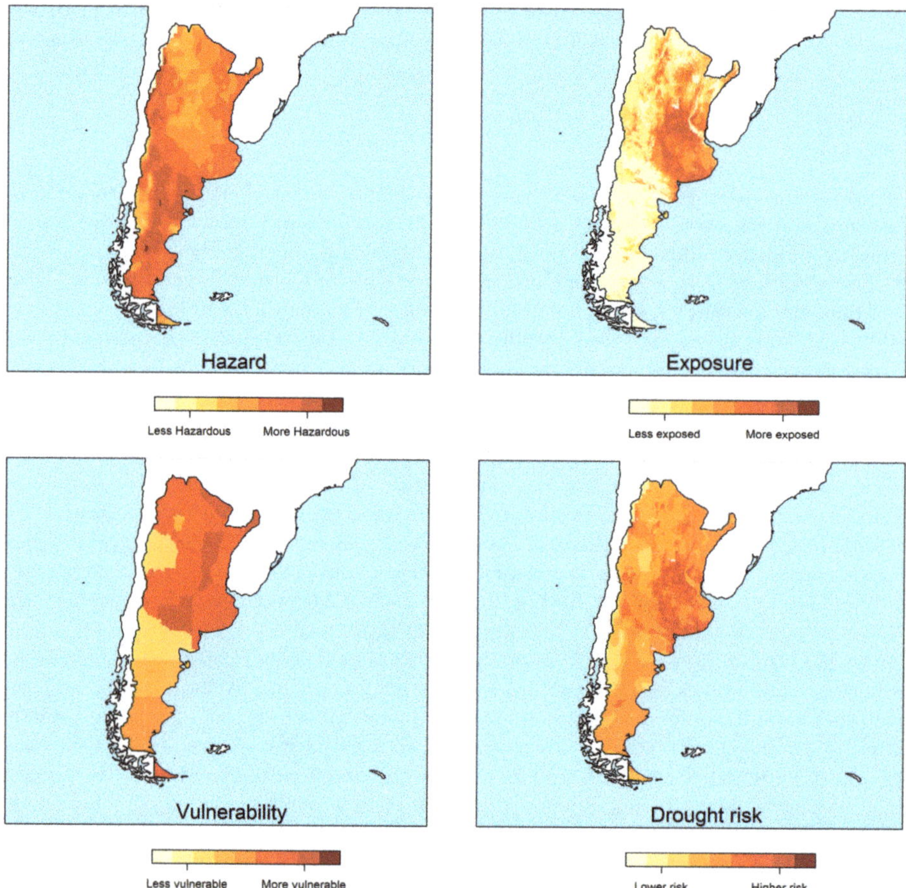

Figure 1. Drought hazard, exposure, vulnerability and the overall drought risk for Argentina.

3.1.4. Risk

As presented in Figure 1, the overall drought risk is lower for remote regions, and higher for populated areas and regions extensively exploited for crop production and livestock farming, such as the Buenos Aires, Córdoba and Santa Fe provinces. Accordingly, regions that have been less exposed are characterized by a lower drought risk. On the areas where there is at least on sector exposed, and as these regions are still subject to severe drought events as well, then their risk increases as a function of the total exposed entities and their lack of coping capacity. As presented in [4], drought risk on the global level is mainly driven by regional exposure, while hazard and vulnerability exhibit a weaker relationship with the geographic distribution of risk values. However, it is important to note that the approach to compute drought risk corresponds to a spatial relative measure and results might change according to changes in the domain or spatial resolution of the data.

3.2. Drought Events, Exposed Assets and Recorded Impacts

As a summary of the drought events observed during the study period, we show the percentage of total land area affected by moderate and severe droughts since the year 2000 (Figure 2). The widest area affected was observed during spring 2009, when more than 50% of the country was affected by a moderate drought and 20% was under severe drought conditions. This event, according to its intensity and extension was ranked as one of the most severe event that occurred in the globe during the past 60 years [47]. Other widespread events were observed during 2000, 2006, and 2011–2012. This trend towards drier conditions is supported also by [48], who detected a significant reduction in the annual precipitation. The authors indicated that during the first decade of the 21st century in Argentina a reduction in the mean annual precipitation of around 200–400 mm was observed centred in Santa Fe, Corrientes, Entre Ríos and Buenos Aires. This reduction is linked to the variability in the equatorial pacific sea surface temperature, but also to changes of the tracks and strength of both, the southern Atlantic and southern Pacific anticyclones.

These dry periods led to impacts in very diverse sectors including agriculture and livestock production but also inland river transportation, hydropower production (Salto Grande Dam located in the Uruguay River reported reductions in the production during the three events), freshwater availability and tourism (Table 1). Even though, the information contained in the media reports could be biased, they represent a fair description and can be extremely comprehensive, and usually similarly reliable as inventories made from official sources [49].

Table 1. Main drought characteristics and impacts of the 2006, 2009, and 2011 droughts in Argentina. Sources detailed in Tables 2, A1 and A2.

	2006–2007	2009–2010	2011–2012
Areas affected	Central and Northeast Argentina	Central Argentina	Central Argentina, Northern Patagonia
Peak month	November	September	September
Maximum area affected	27% of the country	42% of the country	21% of the country
Number of farming Emergencies declared due to droughts and location	32 (in, Central, North East Argentina and Mesopotamia)	53 (in Central, North East, North West and Patagonia)	1 (Northwest)
Main sectors affected	Hydropower, tourism, inland navigation, crops, cattle	Crops, cattle, groundwater, inland navigation, hydropower	Sheep, fresh water availability, hydropower

Table 2. Reported impacts, persons affected and relocated, damages in crops (ha.) and cattle lost during the 2005–2006 and 2008–2009 droughts (Source: Desinventar; http://desinventar.net/).

Event	Reported Impacts	Affected Persons	Relocated	Damages in Crops [ha]	Lost Cattle
2005–2006	36	41,157	0	4,500,000	4000
2008–2009	135	916,500	200	1,050,100	21,280

In many cases, the severity of the impacts led to the declaration of farming emergencies according to law 26.509 [50] (Table A2). One of the main benefits for the producers linked with this regulation is the suspension of the tax obligations until the end of the crop cycle and/or access to low rate credits. To be eligible for the benefits under this law, producers must be affected in their production or production capacity by at least 50%. In that sense, the declaration of a regional emergency is a reliable report of severe impacts as the producers have to proof that the disaster was linked to the respective event. The blue bars in Figure 2 show the number of emergencies declared due to droughts during the period analysed.

As illustrated in [51], during the period 2009–2012, 110 agricultural emergencies were declared in Argentina, out of which 70 were related to drought events. In most of these declared emergencies the farmers were granted with tax reductions or small-scale credits. According to official information (Table A2) during the 2000–2015 period, 118 farming emergencies were declared by provincial or national authorities in response to severe droughts. The 2009–2010 event triggered by far the most emergencies, with 30 declarations in 2009 and 23 in 2010 (Figure 2). During this period, as the drought was widespread around the country, 17 out of 23 provinces declared emergencies. On the other hand, during the 2011–2012 event only one emergency was declared in the province of Salta (Northwestern Argentina).

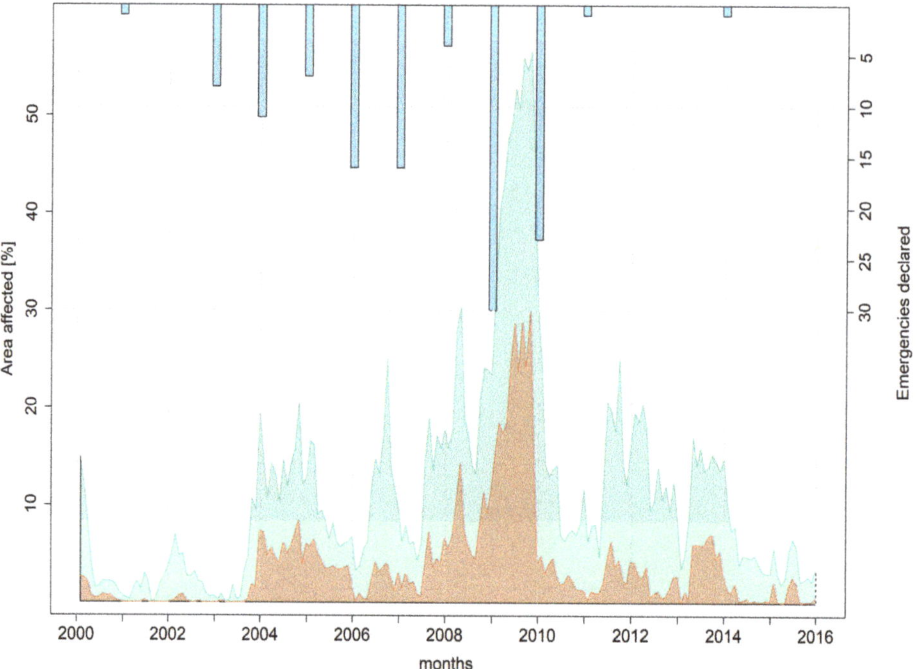

Figure 2. Percent of total land area in Argentina affected by moderate and severe droughts and number of farming emergencies officially declared (blue bars).

Figure 3 shows the spatial distribution of drought severity for the selected events together with the areas that reported impacts or triggered farming emergencies during each period. Overall, there is a good visual match between the most affected areas and the incidence of reported impacts and declaration of emergencies. This is more evident for the 2009 drought, were almost all provinces in the country reported impacts in different sectors.

The dynamic vulnerability and exposure layers are computed for each month. For its computation, the dynamic layer of drought hazard, in this case the monthly SPEI-12 and the structural layers of exposure and vulnerability were taken into account. Figure 4 shows the population, croplands and livestock exposed to droughts for each month together with the main reported impacts during the 2000–2015 period. In addition, in the Tables 1 and 2, summaries of the three main events and related impact statistics and emergency declarations are presented.

As depicted previously, exposed assets were defined as the outcome of the intersection between areas under drought and exposure information. Henceforth, the measure of exposure is the number of assets-months under drought, i.e., persons, livestock or crop land affected represented as the percentage of the total number of exposed assets. When information related to local vulnerability is taken into account (proxies from Table A3) the total number of exposed assets is reduced in proportion of the coping capacity of each region. This reduction is shown in Figure 4 by the broken lines. Even at this relative coarse resolution, the adaptive capacity results in a significant reduction of the total potential assets affected. For instance, during the peak of the 2009 drought, the percentage of population exposed is reduced from 60% to 40% of the total population when adaptation measures are taken into account. Similarly, agriculture and livestock exposed were reduced by about 20% during the same event. This might demonstrate that improvements to the infrastructure like the use of irrigation and/or fertilizers, and improved roads, can lead to a reduction in the overall impacts.

In the following we present a short summary of selected events and related impacts.

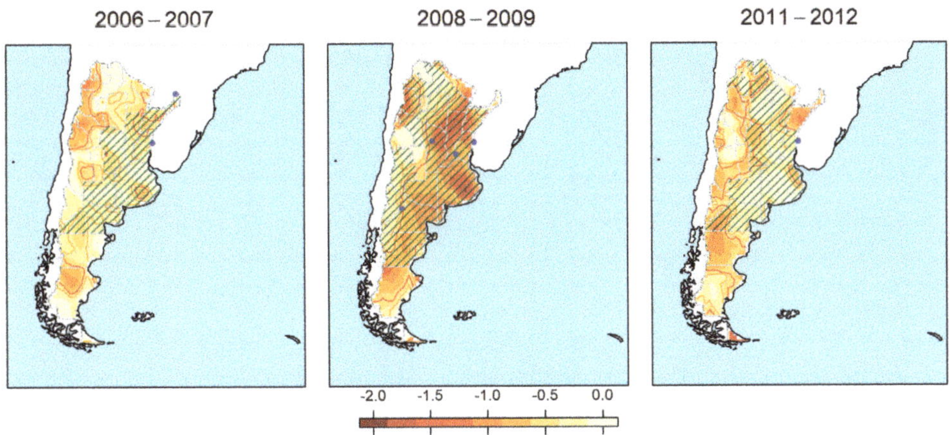

Figure 3. Intensity of drought conditions according to SPEI-12 in Argentina during the events 2006–2007, 2008–2009, and 2011–2012. Orange (red) contours represents the level of severity where at least half year (one year) was affected by a magnitude of SPEI ≤ -1. Hatched lines represent districts were impacts on agriculture and livestock production were reported or emergencies were triggered, while the blue points represent registered hydrological impacts.

3.2.1. 2006–2007 Drought

During this period, drought peaks were centred in central and northeastern Argentina and a prolonged but less intense drought combined with a heat wave affected producers mainly in Buenos Aires, Córdoba, Entre Ríos and Santa Fe. The lack of rain affected mainly maize and soybean production. Losses in the agriculture sector were estimated to around US$2 billion.

Reductions in river discharges were observed in La Plata Basin. Reduction of hydropower generation in Salto Grande Dam located in the Uruguay River and related losses in Buenos Aires of around US$100 million were reported. The reduction in flows from 1500 m^3/s to 350 m^3/s in the

Iguazu falls (Northeastern Argentina) led to the cancelation of several touristic tours producing estimated losses of around US$10 million.

Coincident with these impacts, peaks in exposed population and croplands are evident during spring 2006 (Figure 4). Around 20% of the total population and 30% of the croplands were exposed to moderate droughts.

3.2.2. 2008–2009 Drought

According to Figure 4, during the 2008–2009 event more than half of the total population of Argentina and almost 30% of the agricultural land was affected by a moderate drought. *DesInventar* reported 1 million ha crop damages during the event and almost 1 million persons affected.

More than 40 million head of cattle were exposed to moderate droughts, while 25 million head of cattle were exposed to severe droughts (65% and 40% of total livestock, respectively). According to *DesInventar* entries, 21,280 head of cattle were reported dead. This is the maximum in the analysed period and the third highest loss after the 1989 drought (160,000 head) and 1996 drought (120,000 head). Moreover, media reports stated that around 800,000 head of cattle have been lost [52].

During the 2009 drought farmers called for action to mitigate drought impacts, which are estimated to have caused losses of at least US$4 billion [52]. Among the effects, around 800,000 head of cattle have been lost, while in the Entre Rios province almost 90% of the wheat crop was affected. According to the national weather service, several regions in the country, including the provinces of Buenos Aires, Córdoba, La Pampa and Entre Ríos, were hit by the worst drought since at least 1971. During this period the worst affected area was the Pampas region, where extreme dry conditions were produced by a combination of high evaporative demand and persistent winds.

Significant reductions in the flow rates of the Parana and Uruguay rivers were observed. In Rosario, this prevented the use of small boats, mainly for recreation and forced the international cargo ships to reduce their loads to be able to depart from the port. Some intercontinental cargo ships reported a reduction of their load up to 1.500 tonnes. The most extreme case reported was regarding a cargo ship with 50,000 tonnes of minerals that had to stay in the Port of Rosario for more than one week because the levels of the river were below the minimum depth of water this type of ship can safely navigate. Furthermore, some locations in northeastern Argentina and Paraguay suffered from fuel shortages due to delays in ships transporting these goods from the south, mainly from the Campana and Zárate ports, located in northeastern Buenos Aires.

Starting in 2008, joint droughts affected hydropower generation in the Comahue region and Salto Grande in the Uruguay River forcing to increase the production of electricity from thermal power plants up to 70%. El Chocón was reported to produce energy under minimum flow conditions (140 MW against an installed capacity of 1200 MW with the full reservoir). The minimum of 170 m^3/s was assured to support irrigation and maintain fresh water availability in Neuquén y Río Negro. A shutdown of the hydropower generation was implemented at El Chocón during selected holidays or for specific hours per day to refill the reservoir.

Regarding the persons affected, this is the absolute maximum reported in the DesInventar record (1970–2009). For instance, water restrictions and a low quality of water affected Córdoba City and its surroundings (around 1.5 million inhabitants) due to a deficit of around 1 million m^3 in the San Roque reservoir.

During the period between spring 2008 and the end of 2009, the exposed population, croplands and livestock were more than 40% of the countries' totals (Figure 4). All the observed peaks were related to the reported impacts and declarations of several agricultural emergencies.

3.2.3. 2011–2012 Drought

According to estimates from Argentina's Agro-industry ministry in the eight years starting in 2007, an estimated 1.8 million sheep were lost in Chubut and Río Negro, which is equivalent to 12% of the country's total flock of 14.8 million. These two provinces house 43% of Argentina's sheep population.

During this period, apart from the sheep mortality, the changes in the environmental circumstances promoted by the ash deposits and drought significantly affected Merino wool production and fibre traits in farms of Northwestern Patagonia [53].

During this event, the strongest impacts were recorded in northern Patagonian provinces; however, a longer dry period lasted from 2007 to 2011, when two of the worst droughts in six decades happened [47]. Even if the lack of humidity is recurrent in northern Patagonian provinces, these events were combined with very low summer temperatures and the eruptions of the Chaitén (May 2008) and Puyehue (June 2011) volcanoes.

As represented in Figure 4, during this period the agricultural land exposed to droughts was relatively small and no significant peaks were observed as opposed to the 2005–2006 and 2009–2010 events. However, further noticeable peaks are observed for the population and livestock exposed. All the analysed events presented different areas affected and their linked impacts were accordingly represented by the exposure changes. These changes were verified by a set of impacts reported and the declaration of farming emergencies.

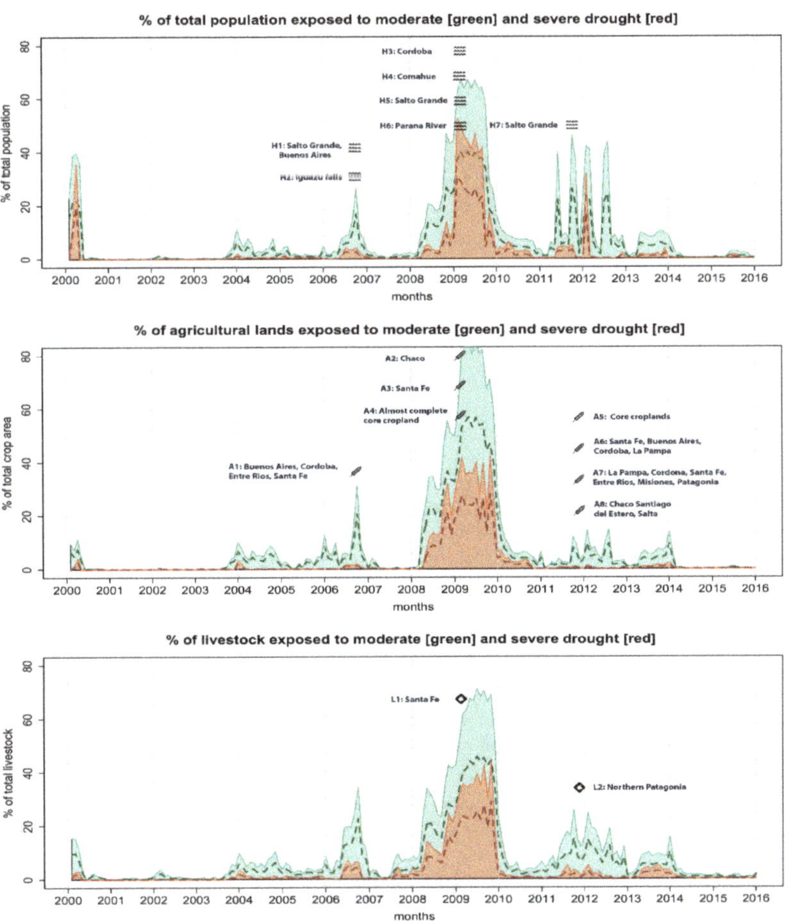

Figure 4. Percent of the total population exposed and hydrological impacts reported; croplands exposed and agricultural impacts reported; and livestock exposed and impacts on livestock reported (period 2000–2015). Dotted lines represent the exposed reduction due to adaptive capacity. Legend abbreviation refers to Appendix A, Table A1.

4. Conclusions

Argentina has committed itself to increasing its food supply to the world, so that it can move from feeding 400 million people in 2015 to 680 million people in 2020, consequently substantially contributing to the achievement of world food security [54]. This increase in food production has to be done under a changing drought risk and thus requires a comprehensive monitoring of drought risk dynamics in order to minimize the related impacts. In the present analysis we explored the sources of drought risk in Argentina by combining recent layers of drought hazard, exposure and vulnerability. According to this information, the key drivers of drought risk for the main economic sectors (i.e., agriculture and livestock production) in Argentina were identified. To do so, the characteristics of the most severe droughts in Argentina during the 2000–2015 period were assessed and their impacts analysed through the incorporation of different information sources (media, official reports, the DesInventar database). The collected impact data enabled for setting up narratives of past drought events that helped us to identify specific sectors that are locally affected by droughts.

During 2006 the drought peaked in the northeast of Argentina, while precipitation deficits indicate the epicenter of the 2009 drought in central Argentina, and in 2011 the most affected areas were observed in the region of Northern Patagonia. The main sectors affected were the ones linked to agriculture (maize, soybean and sunflower production) and livestock production (mainly cattle and sheep). In addition, sectors like hydroelectric power generation, inland water transportation, urban water supply and tourism were severely affected during these events. Impacts in the different sectors became gradually noticeable during the propagation of the drought through the hydrological cycle and due to their interrelations, which led to impacts that are the result of cascading effects and indirect damages. Locally, drought events combined with the exposed assets and the local socioeconomic vulnerabilities (e.g., lack of infrastructure, poverty, etc.), and potentially combined with other natural hazards (e.g., heat waves or volcanic eruptions), can trigger disasters that can affect the economy of the entire country and beyond.

The coincidence of the reported drought peaks and the level of exposed assets was confirmed by documented records of drought impacts and by the number of declared farming emergencies. We could demonstrate that by combining information on exposure and vulnerability with "real-time" drought intensity, it is possible to identify the occurrence of regional impacts across different sectors. This information can be helpful to monitor the probability of impact occurrence from the onset of a drought, or even before, if suitable medium-range forecasts are available. This information is useful to trigger pro-active measures in order to cope with and mitigate the potential impacts of droughts instead of relying on reactive measures to confine the damage when impacts are already manifested and propagated through the entire economic system.

While it is not possible to control the occurrence of droughts, the resulting impacts may be mitigated to a certain degree, namely through appropriate surveillance and management strategies previously agreed and laid out in a Drought Management Plan. These strategies have to be focused on to reduce the local dimensions of vulnerability and increase adaptive capacity.

Author Contributions: Conceptualization, G.N., W.M.V., V.B. and J.V.V.; Methodology, G.N., P.B. and J.V.V.; Data computation, G.N., J.S.; Validation, all authors; Writing-Original and Draft Preparation, All authors; Writing-Review & Editing, all authors.

Funding: This research received no external funding.

Conflicts of Interest: The authors declare no conflict of interest.

Appendix A

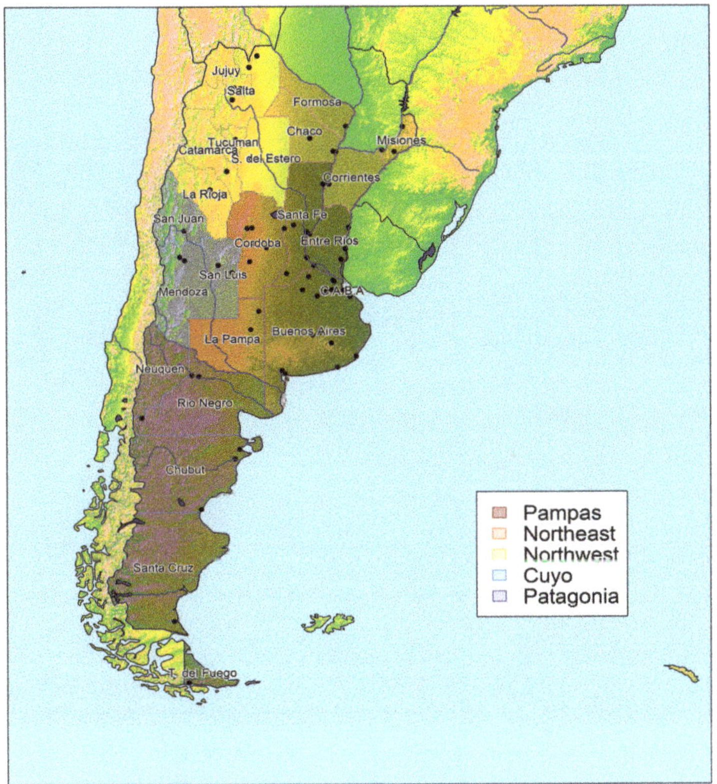

Figure A1. Description of the study area. Digital elevation model, provinces, location of cities with population greater than 50,000 inhabitants and regions used in this study.

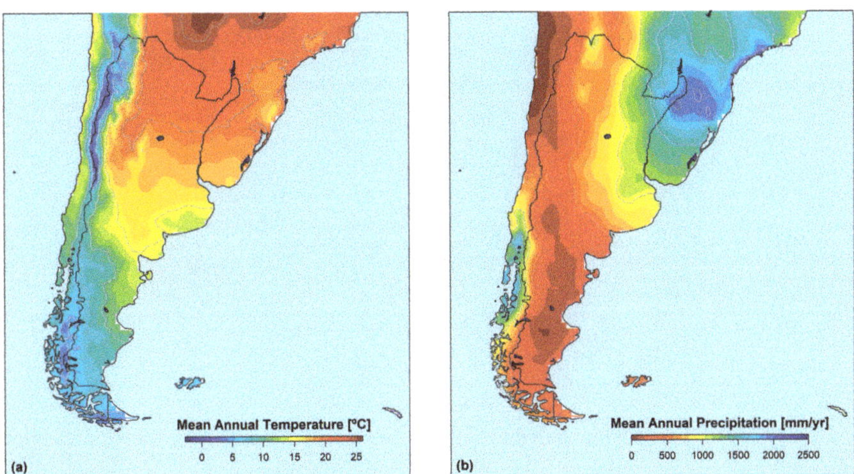

Figure A2. Mean annual temperature (**a**) and mean annual precipitation (**b**) based on CRU v.4.01 dataset.

Table A1. Summary of the selected media reports reviewed in this study. Impacts were classified in three main categories: Agricultural (A), Hydrological (H) and Livestock (L).

Event	Class	Description	Source
2006	A1	Estimated losses in the agriculture sector of around US$2 billion. A prolonged drought combined with a heat wave affected producers in Buenos Aires, Córdoba, Entre Ríos and Santa Fe. The lack of rain mainly affected maize and soybean production.	1
2006	H1	Reduction of hydropower generation in Salto Grande and estimated losses in Buenos Aires of around AR$315 million (circa US$100 million).	2
2006	H2	Flows reduced in the Iguazu falls from 1.5 million L to 350,000 L. Several touristic tours were cancelled, producing estimated losses of around €10 million.	3
2009	H3 A2 A3	Water restrictions in Córdoba City and surroundings. In Chaco province the area sown with wheat was reduced from 40,000 to 10,000 hectares, while that of sunflower was reduced from 300,000 to 60,000 hectares. In Santa Fe 12,500 farmers were assisted with AR$59 million (US$15 million) and losses were estimated at AR$2 billion (US$0.5 billion)	4–5
2009	A4 L1	The drought extends from the southern province of Río Negro through the central provinces of La Pampa and Córdoba and east and north to the provinces of Buenos Aires, Entre Ríos, Santa Fe, Corrientes, Chaco, Formosa and Santiago del Estero. Rural associations estimate that grain production will drop 39 percent and that 1.5 million head of livestock will be lost. In Chaco province agricultural output will be half of precedent year. Corn producers in Entre Ríos estimated losses of over 80 percent. The drought has caused a seven to eight million tonne drop in wheat production, from 16 million tonnes in the last harvest to around eight million with lower levels in 30 years, INTA reported. Estimated losses of at least US$7 billion. In the northern part of the province of Santa Fe alone, 300,000 head of cattle have been lost.	6
2009	H4	Reductions in hydropower generation in the Comahue region and Salto Grande forced production of up to 70% of electricity from thermal power plants. El Chocón was reported to be producing energy under minimum flow conditions. A shutdown of the electric generation was proposed at El Chocón during selected holidays or for specific hours per day to refill the reservoir.	7
2009	H5	Significant reduction of the level of the Paraná River near Rosario that prevented to practice any water sport and navigation. Reductions in wheat production and depletion of water reservoirs.	8
2009	H6	Reductions in the level of the Paraná River. Due to low flows, some intercontinental cargo ships reduced their capacity by 1.500 tonnes. A cargo ship carrying 50,000 tonnes of minerals had to stay in the Port of Rosario for more than one week.	9
2011	A5	60% of maize production affected in the core zone and a reduction in soybean production was observed.	10
2011	A6	Due to the severe drought, the government granted AR$3.5 million (US$0.8 million) for small farmers in Santa Fe, Buenos Aires, Córdoba and La Pampa.	11
2011	A7	Estimations indicated a reduction in 40% maize and around 20% in soybean production.	12

Table A1. *Cont.*

Event	Class	Description	Source
2011	**A8**	80% to 90% reduction in soybean production in some areas in the northern provinces (Chaco, Santiago del Estero and Salta).	**13**
2011	**L2**	The current drought in combination with the volcanic ashes produced by the eruption of volcano Puyehue led to the death of around 800,000 sheep in Northern Patagonia.	**14**
2011	**H8**	Salto Grande hydropower production was reduced five times due to the low flows. Only one out of 14 turbines was operational.	**15**

1 http://www.emol.com/noticias/internacional/2006/01/07/207153/ola-de-calor-y-sequia-afecta-a-zona-centro-de-argentina.html. 2 https://www.urgente24.com/138907-sequia-ya-complica-la-generacion-de-energia-en-salto-grande-y-provoca-perdidas-millonarias. 3 https://www.20minutos.es/noticia/145327/0/Iguazu/cataratas/sequia. 4 http://www.elmundo.es/america/2009/11/05/argentina/1257433211.html. 5 https://www.lanacion.com.ar/1089198-la-produccion-agropecuaria-asfixiada-por-la-sequia. 6 http://www.ipsnews.net/2009/01/agriculture-argentina-worst-drought-in-100-years/. 7 http://www.ambito.com/393351-por-la-sequia-se-agrava-la-falta-de-electricidad. 8 http://eleco.com.ar/interes-general/la-peor-sequia-argentina-de-los-ultimos-tiempos-arrasa-con-los-cultivos-y-el-ganado/. 9 https://www.lacapital.com.ar/la-ciudad/por-la-sequiacutea-el-nivel-del-paranaacute-ya-llegoacute-la-peor-bajante-del-antildeo-n309487.html. 10 https://www.infobae.com/2012/01/07/1041682-argentina-enfrenta-la-que-podria-ser-la-peor-sequia-los-ultimos-46-anos/. 11 http://www.univision.com/noticias/noticias-de-latinoamerica/emergencia-por-sequia-en-argentina. 12 http://noticias.universia.com.ar/vida-universitaria/noticia/2012/01/18/905650/sequia-gana-terreno-campo-argentino.html. 13 https://www.americaeconomia.com/negocios-industrias/sequia-arrasa-con-cosecha-de-soja-201112-en-norte-de-argentina. 14 https://www.eldia.com/nota/2011-10-1-la-patagonia-aun-sufre-por-las-cenizas. 15 http://www.informedigital.com.ar/secciones/departamentales/53243-salto-grande-produce-menos-energia-por-la-bajante.htm.

Table A2. Number of farming emergencies declared by provincial and national authorities due to drought events by province and year. Source "Boletín Oficial de la República Argentina" [39].

	Pampas					**Northeast**				**Northwest**						**Cuyo**			**Patagonia**					
year	**B. Aires**	**Córdoba**	**Santa Fe**	**La Pampa**	**E. Ríos**	**Corrientes**	**Misiones**	**Formosa**	**Chaco**	**Santiago del Estero**	**Salta**	**Tucumán**	**Jujuy**	**La Rioja**	**Catamarca**	**San Juan**	**Mendoza**	**San Luis**	**Neuquén**	**Río Negro**	**Chubut**	**Sta. Cruz**	**T. d Fuego**	**Total**
00																								**0**
01	1																							**1**
02																								**0**
03	3										2	2							1					**8**
04	3	1		1		1						1			1	1	1			1				**11**
05			1				1				2	2		1										**7**
06	4	1	1	1	1	1	2		3	1												1		**16**
07	6		3	1	1	2								1						2				**16**
08			1						1												2			**4**
09	2	1	5	4	3	2			3	3					1				1	3	1	1		**30**
10	3	3	3	3				1	1	1	1			1			2			1	1	2		**23**
11											1													**1**
12																								**0**
13																								**0**
14										1														**1**
15																								**0**

Table A3. Indicators of drought vulnerability in detail: corresponding factors, data sources, reference dates and correlation to the overall vulnerability (negatively correlated indicators are related to adaptive capacity).

Factors	Indicator	Scale	Correlation	Year	Source
Economic	Energy consumption per Capita (Million Btu per person)	Country	Negative	2014	U.S. EIA
	Agriculture (% of GDP)	Country	Positive	2000–2014	World Bank
	GDP per capita (current US$)	Country	Negative	2000–2014	World Bank
	Poverty headcount ratio at $1.25 a day (PPP)	Country	Positive	2000–2014	World Bank
Social	Rural population (% of total population)	Country	Positive	2000–2014	World Bank
	Literacy rate (% of people aged 15 and above)	Country	Negative	2000–2014	World Bank
	Improved water source (% of rural population with access)	Country	Negative	2000–2014	World Bank
	Population ages 15–64 (% of total population)	Country	Negative	2000–2014	World Bank
	Government Effectiveness	Country	Negative	2013	WGI
	Disaster Prevention & Preparedness (US$/Year/capita)	Country	Negative	2014	OECD
Infrastructural	Agricultural irrigated land (% of total agricultural land)	5 arc minute	Negative	2008	FAO
	% of retained renewable water Hydrological	catchment	Negative	2010	Aqueduct
	Road density (km of road per 100 sq. km of land area)	Vector	Negative	2010	gROADSv1

References

1. Wilhite, D.A.; Glantz, M.H. Understanding: The drought phenomenon: The role of definitions. *Water Int.* **1985**, *10*, 111–120. [CrossRef]
2. Knutson, C.; Hayes, M.; Phillips, T. *How to Reduce Drought Risk*; Western Drought Coordination Council Preparedness and Mitigation Group: Lincoln, NE, USA, 1998.
3. Bachmair, S.; Svensson, C.; Prosdocimi, I.; Hannaford, J.; Stahl, K. Developing drought impact functions for drought risk management. *Nat. Hazards Earth Syst. Sci.* **2017**, *17*, 1947–1960. [CrossRef]
4. Carrao, H.; Naumann, G.; Barbosa, P. Mapping global patterns of drought risk: An empirical framework based on sub-national estimates of hazard, exposure and vulnerability. *Glob. Environ. Chang.* **2016**, *39*, 108–124. [CrossRef]
5. Naumann, G.; Spinoni, J.; Vogt, J.V.; Barbosa, P. Assessment of drought damages and their uncertainties in Europe. *Environ. Res. Lett.* **2015**, *10*, 124013. [CrossRef]
6. Rowland, L.; da Costa, A.C.L.; Galbraith, D.R.; Oliveira, R.S.; Binks, O.J.; Oliveira, A.A.R.; Pullen, A.M.; Doughty, C.E.; Metcalfe, D.B.; Vasconcelos, S.S. Death from drought in tropical forests is triggered by hydraulics not carbon starvation. *Nature* **2015**, *528*, 119. [CrossRef] [PubMed]

7. Stahl, K.; Kohn, I.; Blauhut, V.; Urquijo, J.; De Stefano, L.; Acácio, V.; Dias, S.; Stagge, J.H.; Tallaksen, L.M.; Kampragou, E. Impacts of European drought events: Insights from an international database of text-based reports. *Nat. Hazards Earth Syst. Sci.* **2016**, *16*, 801–819. [CrossRef]
8. Blauhut, V.; Stahl, K.; Stagge, J.H.; Tallaksen, L.M.; Stefano, L.D.; Vogt, J. Estimating drought risk across Europe from reported drought impacts, drought indices, and vulnerability factors. *Hydrol. Earth Syst. Sci.* **2016**, *20*, 2779–2800. [CrossRef]
9. Zargar, A.; Sadiq, R.; Naser, B.; Khan, F.I. A review of drought indices. *Environ. Rev.* **2011**, *19*, 333–349. [CrossRef]
10. Vicente-Serrano, S.M.; Van der Schrier, G.; Beguería, S.; Azorin-Molina, C.; Lopez-Moreno, J.-I. Contribution of precipitation and reference evapotranspiration to drought indices under different climates. *J. Hydrol.* **2015**, *526*, 42–54. [CrossRef]
11. Svoboda, M.; LeComte, D.; Hayes, M.; Heim, R.; Gleason, K.; Angel, J.; Rippey, B.; Tinker, R.; Palecki, M.; Stooksbury, D. The drought monitor. *Bull. Am. Meteorol. Soc.* **2002**, *83*, 1181–1190. [CrossRef]
12. Gall, M.; Borden, K.A.; Cutter, S.L. When do losses count? Six fallacies of natural hazards loss data. *Bull. Am. Meteorol. Soc.* **2009**, *90*, 799–810. [CrossRef]
13. Forzieri, G.; Bianchi, A.; e Silva, F.B.; Marin Herrera, M.A.; Leblois, A.; Lavalle, C.; Aerts, J.C.J.H.; Feyen, L. Escalating impacts of climate extremes on critical infrastructures in Europe. *Glob. Environ. Chang.* **2018**, *48*, 97–107. [CrossRef] [PubMed]
14. Wilhite, D.A.; Svoboda, M.D.; Hayes, M.J. Understanding the complex impacts of drought: A key to enhancing drought mitigation and preparedness. *Water Resour. Manag.* **2007**, *21*, 763–774. [CrossRef]
15. INDEC. Available online: https://www.indec.gob.ar/ (accessed on 24 July 2018).
16. Bettolli, M.L.; Penalba, O.C.; Vargas, W.M. Synoptic weather types in the south of South America and their relationship to daily rainfall in the core crop-producing region in Argentina. *Aust. Meteorol. Oceanogr. J.* **2010**, *60*, 37–48. [CrossRef]
17. Minetti, J.L.; Vargas, W.M.; Poblete, A.G.; de la Zerda, L.R.; Acuña, L.R. Regional droughts in southern South America. *Theor. Appl. Climatol.* **2010**, *102*, 403–415. [CrossRef]
18. Vargas, W.M.; Naumann, G.; Minetti, J.L. Dry spells in the River Plata Basin: An approximation of the diagnosis of droughts using daily data. *Theor. Appl. Climatol.* **2011**, *104*, 159–173. [CrossRef]
19. Wu, F.; Guclu, H. Global maize trade and food security: Implications from a social network model. *Risk Anal.* **2013**, *33*, 2168–2178. [CrossRef]
20. Kakwani, N.; Son, H.H. Measuring Food Insecurity: Global Estimates. In *Social Welfare Functions and Development*; Palgrave Macmillan: London, UK, 2016; pp. 253–294. ISBN 978-1-137-58324-6.
21. World Bank. *Improving the Assessment of Disaster Risks to Strengthen Financial Resilience: A Special Joint G20 Publication by the Government of Mexico and the World Bank*; World Bank: Washington, DC, USA, 2012.
22. Peduzzi, P.; Dao, H.; Herold, C.; Mouton, F. Assessing global exposure and vulnerability towards natural hazards: The Disaster Risk Index. *Nat. Hazards Earth Syst. Sci.* **2009**, *9*, 1149–1159. [CrossRef]
23. Vicente-Serrano, S.M.; Beguería, S.; López-Moreno, J.I. A multiscalar drought index sensitive to global warming: The standardized precipitation evapotranspiration index. *J. Clim.* **2010**, *23*, 1696–1718. [CrossRef]
24. Beguería, S.; Vicente-Serrano, S.M.; Reig, F.; Latorre, B. Standardized precipitation evapotranspiration index (SPEI) revisited: Parameter fitting, evapotranspiration models, tools, datasets and drought monitoring. *Int. J. Climatol.* **2014**, *34*, 3001–3023. [CrossRef]
25. Blauhut, V.; Gudmundsson, L.; Stahl, K. Towards pan-European drought risk maps: Quantifying the link between drought indices and reported drought impacts. *Environ. Res. Lett.* **2015**, *10*, 014008. [CrossRef]
26. Sepulcre-Canto, G.; Horion, S.; Singleton, A.; Carrao, H.; Vogt, J. Development of a Combined Drought Indicator to detect agricultural drought in Europe. *Nat. Hazards Earth Syst. Sci.* **2012**, *12*, 3519–3531. [CrossRef]
27. Trambauer, P.; Maskey, S.; Werner, M.; Pappenberger, F.; Van Beek, L.P.H.; Uhlenbrook, S. Identification and simulation of space–time variability of past hydrological drought events in the Limpopo River basin, southern Africa. *Hydrol. Earth Syst. Sci.* **2014**, *18*, 2925–2942. [CrossRef]
28. McKee, T.B.; Doesken, N.J.; Kleist, J. The relationship of drought frequency and duration to time scales. In Proceedings of the 8th Conference on Applied Climatology, Anaheim, CA, USA, 17–22 January 1993; American Meteorological Society: Boston, MA, USA, 1993; Volume 17, pp. 179–183.

29. Harris, I.; Jones, P.D.; Osborn, T.J.; Lister, D.H. Updated high-resolution grids of monthly climatic observations–the CRU TS3. 10 Dataset. *Int. J. Climatol.* **2014**, *34*, 623–642. [CrossRef]
30. Lavell, A.; Oppenheimer, M.; Diop, C.; Hess, J.; Lempert, R.; Li, J.; Muir-Wood, R.; Myeong, S. Climate change: New dimensions in disaster risk, exposure, vulnerability, and resilience. In *Managing the Risks of Extreme Events and Disasters to Advance Climate Change Adaptation*; Field, C.B., V. Barros, T.F., Stocker, D., Qin, D.J., Dokken, K.L., Ebi, M.D., Mastrandrea, K.J., Mach, G.-K., Plattner, S.K., Allen, M., et al., Eds.; A Special Report of Working Groups I and II of the Intergovernmental Panel on Climate Change (IPCC); Cambridge University Press: Cambridge, UK; New York, NY, USA, 2012; pp. 25–64.
31. Ramankutty, N.; Evan, A.T.; Monfreda, C.; Foley, J.A. Farming the planet: 1. Geographic distribution of global agricultural lands in the year 2000. *Glob. Biogeochem. Cycles* **2008**, *22*. [CrossRef]
32. Center for International Earth Science Information Network (CIESIN), European Commission. *GHS Population Grid, derived from GPW4, Multitemporal (1975, 1990, 2000, 2015)*; European Commission: Brussels, Belgium, 2015.
33. Robinson, T.P.; Wint, G.W.; Conchedda, G.; Van Boeckel, T.P.; Ercoli, V.; Palamara, E.; Cinardi, G.; D'Aietti, L.; Hay, S.I.; Gilbert, M. Mapping the global distribution of livestock. *PLoS ONE* **2014**, *9*, e96084. [CrossRef] [PubMed]
34. United Nations International Strategy for Disaster Reduction Secretariat. *Global Assessment Report on Disaster Risk Reduction*; United Nations International Strategy for Disaster Reduction Secretariat: Geneva, Switzerland, 2009.
35. Naumann, G.; Barbosa, P.; Garrote, L.; Iglesias, A.; Vogt, J. Exploring drought vulnerability in Africa: An indicator based analysis to be used in early warning systems. *Hydrol. Earth Syst. Sci.* **2014**, *18*, 1591–1604. [CrossRef]
36. Brooks, N.; Adger, W.N.; Kelly, P.M. The determinants of vulnerability and adaptive capacity at the national level and the implications for adaptation. *Glob. Environ. Chang.* **2005**, *15*, 151–163. [CrossRef]
37. Alkire, S.; Santos, M.E. Measuring acute poverty in the developing world: Robustness and scope of the multidimensional poverty index. *World Dev.* **2014**, *59*, 251–274. [CrossRef]
38. Cook, W.D.; Tone, K.; Zhu, J. Data envelopment analysis: Prior to choosing a model. *Omega* **2014**, *44*, 1–4. [CrossRef]
39. BORA. Available online: https://www.boletinoficial.gob.ar/ (accessed on 9 August 2018).
40. Penalba, O.C.; Rivera, J.A. Precipitation response to El Niño/La Niña events in Southern South America–emphasis in regional drought occurrences. *Adv. Geosci.* **2016**, *42*, 1–14. [CrossRef]
41. Barros, V.R.; Boninsegna, J.A.; Camilloni, I.A.; Chidiak, M.; Magrín, G.O.; Rusticucci, M. Climate change in Argentina: Trends, projections, impacts and adaptation. *Wiley Interdiscip. Rev. Clim. Chang.* **2015**, *6*, 151–169. [CrossRef]
42. Cardona, O.-D.; van Aalst, M.K.; Birkmann, J.; Fordham, M.; McGregor, G.; Mechler, R. *Determinants of Risk: Exposure and Vulnerability*; Cambridge University Press: Cambridge, UK; New York, NY, USA, 2012.
43. World Bank Open Data. Available online: https://data.worldbank.org/ (accessed on 24 July 2018).
44. FAOSTAT. Available online: http://www.fao.org/faostat/en/#home (accessed on 24 July 2018).
45. González, L.M. Vulnerabilidad social y dinámica demográfica en Argentina, 2001–2007. *Cuadernos Geográficos* **2009**, *45*, 209–229.
46. Murgida, A.M.; González, M.H.; Tiessen, H. Rainfall trends, land use change and adaptation in the Chaco salteño region of Argentina. *Reg. Environ. Chang.* **2014**, *14*, 1387–1394. [CrossRef]
47. Spinoni, J.; Barbosa, P.; Cammalleri, C.; De Jager, A.; McCormick, N.; Naumann, G.; Magni, D.; Masante, D.; Mazzeschi, M. Global Collection of meteorological drought events from 1951 to 2016. 2018; in press.
48. Minetti, J.L.; Naumann, G.; Vargas, W.M.; Poblete, A.G. Las sequías en el largo plazo en Argentina y sus precursores invernales. *Rev. Geogr.* **2008**, *10*, 26–37.
49. Desinventar Documentation—Global Disaster Loss Collection Initiative. Available online: https://www.desinventar.net/documentation.html (accessed on 8 August 2018).
50. InfoLeg. Available online: http://servicios.infoleg.gob.ar/infolegInternet/verNorma.do?id=157271 (accessed on 30 July 2018).
51. Casparri, M.T.; Fusco, M.; Fronti, V.G. Ley de emergencia agropecuaria y su impacto sobre los pequeños productores. *Revista de Investigación en Modelos Financieros* **2014**, *3*, 51–67.
52. BBC. Available online: http://news.bbc.co.uk/2/hi/americas/7852886.stm (accessed on 24 July 2018).

53. Easdale, M.H.; Sacchero, D.; Vigna, M.; Willems, P. Assessing the magnitude of impact of volcanic ash deposits on Merino wool production and fibre traits in the context of a drought in North-west Patagonia, Argentina. *Rangel. J.* **2014**, *36*, 143–149. [CrossRef]
54. UNFCCC. *Views on Issues Relating to Agriculture. Submissions from Parties and Admitted Observer Organizations*; UNFCCC: Bonn, Germany, 2017; p. 85.

Review

On the Predictability of 30-Day Global Mesoscale Simulations of African Easterly Waves during Summer 2006: A View with the Generalized Lorenz Model

Bo-Wen Shen

Department of Mathematics and Statistics, San Diego State University, 5500 Campanile Drive, San Diego, CA 92182, USA; bshen@mail.sdsu.edu

Received: 23 April 2019; Accepted: 11 June 2019; Published: 26 June 2019

Abstract: Recent advances in computational and global modeling technology have provided the potential to improve weather predictions at extended-range scales. In earlier studies by the author and his coauthors, realistic 30-day simulations of multiple African easterly waves (AEWs) and an averaged African easterly jet (AEJ) were obtained. The formation of hurricane Helene (2006) was also realistically simulated from Day 22 to Day 30. In this study, such extended predictability was further analyzed based on recent understandings of chaos and instability within Lorenz models and the generalized Lorenz model. The analysis suggested that a statement of the theoretical predictability of two weeks is not universal. New insight into chaotic and non-chaotic processes revealed by the generalized Lorenz model (GLM) indicated the potential for extending prediction lead times. Two major features within the GLM included: (1) three types of attractors (that also appeared in the original Lorenz model) and (2) two kinds of attractor coexistence. The features suggest a refined view on the nature of weather, as follows: The entirety of weather is a superset that consists of chaotic and non-chaotic processes. Better predictability can be obtained for stable, steady-state solutions and nonlinear periodic solutions that occur at small and large Rayleigh parameters, respectively. By comparison, chaotic solutions appear only at moderate Rayleigh parameters. Errors associated with dissipative small-scale processes do not necessarily contaminate the simulations of large scale processes. Based on the nonlinear periodic solutions (also known as limit cycle solutions), here, we propose a hypothetical mechanism for the recurrence (or periodicity) of successive AEWs. The insensitivity of limit cycles to initial conditions implies that AEW simulations with strong heating and balanced nonlinearity could be more predictable. Based on the hypothetical mechanism, the possibility of extending prediction lead times at extended range scales is discussed. Future work will include refining the model to better examine the validity of the mechanism to explain the recurrence of multiple AEWs.

Keywords: African easterly wave; attractor coexistence; chaos; hurricane; limit cycle; Lorenz model; predictability; recurrence; extended range weather prediction

1. Introduction

Due to the pioneering studies of Lorenz [1–3], the finite predictability of weather is well accepted. Subsequent studies have focused on how to estimate the limit of predictability and reveal the fundamental mechanisms responsible for limited predictability. In those studies, a theoretical predictability limit of two weeks was suggested ([4] and references therein; [5]). As a result, some researchers have interpreted this limit as an upper bound for the intrinsic predictability of all weather systems at various scales and, thus, have determined that the practical predictability of all dynamic models cannot be longer than two weeks. Based on recent advances in supercomputing and global modeling technology, as discussed below, promising extended-range (15–30 days) simulations have been reported. For example, as shown in Figure 1a,b, Shen et al. [6] discussed realistic simulations of tropical cyclone (TC) formation, intensification, and movement in 30-day simulations using a global mesoscale model. The simulated TC resembled the real Hurricane Helene (2006) in terms of its movement and intensification. One may wonder how such a surprising result could be obtained while the inherent limits for long-range forecasting that developed within the scientific literature remain (e.g., [4,7]). To address this question, the predictability problem that has been outstanding for decades and only partially resolved must be revisited [8,9]. Since this is not an easy task, we attempt to address this problem based on recent global mesoscale modeling studies [6,10–15], a 10-year analysis of global reanalysis data [16], and generalized Lorenz modeling studies [17–23].

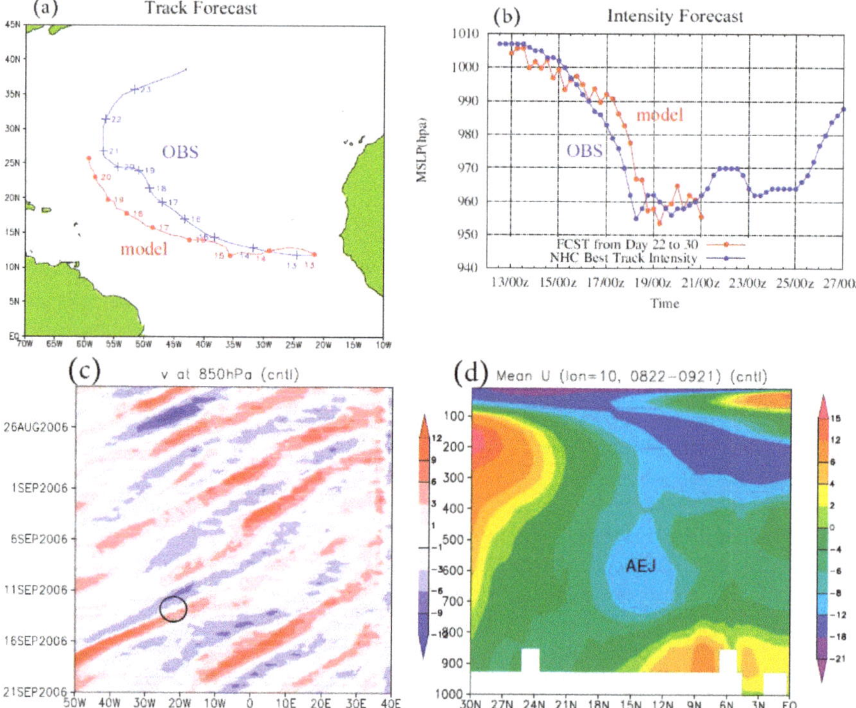

Figure 1. (**a**) Track and (**b**) intensity forecasts for Hurricane Helene (2006) from Day 22 to 30 in the control run initialized at 0000 UTC 22 August 2006. The red and blue lines indicate model predictions and best tracks, respectively. (**c**) Time-longitude diagrams of meridional winds averaged over latitudes 5–20° N. A black circle roughly indicated the time and longitude for the formation of Helene. (**d**) Height-latitude cross sections of time-averaged zonal winds along longitude 10° E. (courtesy of [6]).

While short-term hurricane predictions have been extended to produce promising results at 15–30 days scales, long term climate TC simulations have been shown to provide improved temporal and spatial locality for simulated TCs. For example, recent studies using numerical models and prescribed sea surface temperature have demonstrated the potential of simulating hurricane climate (such as hurricane frequency at a seasonal scale) (e.g., [24] and references therein). All of the above results are encouraging in regards to improving the accuracy of predicting hurricane formation (e.g., at a sub-seasonal scale) in long-term climate simulations. In this study, we revisit the previously published predictability problem for 30-day simulations by discussing the role of large-scale flows, such as African easterly waves (AEWs), and small scale processes in determining the formation of hurricanes at extended-range scales.

As discussed in [25], intrinsic predictability is dependent only on the flow itself, and practical predictability is limited by imperfect initial conditions or dependence on (mathematical) formulas (e.g., [17]). Based on multi-year data analyses (e.g., [16,26]) and global mesoscale modeling simulations [6,10–14], recent studies have been conducted to understand to what extent high intrinsic predictability may exist and how the corresponding practical predictability can realistically be obtained. A conceptual model for discussing the role of hierarchical multiscale processes in the predictability of TCs and large scale waves was proposed, including: (1) downscaling processes associated with modulation due to large-scale flows such as African easterly waves (AEWs; [27]) or Madden–Julian oscillations [28] and (2) upscaling processes associated with feedbacks from small-scale processes such as convection or precipitation. Here, downscaling (or upscaling) processes indicate the transfer of energy from a large (or small) scale system to a small (or large) scale system. The conceptual model suggests the possibility of extending the lead time of TC genesis prediction by realistically simulating the evolution of large-scale processes and their modulation on TC activities, as well as feedbacks by small-scale resolved and parameterized processes.

By conducting and analyzing global mesoscale simulations for selected cases, as previously discussed, the potential impact of downscaling processes on TC simulations was discussed by revealing the relationship between: (i) TC Nargis (2008) and an equatorial Rossby wave [12]; (ii) Hurricane Helene (2006) and an intensifying AEW [6]; (iii) Twin TCs (2002) and a mixed Rossby gravity wave [13] during an active phase of a Madden–Julian oscillation; and (vi) Hurricane Sandy (2012) [29] and upper-level tropical waves associated with a Madden–Julian oscillation [14]. The above studies collectively support the view that a large-scale system (e.g., tropical waves) can provide determinism on the prediction of TC genesis, making it possible to extend the lead time of genesis predictions. For example, analyses using traditional methods found that 30-day runs in [6] produced realistic initiation, as well as the propagation of six consecutive AEWs, between late August and late September 2006 and the mean state of an African easterly jet (AEJ) over Africa and downstream within the tropical Atlantic (Figure 1c,d). The results suggested a relationship between the improved simulations for an individual AEW as well as its interaction with local environments and a very impressive simulation of Helene's formation downstream from Day 22 to Day 30.

Since global simulations possess multiscale processes and interactions, the immediate question was how to reveal time varying scale interactions crucial for the formation of the above TCs? To address the above question, we developed tools for multiscale analysis and visualizations. In the multiscale analysis package, the core technology is empirical mode decomposition (EMD) and ensemble EMD (EEMD) methods. Both have shown remarkable performance in revealing multiscale processes of non-stationary and nonlinear data. The EMD method was originally developed by Huang et al. (1998) [30] and was extended to the EEMD by Wu and Huang (2009) [31] through the addition of ensemble members. The inclusion of ensemble computations into the EEMD was executed in order to overcome the scale (or mode) mixing problem of the EMD. Both methods decompose one set of observational data into so-called intrinsic mode functions (IMFs) that represent various oscillatory components of the data. One of the unique features of both methods is the possession of a filter bank property (e.g., [15,32,33]) (i.e., decomposed mean IMFs staying within natural filter period windows, to be discussed in Section 3.6). We implemented a three-level parallelism into the EEMD, referred to as the parallel EEMD (PEEMD), to improve parallel performance [15,34]. The newly developed PEEMD was first used to perform a multiscale analysis of Hurricane Sandy (2012) [15]. We then utilized it to perform a 10-year analysis on global reanalysis data (e.g., [16,35]), showing the statistics of cases with energy transfer between developing systems and environmental flows. To focus on the performance of the PEEMD in revealing the intensification of an AEW and its association with the formation of Hurricane Helene (2006), a brief summary is provided in Section 3.6. Such an analysis provides additional support to the view that large scale processes provide determinism for the time evolution and spatial location of tropical cyclones.

The simulations of large-scale flows and mesoscale TCs were satisfactorily compared with the observations (e.g., reanalysis and best track). The aforementioned studies called for revisiting the predictability problem in order to understand whether or not such simulations were consistent with chaos theory. Specifically, we asked whether or not the so-called predictability limit of two weeks applied to this case. And if not, we asked why. Based on a comprehensive literature review and the development of a generalized Lorenz model, insightful understanding regarding chaos, butterfly effects, and predictability were obtained, as summarized in the following discussion. While the Lorenz 1963 model [1] was developed in order to determine the sensitive dependence of solutions on the initial conditions, which was then used to define "chaos", the Lorenz 1969 model [2] was proposed in order to reveal the instability of basic wind, showing the dependence of growth rates and predictability on spatial scales. Deterministic chaos within the Lorenz 1963 model suggested finite predictability, which was fundamentally different from the Laplacian view of deterministic predictability. By comparison, the appearance of instability within the Lorenz 1969 model indicated a finite practical predictability. The degree of chaotic responses displayed a dependence on system parameters (representing a system's heating or dissipation). The growth rates of solutions associated with system instability displayed dependence on the selection of basic winds (e.g., with a different slope for the wind spectrum, [36]) and physical processes (e.g., with or without dissipation, [37]) in the governing equations. Thus, an estimate of the predictability limit using the above simplified models should be interpreted with caution, as discussed in Sections 3.1–3.3.

Progress in the aforementioned global modeling has been enabled by recent advances in both supercomputing and visualization technologies over the past 15 years or so, in particular after the birth of the Japan Earth Simulator and NASA's Columbia supercomputer (e.g., [38,39]). High-resolution, high-fidelity global simulations of TCs and large-scale waves effectively revealed the role of hierarchical multiscale processes, as well as butterfly effects, in TC simulations. For example, a TC whose vertical structure varies with height possesses two major features, such as low-level cyclonic circulation and upper-level anticyclonic circulation. Such features of a TC and its interaction with large-scale environmental processes can better be displayed using quasi, three-dimensional streamline packages and concurrent visualization technology (e.g., [40,41]), which enable both high spatial and temporal resolution solutions. A summary of the technical details and visualizations are provided in Section 2.4, Section 3.3, and Section 3.4.

Sections 2.1 and 2.2 briefly introduce the global mesoscale model (GMM) and global reanalysis data, respectively. The PEEMD and its application for the multiscale analysis are discussed in Section 2.3. Section 2.4 applies 2D or 3D concurrent visualizations in order to reveal multiscale processes and initial noise within the model simulations. Section 2.5 presents the generalized Lorenz model (GLM). Section 3.1 discusses various types of solutions within the Lorenz 1963 model and the GLM, including nonlinear periodic solutions, also known as limit cycle solutions, and attractor coexistence. Section 3.2 provides a brief review of Lorenz 1969 study [2] for a comparison. Impact of errors of small scale processes and 30-day global simulations of large scale systems are discussed in Sections 3.3 and 3.4, respectively. Section 3.5 briefly comments the simulation of two hurricanes with larger errors. Section 3.6 summaries the 10-year analysis using the PEEMD. By comparing AEW simulations obtained using the GMM and limit cycle solutions of the GLM, Section 3.7 proposes a hypothesis for the recurrence (or periodicity) of AEWs during summer. Concluding remarks are provided at the end.

2. Numerical Model, Global Data, Visualization, and Analysis Methods

2.1. The Global Mesoscale Model

Enabled by advanced computational technologies, a global mesoscale model (GMM) at the highest horizontal resolution of 1/12 degrees (approximately 9 km) was deployed in 2005 [38], producing encouraging forecasts of intense hurricanes [6,10–12]. In this study, from a perspective of the generalized Lorenz model, we performed an analysis of global mesoscale simulations reported in earlier studies by myself and other colleagues. These simulations were obtained using the GMM that consists of three major components: finite-volume dynamics [42], the National Center for Atmospheric Research (NCAR) Community Climate Model physics, and the NCAR Community Land Model (CLM). Control runs were performed using basic model configurations including a 1/4 or 1/8 degree resolution and a large-scale condensation scheme. Such modeling settings enabled latent heat release from grid-scale condensation processes when cumulus parameterizations were disabled. The model configurations systematically produced reliable results. Our analysis indicated that a quasi-equilibrium assumption within cumulus parameterizations may limit the scale interaction between convection and large-scale flows and, thus, cause uncertainties in simulations of AEWs as well as hurricanes. Additional discussions regarding the impact of including or not including cumulus parameterizations in high-resolution simulations were provided in [10]. The best tracks for TCs were made available by the National Hurricane Center (NHC).

Additional parallel runs were performed using varying physics (e.g., different cumulus parameterizations) in order to understand the underlying dynamics and to examine the sensitivity of solutions to the initial conditions and/or physical processes. In this study, the control and three parallel runs, listed in Table 1, were further analyzed. Two parallel experiments containing different dynamic conditions during different months, referred to as experiments P1 and P2, respectively, were designed in order to reveal the dominant impact of land and physical processes. Experiments P1 and P2 used dynamic initial conditions (ICs) from 22 April and 22 June, respectively. To maintain the same model physics (e.g., radiation) and land model configurations as the control run, timestamps in the dynamic ICs were changed to 22 August. The goal was to show whether or not and how an AEJ that does not appear within the IC may be simulated using realistic land and physics conditions.

Table 1. Sensitivity experiments for studying the dependence of AEW simulations on different dynamic ICs and modified Guinea highlands. Initial conditions for physics and land processes remain the same in all experiments, as shown in the third column.

Case ID	Dynamic IC	CLM and Physics IC	Guinea Highlands
CNTL	08/22	08/22	-
P1	04/22	08/22	-
P2	06/22	08/22	-
P3	08/22	08/22	A factor of 0.6 in heights

In the third experiment, P3, mountain heights were reduced in order to test whether or not and how lowering mountain heights (e.g., resolved in coarser resolution models) in the downstream impacts the simulation (and the successive initiation) of AEWs in the upstream. This was achieved as follows: within the longitude 16.5° W to 6° W and latitude 5° N to 13° N, mountain heights were multiplied by a factor of 0.6. The reduced mountains could have a "direct" impact on interactions with an approaching AEW, and an indirect impact that changed environmental flows and the subsequent initiation of AEWs upstream. Both could have an impact on a simulation of downstream hurricane formation.

2.2. Global Reanalysis Data

For this study, we selected the latest European Centre for Medium-Range Weather Forecasts (ECMWF) global reanalysis, i.e., European Reanalysis Interim (ERA-Interim) dataset [35]. The data spans the time period from January 1979 to the present day. The dataset has a sampling rate of six hours and a horizontal grid spacing of 0.75 degrees, yielding a spatial resolution of approximately 78 km. For multiscale analysis using the PEEMD [16], zonal and meridian winds at 700 hPA were analyzed over a 10-year period from 2004 to 2013. A brief summary is provided below.

2.3. Parallel Ensemble Empirical Mode Decomposition (PEEMD)

As mentioned, the EEMD was developed to include sufficient ensemble members in order to obtain an ensemble average of IMFs, minimizing the issue of mode mixing. Depending on the required level of accuracy within the decomposed IMFs, 200–400 ensemble members are often used to obtain averaged IMFs, significantly increasing computational demands. Therefore, a three-level parallelism was implemented into the EEMD, referred to as the parallel EEMD (PEEMD) method, to efficiently and effectively reveal multiscale processes from high-resolution, global, multi-dimensional Earth science data [15]. The PEEMD achieved promising scalability with a parallel speedup and efficiency of 52.8 and 63 percent, respectively, by increasing the number of cores from 60 to 5000 [34]. In Section 3.6, we discuss a 10-year multiscale analysis using the PEEMD to reveal the scale interactions that contributed to the formation of Hurricane Helene (2006).

2.4. Streamline Package and Concurrent Visualizations

To visualize the multiscale interactions of a TC with its environmental flows, a time series of 2D frames of streamlines for TC circulations were produced at each vertical level, and the frames of several (e.g., 3–5) contiguous levels were grouped into one layer. The streamlines in the three different layers, low, middle and upper, are shown in blue, green, and pink, respectively. With other features such as opacity, used to control the (vertical) "transparency" of streamlines at different heights, the evolution of streamline density at a specific level qualitatively depicted the evolution of average wind speeds (i.e., denser streamlines signified stronger average wind speeds). See the technical details in [41]. To effectively process the simulations into visualizations, concurrent visualization technology was deployed and integrated within the model (e.g., [43]). Such an integrated system generated simulation data and, concurrently, produced animations. Specifically, when the model completed integration at each time step, the integrated system sent the model's outputs from the computing nodes to the visualization nodes via direct remote memory access, and a visualization module started processing the data and generating image frames for display. As a result, the simulations at ultra-high spatial and temporal resolutions were feasible. We showcase both 2D and 3D visualizations in Sections 3.3 and 3.4, respectively.

2.5. The Generalized Lorenz Model (GLM)

Over the past several years, a series of papers on high-dimensional Lorenz models [17–20] have yielded the following generalized Lorenz model (GLM) [22]:

$$\frac{dX}{d\tau} = \sigma Y - \sigma X \tag{1}$$

$$\frac{dY}{d\tau} = -XZ + rX - Y \tag{2}$$

$$\frac{dZ}{d\tau} = XY - XY_1 - bZ \tag{3}$$

$$\frac{dY_j}{d\tau} = jXZ_{j-1} - (j+1)XZ_j - d_{j-1}Y_j, \ j \in \mathbb{Z} \ : \ j \in [1,\ N] \tag{4}$$

$$\frac{dZ_j}{d\tau} = (j+1)XY_j - (j+1)XY_{j+1} - \beta_j Z_j, \ j \in \mathbb{Z} \ : \ j \in [1,\ N] \tag{5}$$

$$N = \frac{M-3}{2};\ d_{j-1} = \frac{(2j+1)^2 + a^2}{1+a^2};\ \beta_j = (j+1)^2 b \quad (6)$$

Here, τ is dimensionless time. The three integers are j, M, and N. While M represents the total number of modes (or equations), N indicates the total number of pairs $(Y_j,\ Z_j)$ for higher wavenumber modes that do not appear within the original three-dimensional Lorenz Model (3DLM, [1]). The time-independent parameters include σ, r, a, b, d_{j-1}, and β_j. The first two represent the Prandtl number and the normalized Rayleigh number (or the heating parameter), respectively (e.g., [17]). The parameter a is defined as the ratio of the vertical scale of the convection cell to its horizontal scale, which is equal to $1/\sqrt{2}$. The last three are coefficients for the dissipative terms. Detailed discussions on each of the above terms can be found in [22]. The GLM with many modes was derived from the partial differential equations (PDEs) for Rayleigh Benard convection, while different chaotic systems with many modes proposed by Lorenz in 2005 [44] were not derived from physically-based PDEs. Variable X denotes the amplitude of the Fourier mode for stream function that defines horizontal and vertical velocities. The variables (Y, Z), (Y_1, Z_1), (Y_2, Z_2), and (Y_3, Z_3), referred to as the primary, secondary, tertiary, and quaternary modes, respectively, represent the amplitudes of the Fourier modes at different wavenumbers for temperature. When the nonlinear term $-XY_1$ is ignored, Equations (1)–(3) become the classical 3DLM. The results obtained from the GLM with M = 5, 7, and 9, referred to as the 5DLM, 7DLM, and 9DLM, respectively, are presented in the following sections.

3. Discussion

3.1. New Insights into Predictability and Chaos

Chaotic solutions using the 3DLM have been a focal point and have led to the statement of "weather is chaotic". In fact, depending on the relative strength of heating, chaotic solutions appear as one of three types of solutions within the 3DLM, including:

1. Steady-state solutions with small heating parameters (i.e., $r < r_c$; $r_c = 24.74$);
2. Chaotic solutions with moderate heating parameters (i.e., $r_c < r < R_c$; $R_c = 313$);
3. Limit cycle solutions with large heating parameters (i.e., $R_c < r$, e.g., [45,46]).

Additionally, the coexistence of chaotic and steady-state solutions may appear within a small range for the heating parameter, (i.e., $24.06 < r < 24.74$). In this paper, we briefly discuss the second and third types of solutions within the 3DLM and the coexistence of two types of solutions within the GLM.

We first present two features of chaotic solutions, including the divergence of nearby trajectories and solution boundedness. The sensitive dependence of solutions on ICs has been illustrated using the divergence of two initial nearby trajectories within the phase space of the 3DLM. For example, using the 3DLM with typical parameters (e.g., $\sigma = 10$, $b = 8/3$, and $r = 28$), Figure 2a–c displays a very different time evolution for two solution orbits whose starting points are very close to one another. In addition to the divergence of nearby trajectories, the solutions or orbits are bounded. Solution boundedness is indeed indicated by the finite size of the butterfly pattern for a chaotic solution. For such a system, the divergence of two orbits, which may be viewed as an error between a control and parallel run, should be bounded (e.g., Figure 2d).

When the Rayleigh parameter becomes large (say, $r > R_c$; $R_c = 313$), nonlinear oscillatory solutions appear (e.g., [45–47]). As show in Figure 3a, 200 runs with different ICs eventually approach the same closed orbit in black. Such a convergence of orbits indicates the isolated nature of a closed stable orbit where nearby trajectories approach the stable orbit. As a result of its isolated and closed nature (in Figure 3b), the nonlinear oscillatory stable orbit is indeed a limit cycle solution [48,49]. One interesting characteristic of a limit cycle is that its orbit is solely determined by the system and independent of the ICs. The "closed" feature associated with periodicity (or a "recurrence" feature with quasi-periodicity for a limit torus) can be illustrated using the non-dissipative model (e.g., [21,22,50]), suggesting that nonlinearity alone or competition between heating and nonlinearity may produce nonlinear periodic or quasi-periodic solutions. By comparison, the isolated feature of limit cycles within the 3DLM indicates the importance of weak dissipation. An important message for the appearance of limit cycle solutions at large Rayleigh parameters is that chaotic solutions only occur over a finite range of Rayleigh parameters.

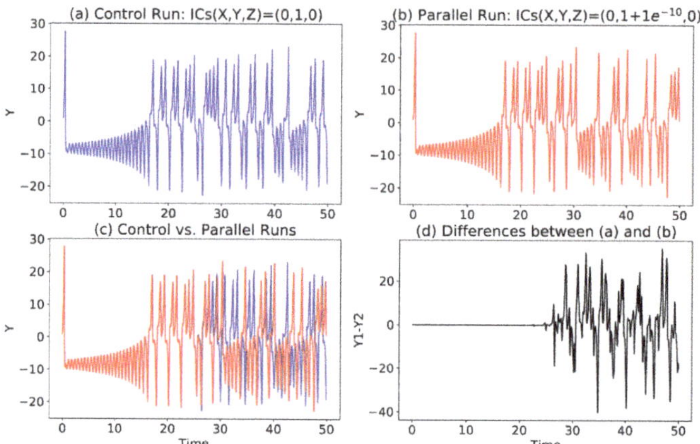

Figure 2. An illustration of the bounded divergence of two nearby trajectories within the 3DLM with $r = 28$ and $\sigma = 10$. Panels (**a**) and (**b**) display solutions from the control and parallel runs, respectively, the latter of which adds a small perturbation ($1 \times e^{-10}$) into the initial value of Y. Panel (**c**) reveals the sensitive dependence of solutions on the initial conditions. Panel (**d**) displays bounded differences in solutions for the control and parallel runs.

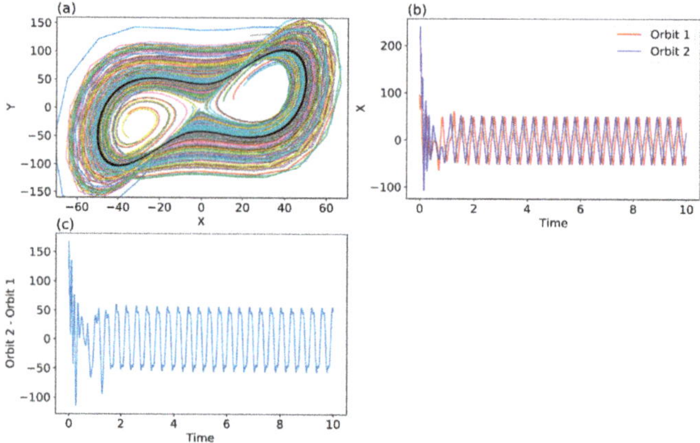

Figure 3. (**a**) A limit cycle (black) as indicated by the convergence of 200 orbits (color) beginning with 200 different initial conditions for $r = 350$ within the 3DLM. Color lines depict orbits during the period of $\tau = 1$ and $\tau = 10$ (i.e., $\tau \in [1, 10]$). "A thick black line" displays 200 orbits for $\tau \in [9, 10]$. Convergence of the 200 orbits into the black orbit indicates the isolated nature of a limit cycle. Panels (**b**) and (**c**) display two orbits and their differences, respectively.

Compared to the 3DLM, the GLM, with any odd number of M greater than three, possesses the following features: (1) three types of solutions (that also appear within the 3DLM); (2) two kinds of attractor coexistence; (3) aggregated negative feedback; and (4) hierarchical scale dependence. Below we briefly discuss (3) and (4) and then illustrate (2). Within the GLM, the negative feedback associated with smaller-scale modes can be aggregated to provide a stronger effective dissipation. As such, a higher critical value for Rayleigh parameter is required for the onset of chaos in a higher dimensional Lorenz mode. This can be seen in Table 2. As indicated by the high Pearson correlation coefficients between the primary and secondary modes (Z and Z_1) and between the secondary and tertiary modes (Z_1 and Z_2), 0.988 and 0.998, respectively, Figure 4 using the 7DLM displays hierarchical scale dependence within the chaotic solutions.

Table 2. The characteristics of various Lorenz models. Values for r_c are determined based on analyses of the ensemble Lyapunov exponents [17,51]. The "Heating terms" column indicates heating terms within the corresponding LM.

Model	r_c	Heating Terms	References
3DLM	23.7	rX	[1]
5DLM	42.9	rX	[17]
6DLM	41.1	rX, rX_1	[18]
7DLM	116.9	rX	[19]
8DLM	103.4	rX, rX_1	[20]
9DLM	102.9	rX, rX_1, rX_2	[20]
9DLM (new)	679.8	rX	[22]

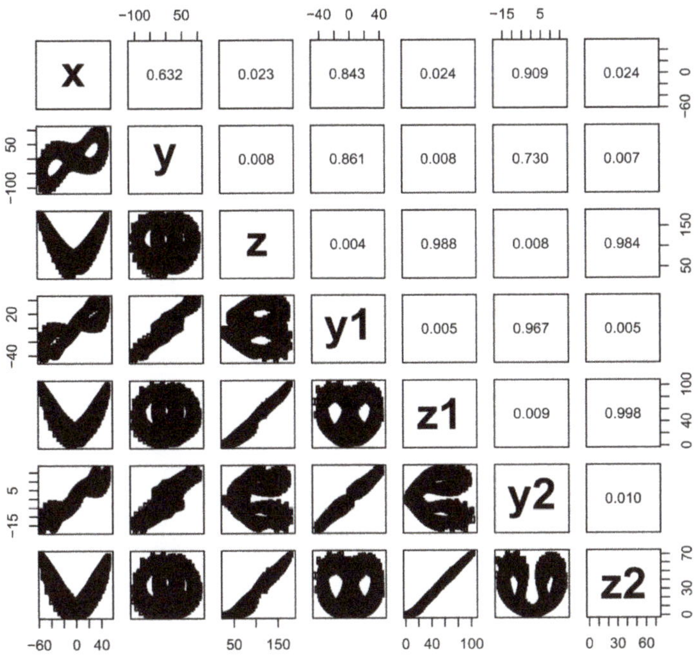

Figure 4. A matrix of scatter plots for the 7DLM with its seven variables, as listed in the principal diagonal. r and σ are 120 and 10, respectively. Each of the cells above (below) the diagonal provides a Pearson correlation coefficient (scatter plot) between the two variables. Scale dependence is indicated by high Pearson correlation coefficients, as well as the linear relationship in scatter plots (courtesy of [20]; See details in [19]).

Here, two kinds of attractor coexistence are discussed. The first kind of attractor coexistence, with chaotic and steady-state solutions, appears in a wider range of Rayleigh parameters in higher-dimensional Lorenz models (i.e., $679.8 < r < 1058$ for the 9DLM) as compared to a smaller range of Rayleigh parameters within the 3DLM. The appearance of either chaotic or non-chaotic solutions depends on the initial conditions. Within the 3DLM or 9DLM, coexisting attractors appear within a system that has stable, non-trivial equilibrium points ([22] and references therein). Therefore, it is reasonable to hypothesize that orbits beginning "near" the stable non-trivial critical points may move toward the non-trivial equilibrium point, at least from a statistical perspective. To show such features using the 9DLM, a large number of runs using different initial conditions were performed. A Gaussian random generator was applied in order to produce N ($N = 256$) data points that were distributed over a hypersphere centered at the non-trivial equilibrium point with a radius of R ($R = 300$). As shown in Figure 5 with $r = 680$, the 256 orbits could be classified as chaotic orbits and steady-state solutions, indicating the coexistence of two types of solutions. A detailed analysis of the dependence of chaotic and non-chaotic solutions on ICs using the ensemble modeling approach with various Ns and Rs can be found in Figure 5 of [23]. The second kind of attractor coexistence with limit cycle and steady solutions was first documented by Shen [22]. Such coexistence was further discussed using the 9DLM with $r = 1600$ and 128 different ICs in Figure 9 of [52]. From a practical perspective, better predictability can be obtained for non-chaotic solutions with insensitivity to the initial conditions.

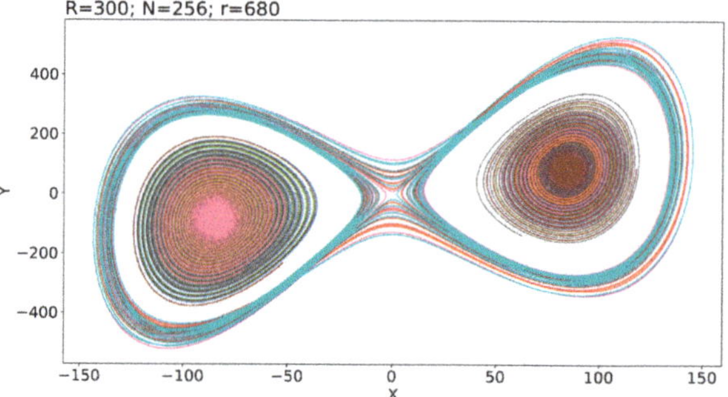

Figure 5. The coexistence of chaotic and non-chaotic orbits within the GLM with $M = 9$; 256 ICs distributed over a hypersphere with a radius of 300. $\tau \in [0.625, 5]$ with $\Delta\tau = 0.001$.

As a result of the above four features within the GLM, a refined view of the nature of weather is proposed [22,23] as follows: The entirety of weather is a superset that consists of both chaotic and non-chaotic processes. In other words, both chaos and order may coexist. Such coexistence suggests better predictability for non-chaotic processes if we can identify them in advance (e.g., [52]).

3.2. *A Brief Review of Lorenz (1969)*

In addition to the 3DLM, Lorenz proposed another model with 21 modes in 1969 [2], showing two major features, the dependence of growth rates on scales and energy transferring across scales. The 3DLM and the 1969 model were used to reveal finite intrinsic predictability for chaotic solutions and finite practical predictability for unstable solutions, respectively. From the perspective of predictability estimates, the calculation of growth rates was more feasible in the 1969 multiscale, linear model than in the 1963 chaotic model since the latter displays sensitivity of solutions to the initial conditions and, thus, produces larger variations of time dependent growth rates. However, as a result of the simplicity in the dynamics and physics of these models as well as other existing models, any quantitative estimate of the predictability should not be viewed as an upper or lower bound of intrinsic predictability.

Lorenz (1969) [2] studied energy transferring across scales and its impact on estimates of growth rates using two experiments illustrating the impact of upscale and downscale transfer for initial error.

While upscale transfer had been a main focus in [2], the two experiments indeed produced comparable results with similar predictability over a range of scales (e.g., Table 3 of [2]). To this end, a recent study in [53] emphasized the role of the downscale transfer of an initial error, originally at a larger (synoptic- or meso-) scale, in producing a reduction of predictability on a smaller scale. These authors further suggested that the upscale transfer of error associated with "the butterfly effect" may not be so crucial for obtaining daily weather prediction (e.g., [54]).

Upscale transfer is a major feature in turbulence models. However, it is not clear whether the 1969 model is a turbulence model because the a conservative partial differential equation with a realistic (turbulent) basic state was used derived the 1969 model. By comparison, as documented on page 139 of Lorenz (1993) [55], the equations of the 3DLM lacked important properties associated with turbulence. Therefore, the 3DLM was not capable of addressing "deterministic turbulence" but illustrated "deterministic non-periodic flow". As a result, detailed similarities and differences in the underlying mechanisms for finite predictability within the 3DLM and the 1969 model of Lorenz should be examined to understand the role of upscale transfer. This will be undertaken in future studies.

3.3. Impact of Errors on Small Scale Processes

Based on the major findings of Lorenz's studies, the following two features have been accepted within the scientific community:

1. Small-scale processes are less predictable than large-scale processes.
2. Errors associated with small-scale processes may "quickly" contaminate simulations of large-scale flows.

Since small errors can easily appear in initial conditions, the above leads to a pessimistic view for extending prediction lead times beyond 5–7 days. For example, such errors appear in my colleagues and my modeling approaches that apply coarser-resolution National Centers for Environmental Prediction (NCEP) analysis data to drive higher-resolution model runs. However, as reported above, encouraging results were still obtained. Why? Here, an analysis on the above two features is provided. While the first feature may be derived from the result of scale dependence of growth rates on scales in [2], the second feature seems to be associated with the sensitive dependence of solutions on the initial conditions in [1]. For the Lorenz (1969) study [2], the model should have included realistic dissipations to allow dissipative small-scale processes with negative growth rates. For the Lorenz 1963 model [1], as well as for the GLM, the coexistence of two types of solutions suggests that a steady-state solution with a negative growth rate may coexist within a chaotic solution. Small-scale processes with negative growth rates are stable and possess deterministic predictability. As a result, the first statement is not very accurate. The coexistence of chaotic and steady-state solutions suggests that tiny errors associated with chaotic solutions should not cause a large impact on steady-state solutions, and vice versa. The second statement is also not accurate. This is qualitatively illustrated in Figure 6 and its corresponding animation for the global simulation of vertical velocity over an initial 5-day period. The animation shows that some noise associated with initial imbalance between the higher-resolution GMM and coarser-resolution NCEP reanalysis data were dampened, yielding no significant impact on the subsequent simulations.

In summary, the two features outlined in the beginning of this subsection may not always occur. Although dissipative small-scale processes should be predictable and may not necessarily contaminate the simulations of large-scale flows, numerical models that contain highly dissipative processes may have a numerical issue with the so-called stiff problem, which can be illustrated using the simplest ODE: $dy/d\tau = \lambda y$, with a negative λ but a large $|\lambda|$ [56]. Note that mathematically and physically, the solution is stable with a large decay rate. However, numerical instability may appear within such a system, thereby requiring a better scheme or a very small time step in order to improve solution stability. For global high-resolution models with many scales, a stiff problem may appear as a result of a large ratio between the largest and smallest eigenvalues and, thus, produce unstable solutions.

Figure 6. A selected frame from a global animation of the vertical velocity in pressure coordinates from a run initialized at 0000 UTC 21 October 2005. The corresponding animation is available as a google document: http://bit.ly/2GS2flD. The animation displays dissipation of the initial noise associated with an imbalance between the model and the initial conditions.

3.4. 30-Day Global Simulations of Large-Scale Systems

We previously discussed a remarkable 30-day simulation with a realistic formation simulation for Hurricane Helene (2006). Hurricane Helene was found to be related to interactions between an observed AEW and the local environment (e.g., [57]). Below we provide a brief summary on simulations of AEWs and an AEJ during the 30-day period. In our study the criteria for the analysis of AEW simulations was similar to those in [58]. The ridge (or the trough) of the AEW was defined as the location with a transition from southerly to northerly flow (or from northerly to southerly flow).

Time–longitude diagrams of 850 hPa meridional winds averaged over 5° to 20° N are shown in Figure 1c. The figure displays the occurrence of six westward propagating AEWs over the 30-day period. The waves had a timescale of 3–5 days, a wavelength of approximately 2000–2500 km, and a propagation speed of roughly 10 m/s. Overall, the model simulations of multiple AEWs were in good agreement with the analysis, especially over the African continent. Spatial and temporal variations existed but were within one characteristic (time and spatial) scale. The strong wind shear along 20° W during 11–13 September, as shown by the black circle in Figure 1c, roughly indicated the formation of Hurricane Helene.

Additionally, Figure 1d shows altitude–latitude cross-sections of zonal winds averaged over the 30-day period along longitude 10° E from the control run. A low-level jet with a maximum of around 10–14 m/s at (14° N, 600 hPa), referred to as the AEJ ([59]), was clearly evident. Overall, the model simulation was in good agreement with the NCEP analysis, but the simulated AEJ was slightly weaker. Below and south of the AEJ, a low-level westerly monsoon flow was simulated (Figure 1d). At roughly 200 hPa and equatorward of the AEJ, the model simulated an upper-level tropical easterly jet that appeared at the right altitude but had a stronger intensity between 9° N and 15° N, as compared to analysis data.

Promising simulations with multiple AEWs and an averaged AEJ support the view that large-scale processes may determine modulation within the simulation of Hurricane Helene (in Figure 1a,b). As shown in Figure 7, such a feature can be illustrated using a quasi, three-dimensional streamline visualization that depicts the evolution of an intensifying AEW into a hurricane from a 30-day simulation.

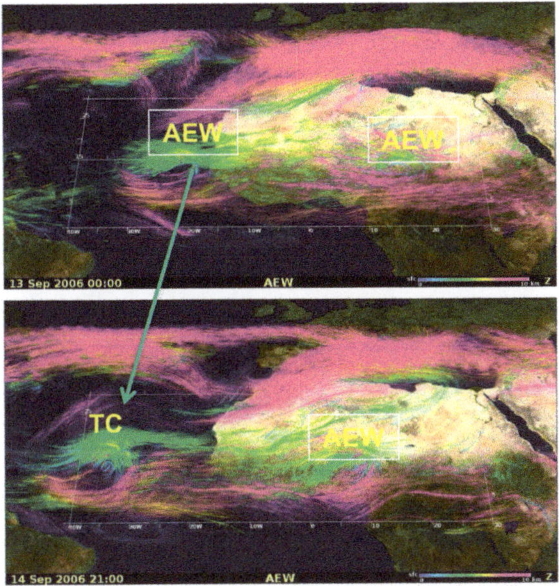

Figure 7. A visualization of formation of Hurricane Helene (2006) and its association with the intensification of an African Easterly Wave (AEW) in a 30-day run initialized at 0000 UTC 22 August 2006. Upper-level winds are shown in pink, middle-level winds in green, and low-level winds in blue. (**top**) Initial formation of Helene as the AEW moves into the ocean, validated at 0000 UTC 13 September 2006 (day 22); (**bottom**) initial intensification associated with intensified low-level inflow with counter clockwise circulation, validated at 2100 UTC 14 September 2006. An animation can be found at: http://tiny.cc/j9ul9 (courtesy of [41]).

3.5. Simulations of Hurricanes Debbie and Florence

In addition to Hurricane Helene, the 30-day control run in [6] also simulated two other hurricanes, Hurricanes Debbie and Florence, for their movement and formation, respectively (as shown in Figure S6 of the supplemental materials in [6]. The control run did not produce any false alarm. Overall, the movement of hurricane Debbie, which appeared in the ICs, was well simulated. Simulated Hurricane Florence first appeared 10° east of the observed hurricane several days earlier, yielding larger errors in location and timing. Different accuracy in the formation simulation for Hurricanes Florence and Helene may be explained as follows. While the model possesses two-way interaction between the surface and atmosphere over land (provided by the land model), it only contained one-way interaction over the ocean (provided by the prescribed weekly sea surface temperature). As a result, the predictability of hurricane formation may be better near the Cape Verde Islands that are closer to the continent (e.g., for Helene) than over the Atlantic Ocean (e.g., for Florence).

3.6. Downscaling Processes Revealed by the PEEMD

Here, we first illustrate that the EMD/EEMD/PEEMD behave as a collection of band-pass filters [32,33] and, thus, are ideal for multiscale analysis. We previously generated a data set with one million points as Gaussian white noise and decomposed the data set into nine IMFs using the EMD. For each of IMFs a spectrum was determined using Fourier (spectral) analysis. As shown in Figure 8, the spectrum of each IMF that displays a Gaussian distribution contained signals within a range of frequencies. In general, a higher order IMF has lower frequencies (or wavenumbers) associated with larger temporal (or spatial) scales. Specifically, the averaged period (or wavelength) of the $(n + 1)$th

IMF doubles the period (or wavelength) of *n*th IMF (e.g., [15]). This feature, similar to that of a dyadic filter, is ideal for multiscale analysis.

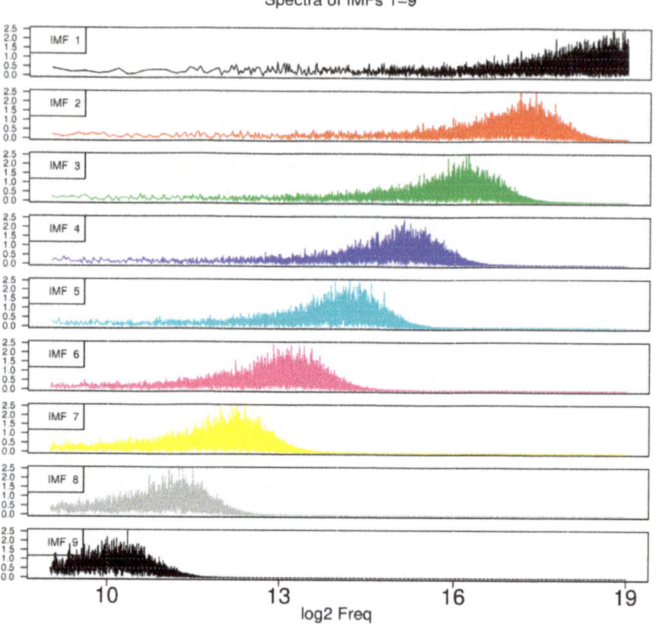

Figure 8. The feature of a collection of band-pass filters within the EMD/EEMD/PEEMD. Panels display the spectra of the first nine intrinsic mode functions (IMFs) that were obtained from the decomposition of Gaussian white noise with one million points. The horizontal represents the logarithm of the frequency (courtesy of [15]).

The PEEMD was applied in order to extract oscillatory IMFs from ECMWF global data. The difference between the raw data and a sum of its oscillatory IMFs is defined as a residual, called a trend mode. In the 10-year data analysis, we found that the trend mode and the third IMF (IMF3) largely represented environmental flows and a mixed AEW and TC, respectively. Below we illustrate energy transfer between the trend mode and IMF3 during the intensification of an AEW and the formation of a TC, including Hurricane Helene (2006). Table 3 summarizes an analysis of the ERA-interim dataset and the NHC best tracks dataset from July 2004–September 2013. In the data, the number of AEWs remained nearly constant and the number of TCs changed significantly from year to year. The former seemed to suggest "stable" large-scale forcing (e.g., seasonal forcing remained nearly unchanged inter-annually), while the latter suggested various mechanisms for the further intensification of an AEW that may or may not have developed into a TC. Since these characteristics were observed during July, August, and September, they may indicate the importance of low-level heating/radiation.

For developing cases, as shown in Figure 9 for Hurricane Helene, the time evolution of wind shear along their tracks was analyzed to reveal downscaling processes during intensification. Storm intensification was indicated by the dropping of sea-level pressure at the storm's center. During storm intensification, the shear of the trend mode decreased with time, although such a tendency was not clear in the shear of total wind. At the same time, the shear of the IMF3 mode gained strength. As a result, the decrease and increase of shear for the trend and IMF3 modes, respectively, indicated the role of downscaling transfer from the trend mode to the IMF3 mode in storm intensification. Below we provide statistics for cases with similar features of downscaling processes.

Table 3. Breakdown by year of AEWs (2nd column), the NHC tracked storms (3rd column), and hurricanes (4th column) for July, August and September from 2004 to 2013 for the main development region. In the 3rd column, numbers outside (or between) the parentheses include tropical depressions (TDs) that were (or were not) associated with AEWs. Numbers in the 4th column contain hurricanes that developed from the AEW associated storms and the numbers between the parentheses are hurricanes with downscaling features. Table was reproduced by permission of the American Meteorological Society [16].

Year	No. of AEWs	No. of TDs	No. of Hurricanes
2004	28	7(8)	4(2)
2005	28	3(6)	2(1)
2006	26	3(4)	1(1)
2007	28	4(5)	3(1)
2008	28	3(4)	2(0)
2009	30	4*(5)	1(1)
2010	27	6(8)	4(2)
2011	27	4(4)	3(2)
2012	27	5(7)	4(2)
2013	24	3(4)	1(1)
total	272	42(56)	25(13)

* The number includes one storm that was already a tropical system (TS) when first classified by the NHC.

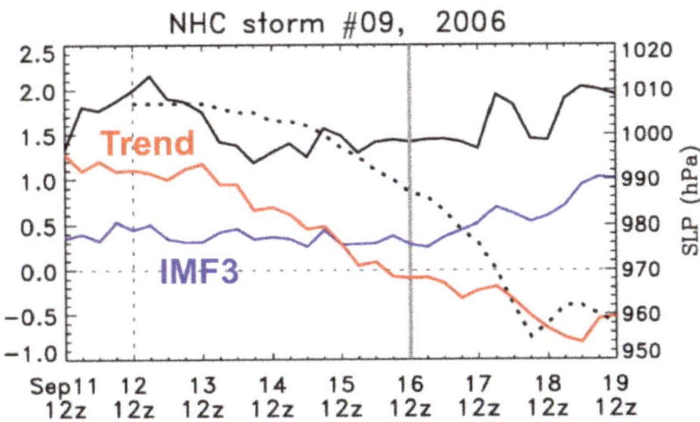

Figure 9. The tendency for horizontal wind shear of the total winds (black line), IMF3 (blue line), and the trend mode (red line) and the minimum sea level pressure (black dotted line) along the track of Helene (2006). The vertical thin dotted line indicates the time when the storm was first classified as a TD by the NHC, while the vertical gray line is provided at 1200 UTC 16 September 2006 as a reference line. The plotted shear of the IMF3 and the trend mode were multiplied by three. Figure was reproduced by permission of the American Meteorological Society [16].

With the goal of understanding the impact of AEWs on the formation of Cape Verde hurricanes, a selected domain for the analysis covered latitudes between 7° and 20° N and longitudes between 15° E and 60° W, the same main development region studied in [60]. Over the 10-year period, 42 AEWs developed into NHC classified storms within the selected domain and 25 further developed into hurricanes (Table 3). Decomposed components for the trend mode and IMF3 suggested that 13 of the 42 developing AEWs exhibited a downscaling feature for shear transfer from the trend mode to the IMF3. All of the 13 cases developed into hurricanes. In other words, 13 of 25 hurricanes were associated with prominent downscaling processes. For these 13 cases, the average drop of minimum sea level pressure was 53 hPa during the intensification phase. The average decrease of the u-component wind shear for the trend mode, representing a decrease in the basic state wind shear, was 0.37×10^{-5} s^{-1}, while the average enhancement of the u-component wind shear (as well as the v-component wind shear) of the IMF3 was 0.28×10^{-5} s^{-1} during storm intensification.

3.7. A Hypothetical Mechanism for Recurrence and Periodicity of Multiple AEWs

Here, the focus of the analyses is on the simulation and consecutive initiation of multiple AEWs. We formally introduce "recurrence" that is defined when the trajectory of a state returns back to the neighborhood of a previously visited state. Thus, recurrence braces quasi-periodicity and chaos and may be viewed as a generalization of periodicity [61]. The recurrence was clearly shown by the consecutive appearance of multiple AEWs (in Figure 1c). For example, Table 3 indicates 27 AEWs per 92 days (for July, August and September) each year. The "recurrence" time is about 92/27 = 3.4 days. The following discussions suggest that the recurrence may contribute to the predictability at extended-range (15–30 days) scales (in Figure 1a,b). Below we first provide a brief review on the key role of land surface processes in contributing to the predictability of both AEWs and the AEJ in extended-range simulations, a feature that can be viewed as a boundary value problem. As a result, extended-range simulations for these AEWs could be the mixed form of an initial value problem and a boundary value (forced) problem. Furthermore, we apply the dynamics of the limit cycle to propose a hypothetical mechanism for the recurrence (or periodicity) of AEWs.

In addition to Figure 1c, the nature of temporal "periodicity" could be qualitatively seen in Figure 10, displaying oscillatory correlation coefficients for the 850 hPa temperatures from the control run and the NCEP analysis over the 30-day period. Results with correlation coefficients above 0.65 suggest that temperatures are simulated with some degree of realism. Such results indicate the advantage of a high-resolution global model that can reduce uncertainties associated with imposed lateral boundary conditions, making it possible to perform longer simulations, as compared to regional models. However, the correlation coefficients oscillate with time. Why? Was this consistent with the divergent feature of chaotic processes? Are correlation coefficients supposed to be a monotonically decreasing function with time? We address this question below.

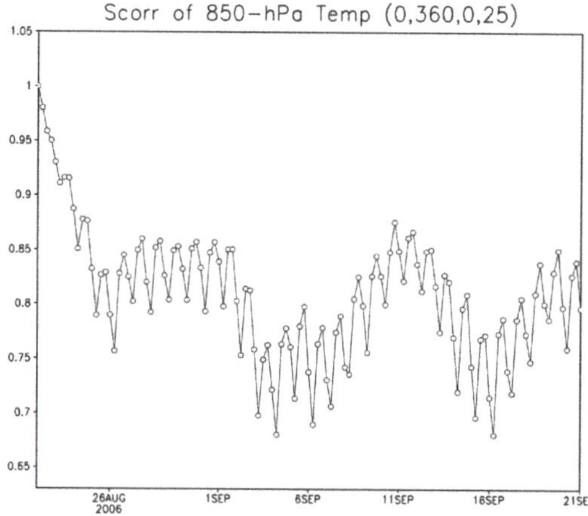

Figure 10. Correlation coefficients between the simulated 850 hPa temperature and the corresponding NCEP analysis over the domain longitude 0° E to 360° E and latitude 0° N to 25° N over the 30-day period. The choice of a global belt domain for verification is due to zonally-moving weather systems. Correlation coefficients are calculated with the scorr function provided by the Grid Analysis and Display System (GrADS). It should be noted that the correlation coefficients for the 850 hPa temperatures above are 0.65 for the entire 30-day period. Data are from the supplemental materials in [6].

In the early study [6], we additionally presented good agreement of the spatial distribution of 30-day averaged 850 hPa temperatures between the control run and the NCEP analysis (e.g., Figure 2 in [6]). The result may suggest the importance of surface forcing (or low-level forcing) on simulations of AEWs as well as hurricanes during the July–September period. As shown in Figure 11a,b via dynamic ICs on 22 April and 22 June, respectively, experiments P1 and P2 with no or a weaker AEJ initially indicated that a realistic AEJ and westward moving AEWs could be simulated using the same model physics (e.g., radiation) and land model configurations as the control run. For example, after 20–25 days of integration, experiment P1 was able to produce AEWs (Figure 11a) and an AEJ (Figure 11c–f), although simulated AEWs and AEJs had large errors. In the two experiments with very different dynamic ICs, large discrepancies in timing and location also existed. However, the two experiments indeed illustrated the importance of accurate land surface and physics ICs on simulations of AEW(s), and a realistic AEJ. Additionally, experiment P3 with reduced mountain heights was used to examine their impacts on AEW simulations, displaying higher impact on the downstream development of AEWs than on upstream initiation. This suggested that it takes time for errors to have impact on simulations.

Here a brief summary of the limit cycle solution is provided as a baseline for proposing a mechanism for the recurrence of the multiple AEWs. The GLM and 3DLM were derived based on partial differential equations for Rayleigh–Benard convection with heating at the bottom. The appearance of nonlinear limit cycle solutions at large Raleigh parameters suggested that: (1) the collective impact of nonlinearity and "strong" heating could lead to periodicity while (2) relatively small dissipation was responsible for the isolated nature of the limit cycle solution. Such an isolated feature was shown in Figure 3a, suggesting no long-term memory of initial conditions. In other words, a limit cycle solution is insensitive to tiny changes in initial conditions. Specifically, a limit cycle and its periodicity and amplitudes are solely determined by the system and thus independent of initial conditions within the idealized 3DLM and GLM that contain constant forcing and dissipative terms. An accurate IC is effective in helping reach the balance between the nonlinearity and heating, playing a role in determining the initial evolution of solution. Based on the analyses above and below, we propose that the periodicity of AEWs may be largely determined by strong surface heating and nonlinearity. The hypothesis was consistent with the fact that AEWs mainly appear during July, August, and September. Since periodic signals produced oscillatory differences between the control and parallel runs, as shown in Figure 3c for the limit cycle solution, the hypothesis was also consistent with the simulated result containing oscillatory forecast scores in Figure 10. Note that, by comparison, chaotic solutions produced irregularly oscillatory "errors" (or divergences) (Figure 2d).

The above discussions suggest the possibility of extending the lead time of prediction at extended-range scales (say to 23 days). For example, when a model has accurate land and physics components, as well as ICs, to simulate the "periodic" nature of AEWs with a period of 5 (or 3.4) days and has the capability of simulating the downscaling processes associated with the fourth (or fifth) AEW for a period of 3 days, it may produce simulations of TCs with a predictability of 23 (or 20) days, computed as follows: $5 \times 4 + 3 = 23$ (or $3.4 \times 5 + 3 = 20$). The 3-day predictability of downscaling processes is not inconsistent with findings using current regional models. As a result of the periodic (or recurrent) nature of large-scale systems (that lead to a high intrinsic predictability), extended-range predictability is possible.

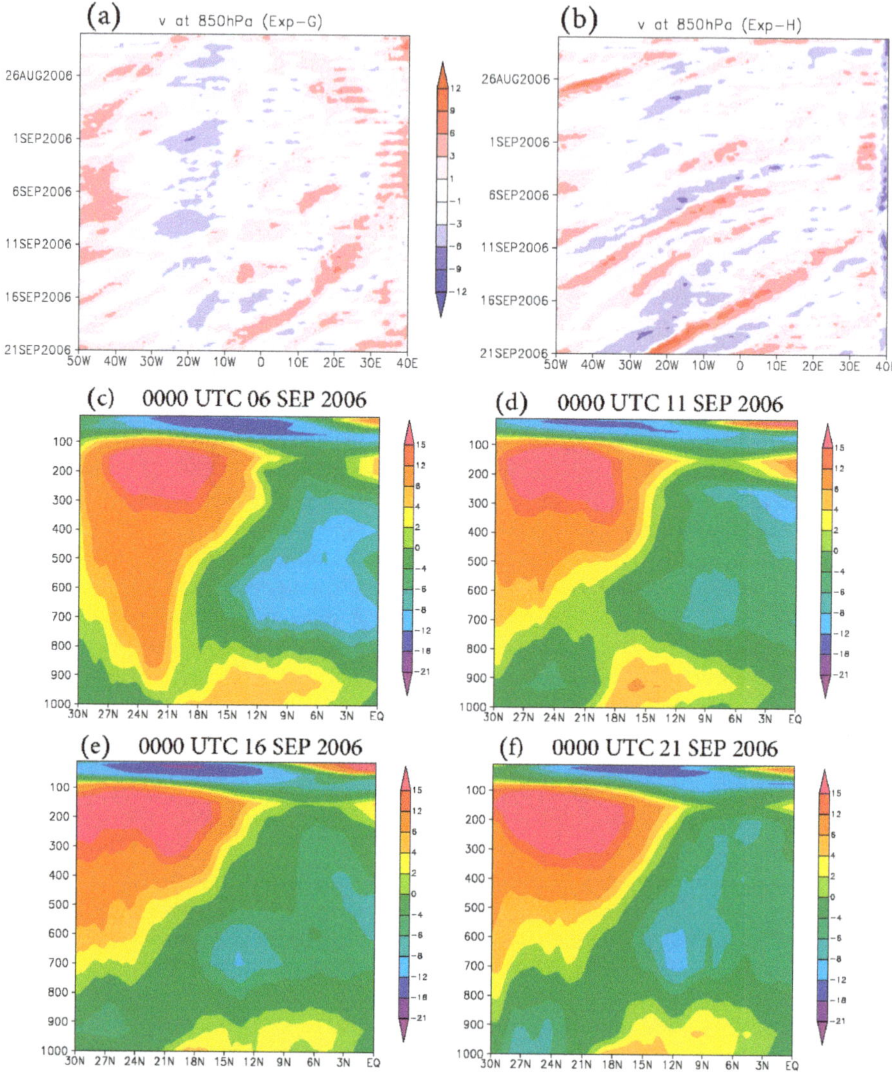

Figure 11. Panels (**a**,**b**) illustrate the sensitivity of AEW simulations to different dynamic ICs in experiments P1 and P2, respectively. Time-longitude diagrams of simulated meridional winds are plotted. Panels (**c**–**f**) display altitude–latitude cross sections of zonal winds averaged over longitude 20° W to 20° E on 6, 11, 16, and 21 September 2006, respectively, showing the development of an AEJ after 25 days in experiment P1. Data are from the supplemental materials in [6].

4. Conclusions

Recent advances in computational and global modeling technology have shown the potential for improving weather predictions at extended-range scales. In some of our past work, establishing remarkable predictability, we proposed a conceptual model that examined the role of (i) the downscaling processes of large-scale tropical systems (e.g., African easterly waves (AEWs) and Madden–Julian oscillations), and (ii) the upscaling processes of small-scale flows (e.g., precipitation) in the formation, intensification, and movement of mesoscale tropical cyclones (TCs). In earlier studies [6], we reported realistic simulations of multiple AEWs and an averaged AEJ in 30-day runs and promising simulations of formation for hurricane Helene (2006) from Day 22 to Day 30. In this study, such extended predictability was further analyzed based on a recent understanding of chaos and instability derived from studies using the Lorenz models [1–3,25] and the generalized Lorenz model [22].

By definition, intrinsic predictability and practical predictability are different. Since numerical models cannot perfectly represent weather, an estimate of practical predictability should be interpreted with caution. While the Lorenz 1963 and 1969 models [1,2] suggest finite predictability, the underlining mechanisms are different. The former focuses on chaos (i.e., sensitive dependence on initial conditions) and the latter associates finite predictability with instability. Both models possess simplified physical processes (i.e., without realistic dissipations), so, in practice, an estimate of predictability using the above models may be a qualitative indicator but not necessarily an upper bound for intrinsic predictability. Therefore, the statement regarding a theoretical predictability of two weeks is not universal. The potential for extending the lead time of predictions was discussed by providing insight into understandings of chaotic and non-chaotic processes within the generalized Lorenz model (GLM).

The GLM possesses the following features: (1) three types of attractors (that also appear within the original Lorenz model), (2) two kinds of attractor coexistence, (3) aggregated negative feedback, and (4) hierarchical scale dependence. When additional realistic dissipative processes at smaller scales are included within the GLM, their negative feedback can be aggregated to produce a stronger dissipation to stabilize the system, leading to stable equilibrium points and steady-state solutions. The feature of hierarchical scale dependence provides a theoretical basis for the role of large-scale processes in modulating small scale processes. All of these features suggest a refined view on the nature of weather as follows: the entirety of weather is a superset that consists of chaotic and non-chaotic processes. Stable, steady-state solutions should have better predictability compared to chaotic solutions. Additionally, errors associated with dissipative, small-scale processes do not necessarily contaminate the simulations of large-scale processes.

In Lorenz systems, chaotic solutions may appear within a subset of the entirety of solutions. In addition to chaotic and steady-state solutions, nonlinear periodic solutions may appear alone in the 3DLM at large Rayleigh parameters or they may coexist with steady-state solutions in the GLM with M = 9 or larger. Based on a simple comparison between Rayleigh–Benard convection and AEW/AEJ problems, we applied the dynamics of the limit cycle solution in order to propose a hypothetical mechanism for the recurrence (or periodicity) of successive AEWs. The recurrence may appear as a result of a balance between the dominant surface heating and nonlinearity. Specifically, a system with strong heating balanced by nonlinearity may produce recurrence and, thus, be more predictable. The characteristics of recurrent signals (e.g., periods) may be less sensitive to initial conditions (as shown in Figure 11), as suggested by the insensitivity of limit cycles to initial conditions. By comparison, the initial evolution and the phase of oscillatory solutions may be influenced by accurate initial conditions. Such an impact by initial conditions can be seen in the simulations of two limit cycle solutions (Figure 3b) as well as in global model simulations of AEWs from the control run and parallel experiments P1 and P2. Therefore, within simulations of oscillatory (or recurrent) signals, forecast scores may be oscillatory. As a result of better predictability for recurrence, near the end of Section 3.7 we discussed the possibility of extending the lead time of predictions at extended range scales. Future work will include refining the idealized Lorenz models (e.g., with realistic parameters) to better examine the validity of the mechanism in explaining the recurrence of multiple AEWs.

Funding: This research received no external funding.

Acknowledgments: We thank reviewers and R. Atlas, R. Anthes, J.-J. Baik, D. Durran, F.D. Marks, Z. Musielak, T. Krishnamurti, C.-D. Lin, R. Rotunno, I. A. Santos, R. Pielke, Sr., X. Zeng, and F. Zhang for valuable comments and discussions. We are grateful for the support from the College of Science at San Diego State University. We thank J. Cui for his help in producing the figures. This paper was completed based on a recent presentation entitled "Butterfly Effects and Chaos within a Generalized Lorenz Model: New Insights and Opportunities" at NOAA/AOML/HRD http://bit.ly/2LuiIAY.

Conflicts of Interest: The author declares no conflict of interest.

References

1. Lorenz, E.N. Deterministic nonperiodic flow. *J. Atmos. Sci.* **1963**, *20*, 130–141. [CrossRef]
2. Lorenz, E.N. The predictability of a flow which possesses many scales of motion. *Tellus* **1969**, *21*, 289–307. [CrossRef]
3. Lorenz, E.N. Predictability: Does the flap of a butterfly's wings in Brazil set off a tornado in Texas? In Proceedings of the 139th Meeting of AAAS Section on Environmental Sciences, New Approaches to Global Weather: GARP, Cambridge, MA, USA, 1972; p. 5. Available online: http://eaps4.mit.edu/research/Lorenz/Butterfly_1972.pdf (accessed on 17 June 2019).
4. Lewis, J.M. Roots of ensemble forecasting. *Mon. Weather Rev.* **2005**, *133*, 1865–1885. [CrossRef]
5. Anthes, R. Turning the Tables on Chaos: Is the Atmosphere More Predictable than We Assume? UCAR Magazine, Spring/Summer. 2011. Available online: https://news.ucar.edu/4505/turning-tables-chaos-atmosphere-more-predictable-we-assume (accessed on 26 November 2018).
6. Shen, B.-W.; Tao, W.-K.; Wu, M.-L.C. African Easterly Waves in 30-day High-resolution Global Simulations: A Case Study during the 2006 NAMMA Period. *Geophys. Res. Lett.* **2010**, *37*, L18803. [CrossRef]
7. Palmer, T.N.; Doring, A.; Seregin, G. The real butterfly effect. *Nonlinearity* **2014**, *27*, R123–R141. [CrossRef]
8. Lorenz, E.N. *"Predictability—A Problem Partly Solved" (PDF)*; Seminar on Predictability; ECMWF: Reading, UK, 1996; Volume I.
9. Lorenz, E.N. *Predictability—A Problem Partly Solved*; Predictability of Weather and Climate; Palmer, T., Hagedorn, R., Eds.; Cambridge University Press: Cambridge, UK, 2006; pp. 40–58.
10. Shen, B.-W.; Atlas, R.; Reale, O.; Lin, S.-J.; Chern, J.-D.; Chang, J.; Henze, C.; Li, J.-L. Hurricane Forecasts with a Global Mesoscale-Resolving Model: Preliminary Results with Hurricane Katrina (2005). *Geophys. Res. Lett.* **2006**, *33*, L13813. [CrossRef]
11. Shen, B.-W.; Atlas, R.; Chern, J.-D.; Reale, O.; Lin, S.-J.; Lee, T.; Chang, J. The 0.125 degree finite-volume General Circulation Model on the NASA Columbia Supercomputer: Preliminary Simulations of Mesoscale Vortices. *Geophys. Res. Lett.* **2006**, *33*, L05801. [CrossRef]
12. Shen, B.-W.; Tao, W.-K.; Lau, W.K.; Atlas, R. Predicting Tropical Cyclogenesis with a Global Mesoscale Model: Hierarchical Multiscale Interactions During the Formation of Tropical Cyclone Nargis (2008). *J. Geophys. Res.* **2010**, *115*, D14102. [CrossRef]
13. Shen, B.-W.; Tao, W.-K.; Lin, Y.-L.; Laing, A. Genesis of Twin Tropical Cyclones as Revealed by a Global Mesoscale Model: The Role of Mixed Rossby Gravity Waves. *J. Geophys. Res.* **2012**, *117*, D13114. [CrossRef]
14. Shen, B.-W.; DeMaria, M.; Li, J.-L.F.; Cheung, S. Genesis of Hurricane Sandy (2012) Simulated with a Global Mesoscale Model. *Geophys. Res. Lett.* **2013**, *40*. [CrossRef]
15. Shen, B.-W.; Cheung, S.; Li, J.-L.F.; Wu, Y.-L.; Shen, S.S. Multiscale Processes of Hurricane Sandy (2012) as Revealed by the Parallel Ensemble Empirical Mode Decomposition and Advanced Visualization Technology. *Adv. Data Sci. Adapt. Anal.* **2016**, *8*, 1650005. [CrossRef]
16. Wu, Y.-L.; Shen, B.-W. An Evaluation of the Parallel Ensemble Empirical Mode Decomposition Method in Revealing the Role of Downscaling Processes Associated with African Easterly Waves in Tropical Cyclone Genesis. *J. Atmos. Ocean. Technol.* **2016**, *33*, 1611–1628. [CrossRef]
17. Shen, B.-W. Nonlinear Feedback in a Five-dimensional Lorenz Model. *J. Atmos. Sci.* **2014**, *71*, 1701–1723. [CrossRef]
18. Shen, B.-W. Nonlinear Feedback in a Six-dimensional Lorenz Model. Impact of an additional heating term. *Nonlin. Process. Geophys.* **2015**, *22*, 749–764. [CrossRef]
19. Shen, B.-W. Hierarchical scale dependence associated with the extension of the nonlinear feedback loop in a seven-dimensional Lorenz model. *Nonlin. Process. Geophys.* **2016**, *23*, 189–203. [CrossRef]
20. Shen, B.-W. On an extension of the nonlinear feedback loop in a nine-dimensional Lorenz model. *Chaotic Model. Simul. CMSIM* **2017**, *2*, 147–157.
21. Shen, B.-W. On periodic solutions in the non-dissipative Lorenz model: The role of the nonlinear feedback loop. *Tellus A* **2018**, *70*, 1471912. [CrossRef]
22. Shen, B.-W. Aggregated Negative Feedback in a Generalized Lorenz Model. *Int. J. Bifurc. Chaos* **2019**, *29*, 1950037. [CrossRef]

23. Shen, B.-W.; Reyes, T.A.L.; Faghih-Naini, S. Coexistence of Chaotic and Non-Chaotic Orbits in a New Nine-Dimensional Lorenz Model. In *CHAOS 2018: 11th Chaotic Modeling and Simulation International Conference*; Skiadas, C.H., Lubashevsky, I., Eds.; Springer Proceedings in Complexity; Springer: Cham, Switzerland, 2019.
24. Chen, J.-H.; Lin, S.-J. Seasonal predictions of tropical cyclones using a 25-km-resolution general circulation model. *J. Climate* **2013**, *26*, 380–398. [CrossRef]
25. Lorenz, E. The predictability of hydrodynamic flow. *Trans. N. Y. Acad. Sci.* **1963**, *25*, 409–432. [CrossRef]
26. Frank, W.M.; Roundy, P.E. The role of tropical waves in tropical cyclogenesis. *Mon. Weather Rev.* **2006**, *134*, 2397–2417. [CrossRef]
27. Carlson, T.N. Synoptic histories of three Africa disturbances that developed into Atlantic hurricanes. *Mon. Weather Rev.* **1969**, *97*, 256–276. [CrossRef]
28. Madden, R.A.; Julian, P.R. Detection of a 40–50 day oscillation in the zonal wind in the tropical Pacific. *J. Atmos. Sci.* **1971**, *28*, 702–708. [CrossRef]
29. Blake, E.S.; Blake, E.S.; Kimberlain, T.B.; Berg, R.J.; Cangialosi, J.P.; Beven, J.L., II. *Tropical Cyclone Report: Hurricane Sandy*; Report AL182012; National Hurricane Center: Miami, Fl, USA, 2013.
30. Huang, N.E.; Shen, Z.; Long, S.R.; Wu, M.C.; Shih, H.H.; Zheng, Q.; Yen, N.; Tung, C.C.; Liu, H.H. The empirical mode decomposition and the Hilbert spectrum for nonlinear and non-stationary time series analysis. *Proc. R. Soc. Lond. A* **1998**, *454*, 903–995. [CrossRef]
31. Wu, Z.; Huang, N.E. Ensemble Empirical Mode Decomposition: A noise-assisted data analysis method. *Adv. Adapt. Data Anal.* **2009**, *1*, 1–41. [CrossRef]
32. Flandrin, P.; Rilling, G.; Gonçalves, P. Empirical Mode Decomposition as a Filterbank (pdf). *IEEE Signal Process. Lett.* **2003**, *11*, 112–114. [CrossRef]
33. Wu, Z.; Huang, N.E. Statistical significance test of intrinsic mode functions. In *Hilbert-Huang Transform: Introduction and Applications*; Huang, N.E., Shen, S.S.P., Eds.; World Scientific: Singapore, 2004; pp. 125–148, p. 311.
34. Shen, B.-W.; Cheung, S.; Wu, Y.; Li, F.; Kao, D. Parallel Implementation of the Ensemble Empirical Mode Decomposition (PEEMD) and Its Application for Earth Science Data Analysis. *Comput. Sci. Eng.* **2017**, *19*, 49–57. [CrossRef]
35. Dee, D.P.; Uppala, S.M.; Simmons, A.J.; Berrisford, P.; Poli, P.; Kobayashi, S.; Andrae, U.; Balmaseda, M.A.; Balsamo, G.; Bauer, P.; et al. The ERA-Interim reanalysis: Configuration and performance of the data assimilation system. *Q. J. Royal Meteorol. Soc.* **2011**, *137*, 553–597. [CrossRef]
36. Rotunno, R.; Snyder, C. A generalization of Lorenz's model for the predictability of flows with many scales of motion. *J. Atmos. Sci.* **2008**, *65*, 1063–1076. [CrossRef]
37. Leith, C.E.; Kraichnan, R.H. Predictability of turbulent flows. *J. Atmos. Sci.* **1972**, *29*, 1041–1058. [CrossRef]
38. Biswas, R.; Aftosmis, M.J.; Kiris, C.; Shen, B.-W. Petascale Computing: Impact on Future NASA Missions. In *Petascale Computing: Architectures and Algorithms*; Bader, D., Ed.; Chapman and Hall/CRC Press: Boca Raton, FL, USA, 2007; pp. 29–46.
39. Kerr, R. Sharpening Up Models for a Better View of the Atmosphere. *Science* **2006**, *25*, 1040. [CrossRef] [PubMed]
40. Green, B.; Henze, C.; Shen, B.-W. Development of a scalable concurrent visualization approach for high temporal- and spatial-resolution models. In Proceedings of the AGU 2010 Western Pacific Geophysics Meeting, Taipei, Taiwan, 22–25 June 2010.
41. Shen, B.-W.; Nelson, B.; Tao, W.-K.; Lin, Y.-L. Advanced Visualizations of Scale Interactions of Tropical Cyclone Formation and Tropical Waves. *Comput. Sci. Eng.* **2013**, *15*, 47–59. [CrossRef]
42. Lin, S.-J. A vertically Lagrangian finite-volume dynamical core for global models. *Mon. Weather Rev.* **2004**, *132*, 2293–2307. [CrossRef]
43. Shen, B.-W.; Tao, W.-K.; Green, B. Coupling NASA Advanced Multi-Scale Modeling and Concurrent Visualization Systems for Improving Predictions of Tropical High-Impact Weather (CAMVis). *Comput. Sci. Eng.* **2011**, *13*, 56–67. [CrossRef]
44. Lorenz, E.N. Designing Chaotic Models. *J. Atmos. Sci.* **2005**, *62*, 1574–1587. [CrossRef]
45. Sparrow, C. The Lorenz Equations: Bifurcations, Chaos, and Strange Attractors. *Appl. Math. Sci.* **1982**. [CrossRef]

46. Strogatz, S.H. *Nonlinear Dynamics and Chaos. With Applications to Physics, Biology, Chemistry, and Engineering*; Westpress View: Boulder, CO, USA, 2015; p. 513.
47. Shimizu, T. Analytical Form of the Simplest Limit Cycle in the Lorenz Model. *Physica* **1979**, *97A*, 383–398. [CrossRef]
48. Jordan, D.W.; Smith, P. *Nonlinear Ordinary Differential Equations. An Introduction for Scientists and Engineers*, 4th ed.; Oxford University Press: New York, NY, USA, 2007; p. 560.
49. Nagel, R.K.; Staff, E.; Snider, A. *Fundamentals of Differential Equations*, 7th ed.; Pearson: New York, NY, USA, 2008; p. 912.
50. Faghih-Naini, S.; Shen, B.-W. Quasi-periodic in the Five-dimensional Non-dissipative Lorenz Model: The Role of the Extended Nonlinear Feedback Loop. *Int. J. Bifurc. Chaos* **2018**, *28*, 1850072. [CrossRef]
51. Wolf, A.; Swift, J.B.; Swinney, H.L.; Vastano, J.A. Determining Lyapunov Exponents from a Time Series. *Physica* **1985**, *16D*, 285–317. [CrossRef]
52. Reyes, T.; Shen, B.-W. A Recurrence Analysis of Chaotic and Non-Chaotic Solutions within a Generalized Nine-Dimensional Lorenz Model. *Chaos. Solitons Fractals* **2019**, *125*, 1–12. [CrossRef]
53. Durran, D.R.; Gingrich, M. Atmospheric predictability: Why atmospheric butterflies are not of practical importance. *J. Atmos. Sci.* **2014**, *71*, 2476–2478. [CrossRef]
54. Weyn, J.; Durran, D.R. The dependence of the predictability of mesoscale convective systems on the horizontal scale and amplitude of initial errors in idealized simulations. *J. Atmos. Sci.* **2017**, *74*, 2191–2210. [CrossRef]
55. Lorenz, E.N. *The Essence of Chaos*; University of Washington Press: Seattle, WA, USA, 1993; p. 227.
56. Boyce, W.E.; Diprima, R.C. *Elementary Differential Equations*, 10th ed.; JohnWiley & Sons, Inc.: Hoboken, NJ, USA, 2012; ISBN1 13: 978-0470458327. ISBN2 10: 0470458321.
57. Brown, D.P. *Tropical Cyclone Report: Hurricane Helene*; Report AL082006; National Hurricane Center: Miami, FL, USA, 2006.
58. Sander, N.; Jones, S.C. Diagnostic measures for assessing numerical forecasts of African easterly waves. *Meteorol. Z.* **2008**, *17*, 209–220. [CrossRef]
59. Thorncroft, C.D.; Blackburn, M. Maintenance of the African easterly jet. *Q. J. R. Meteorol. Soc.* **1999**, *125*, 763–786. [CrossRef]
60. Hopsch, S.B.; Thorncroft, C.D.; Hodges, K.; Aiyyer, A. West African storm tracks and their relationship to Atlantic tropical cyclones. *J. Clim.* **2007**, *20*, 2468–2483. [CrossRef]
61. Thompson, J.M.T.; Stewart, H.B. *Nonlinear Dynamics and Chaos*, 2nd ed.; John Wiley & Sons Ltd.: Chichester, UK, 2002; p. 437.

Communication

Insights into the Oroville Dam 2017 Spillway Incident

Aristotelis Koskinas [1], Aristoteles Tegos [1,2,*], Penelope Tsira [1], Panayiotis Dimitriadis [1], Theano Iliopoulou [1], Panos Papanicolaou [1], Demetris Koutsoyiannis [1] and Tracey Williamson [3]

1 Department of Water Resources and Environmental Engineering, National Technical University of Athens, Heroon Polytechniou 9, Zografou, GR-15780 Zographou Athens, Greece; tel9021@yahoo.gr (A.K.); tsira_p@hotmail.com (P.T.); pandim@itia.ntua.gr (P.D.); theano_any@hotmail.com (T.I.); panospap@mail.ntua.gr (P.P.); dk@itia.ntua.gr (D.K.)

2 Arup Group Limited, 50 Ringsend Rd, Grand Canal Dock, D04 T6X0 Dublin 4, Ireland

3 Arup, 4 Pierhead Street, Cardiff CF10 4QP, UK; tbulley1@gmail.com

* Correspondence: tegosaris@yahoo.gr

Received: 9 December 2018; Accepted: 7 January 2019; Published: 11 January 2019

Abstract: In February 2017, a failure occurring in Oroville Dam's main spillway risked causing severe damages downstream. A unique aspect of this incident was the fact that it happened during a flood scenario well within its design and operational procedures, prompting research into its causes and determining methods to prevent similar events from reoccurring. In this study, a hydroclimatic analysis of Oroville Dam's catchment is conducted, along with a review of related design and operational manuals. The data available allows for the comparison of older flood-frequency analyses to new alternative methods proposed in this paper and relevant literature. Based on summary characteristics of the 2017 floods, possible causes of the incident are outlined, in order to understand which factors contributed more significantly. It turns out that the event was most likely the result of a structural problem in the dam's main spillway and detrimental geological conditions, but analysis of surface level data also reveals operational issues that were not present during previous larger floods, promoting a discussion about flood control design methods, specifications, and dam inspection procedures, and how these can be improved to prevent a similar event from occurring in the future.

Keywords: Oroville Dam; spillway; incident; flood control; flood-frequency analysis; dam operation

---

1. Introduction

Dam construction and operation across the centuries has resulted from major and multipurpose human needs and is linked to the development of human wealth, health and the growth of civilization [1–4]. Spillways are a crucial aspect of dams, as their most important function is to discharge excess flows during severe floods to prevent dams from failing due to overtopping [5,6]. The International Committee on Large Dams (ICOLD) has suggested that nearly a third of dam related incidents is linked to this cause of failure. These cases are usually brought about by extreme weather conditions that exacerbate faulty or incomplete spillway designs, leading to significant damage [6,7]. As such, reducing the risk of spillway failures is a topic that promotes continuing research and improvement. Ongoing studies have attempted to investigate these structures both from a hydrologic approach [8,9] and by looking into the various available design methods and materials [10–12]. The former studies found several cases of spillways built using outdated data and formulas which are now obsolete, and proposed alternative methods to calculate design flows taking into account the effects of long-term persistence on floods [2,7,8,13]. On the other hand, works that focused on the more practical aspects of building spillways analyzed the core elements of these structures [5,14], and determined scenarios where they can be undermined even when not under extreme conditions, much like the case of Oroville Dam itself. Studies of previous similar dam failures [9,14,15] reveal multiple aspects of a typical dam failure due to a problem developing in its spillway, and prove that usually when such an incident occurs, several different factors likely contributed to it.

In their publication "Lessons from dam incidents", ICOLD summarizes some 500 incidents from 1800 to 1965. Several of these have been the subject of major investigations and have a substantial literature [16]. Another useful reference in USCOLD's "Lessons from Dam Incidents USA", which lists over 500 incidents in the USA [17]. A report published by the UK's Environment Agency offers insights and information on lessons learnt from over 100 national and international dam incidents and failures [18]. Many of these include incidents caused by overtopping, internal erosion and foundation failures such as those related to the Oroville incident.

Analyzing scenarios like these leads to preventing future disasters, and the Oroville Dam incident is yet another case study that promotes varying topics of dam risk assessment, flood control analysis, as well as offering insights into design and operational procedures regarding these crucial structures.

2. Materials and Methods

2.1. Feather River Basin Characteristics

Oroville Dam's catchment, the Feather River Basin, lies between the north end of the Sierra Nevada range and the east side of the Sacramento River Valley. It is bounded by Mt. Lassen to the northwest, and by the Diamond Mountains to the northeast. Documented results of geological studies in the vicinity [19,20] suggest that the area immediately in and around Lake Oroville composed mostly of what is called the "Bedrock Series". This consists mostly of metavolcanic and pyroclastic rock, such as amphibolite. Above this bedrock lie various younger sedimentary rocks such as shales, dolomites, Quaternary alluvium, playas, terraces, glacial till and moraines, and finally various marine and non-marine sediments [21]. In general, the Feather River Basin is considered an area of low seismicity [20]. The Feather River Basin as well as the city of Oroville are characterized by a Mediterranean climate. Precipitation in the Feather River basin occurs most usually during the cooler months, in rare yet intense events. On average, there are only 57 days of precipitation per year, and 36 of those are liquid. Large floods in the Feather River basin occur due to severe winter rain storms, in some cases augmented by snowmelt. A typical event may last several days, not being a single storm, but a sequence of smaller individual storms in quick succession. In these cases, runoff can produce high-peak intense flows downstream with a variety of flood characteristics [22].

A report [23] contains unregulated, annual maximum flow data for the Feather River at Oroville station resulting from rainfall for 1-day, 3-day, 7-day, 15-day, and 30-day durations as provided by the US Army Corps of Engineers. Each n-day period is useful for different aspects of reservoir management [22–25]. The most intense floods from this analysis can be found in Table 1.

Table 1. Historical maximum 1-day and 3-days floods, Feather River at Oroville [23].

	1-day		3-day	
Water Year	**Date**	**Flow (m^3/s)**	**Date**	**Flow (m^3/s)**
1903–04	24-February	3001.59	18-March	2501.23
1906–07	19-March	5295.25	18-March	4256.87
1908–09	16-January	3879.41	14-January	3643.53
1927–28	26-March	3544.42	25-March	3139.77
1937–38	11-December	4501.81	10-December	3004.98
1939–40	30-March	3815.98	27-February	3055.67
1955–56	23-December	5140.36	22-December	4160.03
1964–65	23-December	5055.97	22-December	4683.32
1979–80	13-January	3896.96	12-January	3032.73
1985–86	17-February	6145.32	17-February	5295.53
1994–95	10-March	3799.78	9-March	3221.98
1996–97	1-January	8860.14	31-December	6923.04

2.2. Oroville Dam Characteristics

Oroville Dam is a zoned earth-fill embankment structure with a maximum height of 235 m above river excavation as shown in Figure 1. The dam embankment has a volume of approximately 61 million m^3 and comprises an inclined impervious core on a concrete foundation, supplemented by zoned earth-fill sections on both sides.

Figure 1. Aerial view of Oroville Dam [26].

This dam has a large catchment, with an area of approximately 9342 km^2 and reservoir surface area of approximately 64 km^2. The reservoir capacity (up to the main spillway sill level) is 3427 hm^3, whereas the maximum operating volume is stated to be 4364 hm^3 (up to the emergency spillway sill level). Further pertinent data on the dam, including a stage-storage capacity curve can be found in related design reports released shortly after the dam's construction [19,22]. An important additional note is that this dam is not the only one that operates in the Feather Basin; it is part of a network that includes several upstream reservoirs and diversion pools [27,28].

Oroville Dam's spillway is located on a natural ridge adjacent to right abutment of the main embankments. It consists of two independent structures, a combined flood control outlet and an emergency weir. The former consists of an unlined approach channel with walls in such a way as to make flows smoothly transit into an outlet passage, a headworks structure, and a concrete lined chute, approximately 929 m in length. The headworks structure is comprised of eight top-seal radial gates, 17.78 cm thick and 5.18 m wide by 10.06 m high. At the end of the lined chute, chute blocks help absorb some of the energy from the outgoing flow before it pours into the Feather River.

The main concept behind designing the flood control outlet was to limit Feather River flow to 5094 m^3/s in the occurrence of a flood event known as the Standard Project Flood (SPF). For Oroville Dam, the peak inflow of the SPF was estimated at 12,700 m^3/s, and is claimed to have a return period of 450 years in related design documents [19]. In order to meet this criterion, the flood control was designed for a 4245 m^3/s controlled release, and a flood control reservation volume of 925.11 hm^3 was deemed necessary. This volume is also mentioned in the official manual for flood control operation of Oroville Dam [19,22].

According to references [19,29], the combined capacity of the main and emergency spillways is 17,472 m^3/s, which corresponds to a peak inflow of 20,160 m^3/s. The event that would cause this inflow corresponds to what has been known as the Probable Maximum Flood (PMF). Given the known design capacity of the main spillway, this would set the design capacity of the emergency spillway to approximately 9900 m^3/s in order to meet the combined outflow required by the PMF.

Blasting was used for almost 90% of the main spillway chute foundation, in order to reach grade. The remaining amount consisted of the removal of several seams of clay located in the foundation, and a few areas where the slope failed [19]. The slopes in the flood control outlet section were of a lower quality rock than initially presumed and several large seams ran parallel with the main spillway chute. The countermeasure that was applied was the replacement of planned anchor bars with grouted rock blots, pigtail anchors and a chain-link covering the area's surface [19].

2.3. Annual Maxima Rainfall Analysis

Until now, known flood control studies for Oroville Dam and the Feather Basin have attempted to determine the Probable Maximum Flood (PMF) for Lake Oroville, based on the Probable Maximum Precipitation (PMP). The most recent existing study available detailing PMP calculations in California is Hydrometeorological Report No. 59 or HMR 59 [30]. In brief, the computational procedure includes tracing an outline of the drainage basin, placing this outline on top of a given PMP 10-m^2, 24-h index map, then determining depth-duration relationships and areal reduction factors, and finally conducting temporal distribution of incremental depths extracted from a given curve.

While this method is simple to use, and the analysis involved in creating these PMP index maps undoubtedly contains valuable information, it would be better to adopt a probabilistic approach to precipitation analysis, where instead of assuming a deterministic, theoretical upper limit, a return period would be assigned to any precipitation and flood value. This would be achieved by studying existing precipitation data and extracting a return period for the already calculated 24-h index depths, for every sub-area of the Feather River Basin, as determined by the California Department of Water Resources [27]. One of the possible methods to achieve this is exposed below.

The 24-h index PMP depth essentially describes a daily maximum precipitation value. If the distribution of daily rainfall for a given area is known, one can assume that the annual maxima of daily rainfall would resemble one of two limiting types: type I, known as Gumbel distribution or type II, known as Fréchet distribution. The Generalized Extreme Value (GEV) distribution, which comprises these types by way of its shape parameter (as well as type III, known as reversed Weibull, which however is not recommended for rainfall maxima [31]) can be fitted to a series of annual maxima of daily rainfall.

In accordance with References [13,32,33], the GEV distribution using the method of L-moments is fitted to various precipitation data gathered from the Feather Basin [34,35]. A map of the basin with the measurement stations used in this analysis can be found in Appendix A [34–36].

To improve accuracy, a filter is applied to the data, i.e., only years with 300 or more daily measurements are taken into account, roughly equivalent to at least 25 days with measurements per month. After discarding stations with data suspected of containing erroneous measurements that could not be cross-referenced with floods around the same time period, four significant precipitation measurement stations were selected for this analysis. Then, annual daily maxima time series are created. The process is simple: First, select the maximum daily precipitation value of every year, then rank them in descending order. Obviously, the highest value is the most important one, so it is imperative that it is cross-referenced with multiple sources to confirm its validity. Finally, the GEV-max distribution with the method of L-moments is fitted using the "Pythia" statistical tool of the HYDROGNOMON open software, which follows the exact principles stated in the related literature [13,32,33].

2.4. The 2017 Event

During the first few days of January 2017, two small rain storms occurred just over Oroville Dam's reservoir [34,37]. The first rain storm was short, lasting only 4 days, peaking at 90 mm on January 3, and the second was a stronger 6-day event, peaking at 136 mm on January 10. These rain storms quickly led into a large increase of inflows into Lake Oroville. Two inflow peaks occurred: The primary one was 4839 m^3/s on January 8 at 21:00 p.m., and a secondary peak of 3079 m^3/s, occurring on January 10 at 22:00 p.m. These inflows are definitely significant, yet expected during a typical wet season. However, outflows from Lake Oroville at the same time were very low, almost zero, as there was a sharp water storage increase in Lake Oroville, as well as a significant rise in its surface elevation. Lake Oroville's surface elevation initially exceeded the flood control minimum on January 12, 2017 at 17:00 p.m. Around that time, outflows from Oroville Dam's main spillway were increased to compensate for this fact and return the surface elevation to below the minimum. Overall, the Oroville Dam operator was able to return the surface level to below the flood control limit on February 3, 2017 at 17:00 p.m, just in time for an upcoming February rain storm.

Thereafter, according to CDEC, a rain storm over the Feather Basin began on February 2, 2017, and ended around February 11. The largest flood value occurred on February 9 at 19:00 p.m., and was 5392 m^3/s. This value is significantly lower than the highest recorded floods to ever occur in the Feather Basin. Under normal circumstances, Oroville Dam should have been able to deal with this event without trouble. On February 6 at approximately 13:00 p.m., outflows from Lake Oroville were raised in order to prepare for incoming inflows to 1500 m^3/s. However, the next day, February 7, at approximately 10:00 a.m., workers at the Oroville Dam site noticed a discoloration in the water flowing through the main spillway. Outflow from the main spillway was immediately halted, in order to detect the source of this discoloration, revealing a large hole in the main spillway chute, seen in Figure 2.

Figure 2. 7 February 2017. Front view of the initial main spillway chute damage [38].

At this point, the main spillway is already severely damaged, and any discharges at that point would rapidly amplify this erosion and move entire parts of the concrete chute and walls downstream. However, Lake Oroville's surface elevation is already past the flood control minimum, and inflows from the February rain storm are imminent. After brief consultation with various dam safety agencies, the operators decided to release test flows into the main spillway and monitor the damage. These small flows ranged hourly from around 300 m^3/s to 900 m^3/s over the course of February 8. On the very next day, February 9, the hole in the main spillway had increased in size.

A worrying aspect of the spillway damage is that it was moving uphill. This is a typical sign of a failure known as headcutting (or undercutting), which is what happens when water flowing across a hard surface falls onto a softer surface below.

With the ever-increasing inflows dangerously raising the reservoir surface level, which is already above the minimum flood control elevation, there was no time to quickly repair the main spillway.

At this point, the Oroville Dam operators were facing a tough dilemma; either continue to release flows through the already damaged chute and cause further erosion, or risk using the untested auxiliary spillway. However, as the latter structure is ungated, if unchecked the dam itself would make that choice for them, as water would flow over the emergency spillway as soon as the surface elevation surpassed its crest, at 274.62 m. As such, a plan was formulated to continue letting small flows pass through the main spillway, while also preparing the area around the auxiliary spillway in case it would have to be put to use. To that end, workers began clearing the area downstream of this secondary structure, as well as placing large rocks at its foot to mitigate possible erosion. At this point, the inflows into Lake Oroville increased tremendously, reaching the aforementioned peak of 5392 m^3/s. On February 11, at 8:00 a.m., surface elevation at Lake Oroville surpassed that of the emergency spillway crest, meaning that for the first time in the dam's history, water would pour over it. According to data from CDEC, water poured over this ogee weir for just over 37 h in total, as the surface level dropped below its crest elevation again on February 12 at 21:00 p.m.

A noticeable fact is that there is a parking lot just next to the emergency spillway, which is at a lower elevation, and thus is flooded by design whenever water pours over the weir. Furthermore, an access road located just below the structure was also subsequently flooded and quickly destroyed, as seen in Figure 3.

Figure 3. 11 February 2017. Image of the flooded parking lot and access road located next to the emergency spillway [39].

Unfortunately, erosion downstream developed much more rapidly than anticipated. While the emergency spillway was only active for a very brief duration, and peak discharge did not exceed 400 m^3/s; large boils occurred downstream, destroying the access road below and threatening to damage the spillway crest itself by failure due to headcutting. The exact extent of the damage was not clearly visible when water was still pouring over the downstream hill on February 12, however, and thus local authorities, fearing the worst outcome, were forced to spring into action and order the evacuation of Oroville and other areas downstream of the dam, including Yuba City and Marysville. The California Department of Water Resources responded to the evacuation order by immediately increasing outflow releases from the main spillway to 2830 m^3/s. This would drastically lower the surface elevation and stop flows over the emergency spillway and any resulting erosions there, at the cost of causing irreparable damages to the main spillway. Luckily, despite the conditions, the upper portion of the main spillway was able to release these discharges without causing further upstream erosion. However, the hill downstream of the initial hole would be quickly eroded away from high velocity flows.

3. Results

Based on the evidence gathered, it is possible to make several hypotheses for the possible causes of failure for both spillways.

3.1. Emergency Spillway

It is much easier to determine the cause of the near failure of the emergency spillway due to the fact that it was actuated for a very brief duration under constant supervision, as authorities were already alerted of the situation. While water was pouring over the concrete weir without a problem, it was the surrounding conditions that posed a threat. Already from the documents describing Oroville Dam's construction, the following facts are known:

- The emergency spillway was untested, even in the model studies conducted by the US Bureau of Reclamation [29].
- The area downstream of the emergency spillway was not cleared [19].
- The emergency spillway foundation excavation continued 3 m—deeper than expected—in order to reach the foundation rock that met the design criteria [19].

While it is known that this concrete overpour weir was built on a solid foundation, no effort was made to secure that the downstream ridge would be able to accommodate flows passing over it without significant erosion occurring as a result. This could have been acceptable if this structure was truly used as an emergency measure (i.e., any outflows from it not being factored into hydrologic design calculations, using only the main structure's design capacity instead), but this is not the case. According to Reference [40], a high risk structure such as Oroville Dam should be able to withstand the PMF. All PMF analyses so far [22,27] have included the emergency spillway in their calculations, and in fact, in the event of the PMF, the emergency spillway is expected to reach outflow discharges of around 10,000 m^3/s. Seeing as erosion threatened to cause structural failure at less than 420 m^3/s, the spillway's ability to withstand PMF-level discharges is questionable. In any case, this warrants the need for the structure to be properly armored with concrete and considered to be an "auxiliary" spillway, not an "emergency" one. This has been repeatedly requested by the community [41–43], and has yet to be fully implemented.

3.2. Revisiting the Minium Flood Control Elevation

When posing the question of why Oroville Dam was capable of withstanding the previous devastating floods of 1986 and 1997, and not the 2017 event, one is prompted to also examine the surface elevation levels prior to each flood. Thus, an attempt is made to compare Oroville Dam reservoir surface levels shortly before and after each of the three recent flood events, occurring in 1986, 1997 and 2017 [34]. In Figure 4, the vertical axis represents surface elevation in meters, whereas the horizontal axis represents time, up to 240 h (10 days) before and after peak inflow. Hour 0 is the hour during which peak inflow occurred for each event.

Naturally, no two flood events are the same and they all impact Oroville Dam in subtly different ways, but this comparison contains clues on what went wrong during the 2017 spillway incident. Notably, while the 2017 peak inflow is the lowest of the three major flood events, its surface elevations are the highest. This is due to two factors. First, as is clear from the graph, shortly prior to peak inflow, the surface elevation during the 2017 event was higher than in previous floods. Already, this has a negative impact on flood management. While this elevation is below the minimum limit specified by the flood control manual [22], the 2017 flood is actually harder to manage than previous events. This is partly why, despite not being a record flood, this event came close to causing severe damages to Oroville Dam's key structures once the main spillway failed.

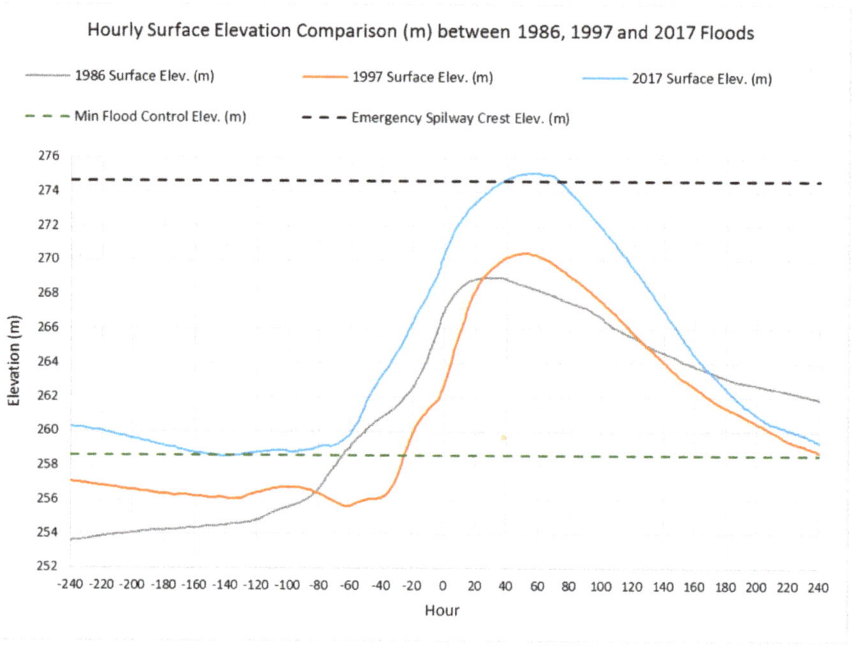

Figure 4. Comparison of Oroville reservoir hourly surface elevations 10 days before and after the peak inflow of the 1986, 1997 and 2017 flood events.

According the flood control manual [22], the additional following restrictions are applied to flows from the main spillway:

- Any water stored in the designated flood control space should be released as quickly as possible, according to a given flood control diagram. Flows from Feather River should not exceed 4245 m^3/s.
- During extreme flood events, releases greater than 4245 m^3/s may be required in order to minimize uncontrolled spillway discharges.
- Releases from Oroville Dam are not to be increased more than 280 m^3/s or decreased more than 140 m^3/s in any given 2-h period.

While levels were within the flood control manual standards [22], the fact that they were close to the limit made dealing with the February 2017 inflows a much more daunting task once the main spillway failed. Thus, it would seem reasonable to request a lowering of the minimum flood control elevation level for Lake Oroville. Lowering the minimum flood control elevation level has its downsides, mainly due to losing reservoir capacity and its valuable resource, but it could be adjusted 5 m lower or even further down at the spillway's sill elevation at 248 m without significant losses in efficiency. Alternatively, it can be maintained even lower by utilizing draw-offs and other outlets to a greater capacity. However, this method of managing the flood risk would need to be considered together with other economic and ecological factors to ensure the benefits are balanced with any other detrimental impacts.

Therefore, taking all of the above into account, it would seem logical to request a small reduction in the minimum flood control level. In their 2006 statement [41], FOR et al., had requested an additional 185 hm^3 of surcharge storage be added to the 925 hm^3 control pool in order to compensate for the never constructed Marysville Dam. This was a project that was included in the flood control pool calculations, yet was never completed. If this measure were to be implemented, according to the flood control manual, the new minimum flood control elevation would be 255 m. Under these conditions, according to Reference [29], the flood control outlet's release capacity is approximately 1274 m^3/s. By chance, this was the Oroville Dam reservoir's surface elevation just before the 1997 flood [44], and the spillway performed adequately even when outflows briefly exceeded the designed discharges. Following additional analyses of other flood events, it may be considered that this small lowering of the flood control level could reduce risks to an acceptable level [5,8,45].

3.3. Main Spillway

Attempting to detect what caused the initial failure of the main spillway is a much more complicated task, as due to the nature of the incident, very few pictures are available showing the initial chute hole that was spotted on February 7. Any physical evidence that could have been gathered from the scene at the time has been likely washed away from the subsequent discharges that eroded away the bottom half of the chute and much of the downstream ridge. Simply looking at pictures of the February 7 chute damage is not enough, and can lead to forming biased conclusions. Thus, prior to studying these pictures, further background research is required.

A dam inspection guide [40] lists potential incidents that can occur on spillway concrete chutes and possible causes based on studies of previous similar events. More specifically, the following defects mentioned in the guide are directly related to the Oroville Dam main spillway chute.

- Cracking of concrete in floor slabs. Visible on casual inspection when concrete is dry, possibly caused by temperature changes or inadequate reinforcement.
- Damaged concrete. Possibly caused by cavitation or erosion due to irregularities or rough surface.
- Lifted slab panels. Indicated by vertical offsets in joints, possibly caused by poor drainage under slabs, and/or inadequate anchoring of slab to foundation.

Furthermore, the geological conditions below the spillway chute are also considered.

3.4. Structural Flaws

Based on previous inspection reports and other sources [19,27,46,47], it is known that cracks had previously occurred in the main spillway chute's floor slabs, just above the herringbone drains. Aside from cracking, removal of joint filler and spalling have been also been documented as mechanisms that cause damages. Attempts to repair the structure took place in 1977, 1985, 2009, and 2013. These efforts usually included a simple removal of spalled concrete and patchwork with the intent of simply restoring flow surface. By 2017, cracks remained present above the drains, which likely allowed water to flow through the chute's concrete slabs whenever the spillway was actuated.

Unfortunately, due to the nature of the incident and the measures that had to be taken to ensure Oroville Dam's safety, if the initial cause of the main spillway chute's failure was slab uplift due to a fault in the drain system, the only available evidence can be found in pictures taken shortly before and after the February 6 chute hole was spotted, as any physical evidence was subsequently eroded away by the February 12 outflows. However, by conducting background research, the following factors are discovered about the main spillway's drain system and the concrete chute slabs [19,46,48,49]:

- The invert slabs have a minimum thickness of 380 mm, are anchored to rock with grouted anchor bars, and are provided with a system of underdrains [19].
- The initial drain system plan ended up being significantly altered during construction. After a recommendation by the Oroville Dam Consulting Board, the original horizontal pipe drains under the chute were enlarged and placed in a herring-bone pattern. The collector system operating in line with the chute was also enlarged and modified so as to enhance its capacity and self-cleaning ability [19].
- The last official inspection of the main spillway chute's full length took place in 2015 [49]. At the time, no structural deficiencies were detected. An additional inspection took place in 2016 [50],

but the spillway chute was only examined from the top of the FCO outlet structure, not up close like in 2015. A reason for this is not specified.

In addition, a comparison of pictures of the spillway shortly before the February 6 hole was discovered yield additional clues. Figure 5 is a comparison of two pictures of the main spillway chute, taken shortly before the February incident. The first was taken on 11 January 2017 and the second on 27 January of the same year.

Figure 5. Views of the Oroville Dam main spillway chute. (**a**) taken on 11 January 2017, and (**b**) taken on 27 January 2017. A red arrow points to the location of the initial chute failure [38,51].

While these pictures were only taken within 16 days of each other, there are significant differences in the spillway chute. A center section of the chute's concrete floor appears dry on the right-hand picture, despite flows passing over the rest of the structure. This indicates possible irregularities among the floor slabs. Furthermore, the fact that this dry patch is not visible in the photo taken earlier, could possibly mean that a possible slab uplift occurred near the red arrow's location, diverting small water flows around it instead of over it.

Furthermore, by looking at the drain system more closely, two clues are revealed: First, water is coming out of the drains under pressure, which is not according to design specifications, and secondly, discharge from these drains significantly increased in a short time, once flows from the January flood filled up the Oroville Dam reservoir. This is a telltale sign of a buildup of excess water occurring beneath the spillway, which could apply significant forces to the concrete slabs from below and cause them to uplift [40]. Additionally, the January 27 photograph shows the drains on the opposite wall operating under pressure as well.

3.5. Possible Cavitation—1-D Water Surface Profile Analysis

One of the possible causes of the initial damage to the concrete chute floor is cavitation. In order to better understand this cause, extensive examination of the USBR hydraulic model study of the main spillway [29] is required. Furthermore, comparing this data to a simple mathematical model of the main spillway chute could help find possible clues. A simple mathematical model is constructed in a spreadsheet software which uses an iterative procedure to simulate 1-D steady open-channel flow, known as the standard step method [52]. In order to construct this model, some additional assumptions must be made, which are analyzed below.

Based on the USBR main spillway chute profile, its main rectangular concrete section is 54.46 m wide, begins at Station +13 00 (past the beginning of the approach channel) and ends at Station +43 00, just before the terminal structure with the concrete chute blocks. As such, this main section is exactly 914.4 m in length, and only this part of the main spillway is modeled. To avoid confusions between the USBR calculations and those of the model, the entire model is constructed using American unit measurements (distance in feet, discharge in cfs, etc.).

To calculate flows, Manning's n coefficient is additionally required. Unfortunately, there is no mention of the specific coefficient used for the hydraulic calculations of the final chute in Reference [29]. However, a profile drawing of an earlier model describes a lined concrete channel with an n value of 0.013. Based on this and the HEC-RAS manual specifications, an n value of 0.014 was selected for the model. Furthermore, in the interest of time and with the intent of keeping the mathematical model as simple as possible, critical flow depth was assumed at the chute's beginning for every discharge profile, instead of the true depth which is partially controlled by the flood control outlet gates. However, as is evident later, this did not have a significant impact on the results.

Four discharge profiles were created, in accordance with those of the USBR model study: 20,000 cfs (566 m^3/s); 50,000 cfs (1416 m^3/s); 100,000 cfs (2832 m^3/s); and finally 277,000 cfs (7484 m^3/s), which is the main spillway's design capacity. A water surface profile view of the chute for the latter discharge is plotted in Figure 6, with an additional data label at the exact point where the 2017 hole occurred (Station +33 00). Figures for the other surface profiles can be found in Appendix B.

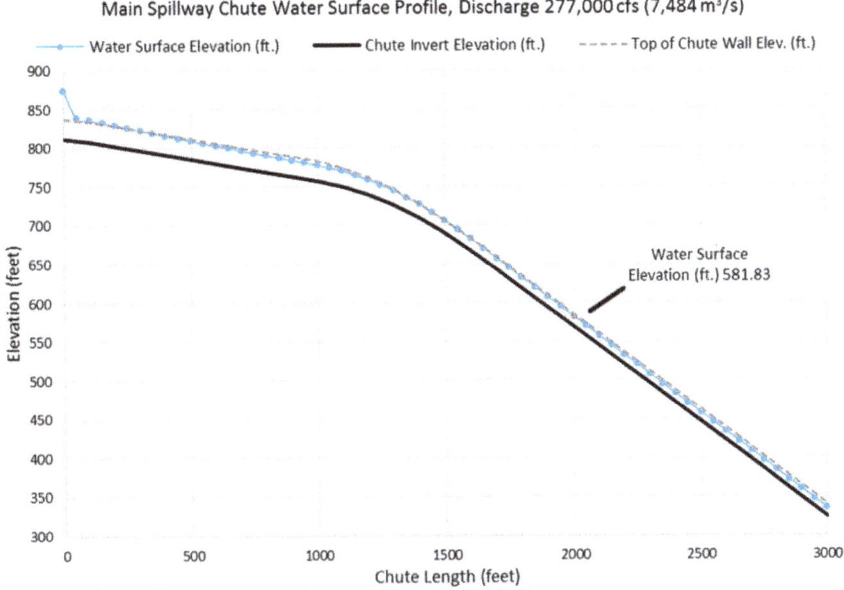

Figure 6. Oroville Dam main spillway chute water surface profile, discharge 7484 m^3/s.

From the chute flow analysis, it is clear that the initial assumption of critical flow depth at the chute's beginning does not negatively impact the results significantly, as due to the chute's design, flow depth quickly approaches normal depth with a standard S2 curve for supercritical flow [52]. For low discharge profiles, normal depth is reached fairly quickly, and only when the spillway is running at maximum capacity, 7484 m^3/s (277,000 cfs), does the flow reach normal depth close to the chute's end. No surface flow irregularities are immediately apparent from this analysis, indicating that cavitation is probably not the initial cause of the of main spillway's failure. However, as this model assumes hydrostatic pressure, to confirm this assumption one could use the method recommended in Reference [53], that allows for the detection of cavitation despite one-dimensional flow assumptions.

A more thorough analysis was carried out as part of the Forensic study and noted that the cavitation was not a contributor to the failure of the service spillway chute. In fact, following our calculations (Appendix C) at the area of failure, one can make an estimate of cavitation number for the 100,000 cfs release of February 12. The channel velocity was around 95.5 fps and the flow depth around 5.9 ft providing a cavitation number around 0.275. This is much higher than 0.15–0.20 which according to Figures 3–8 of Reference [53] could have caused damage after 100 h of operation.

3.6. Geological Conditions Beneath the Main Spillway Chute

The fact that the main spillway chute was built on rock that required blasting to excavate would mean that the rock is suitably hard to serve as a foundation for the concrete chute sections. However, pictures of the initial spillway failure reveal more information about this foundation rock.

Based on Figure 7, it appears that the foundation rock is indeed composed of the metavolcanic materials mentioned previously. However, this particular section of bedrock appears highly fractured and heterogeneous. There is a significant variance of color in the formations, indicating different degrees of weathering. Furthermore, due to the orientation of the seams, the rock is expected to erode away in large chunks, not in sheets. It is also possible that water was able to seep through cracks in the weaker, more weathered sections of rock and undermine the chute from below.

Figure 7. 7 February 2017. Side view of the initial spillway chute failure [38].

3.7. Annual Maximum Rainfall Analysis—Results

After consulting the 24-h PMP index depth maps in Reference [30] and comparing them to those specified in Reference [27] for the subareas of the Feather River Basin, it is possible to use these distribution fits to estimate the annual daily maximum precipitation value with a 10,000 year return period and find the return period of the stated probable maximum precipitation index depths. The results of this analysis are summarized in Table 2.

Table 2. 10,000-year annual daily maximum precipitation forecasts, compared to the 24-h PMP index depths and their return periods, based on the GEV-Max distribution fit.

Station ID	Available Daily Record (years)	10,000-year Daily Maximum Precipitation (mm)	HMR 59 Avg. PMP 24 h Index Depth (mm)	GEV-Max Return Period of PMP (years)
BRS	1986–2017	688.6	800.1	33,333
USC00041159	1959–2016	529.7	647.7	50,000
USC00044812	1913–1967	414.8	635.0	>100,000
QCY	1989–2017	481.1	431.8	4348

The PMP usually has a return period that is extraordinally high, which increases safety. However, it would be an error to assume that designing with the PMP method removes risk entirely simply because it generates large values. This is why assigning a return period to a design precipitation value is better for representing the associated risks, which are inevitable in engineering.

Furthermore, the PMP method evidently does not always generate overly high values. In the case of the Quincy station (QCY), the distribution of the annual maximum rainfall results in a daily maximum precipitation value with a 10,000 year return period that is above the PMP 24-h index depth for the same region. That same probable maximum value has a corresponding return period of only 4348 years, which, while still being very high, leads to the conclusion that the PMP method is not risk-free as some would expect.

3.8. Flood Frequency Analysis

Instead of assuming a fixed "worst case scenario" flood that supposedly cannot be exceeded, which is what the PMF suggests, it is possible to assign a return period to existing design floods by using customary flood frequency analysis methods. The record of unregulated, annual maximum flow data for the Feather River at Oroville station resulting from rainfall for a 1-day duration provided by USGS [23] is an ideal input time series for this purpose, and further cross-examination with known extreme floods such as the 1964, 1986 and 1997 events as mentioned above confirms its accuracy. Using HYDROGNOMON, two distributions are fitted to the data, namely the Log-Pearson III with the method of maximum likelihood estimators and the GEV distribution using the L-Moments method, according to References [13,54,55]. The results of the distribution fitting can be found in Appendix C.

From this analysis, it is possible to extract the 10,000 year floods for each of the distribution fits. For the Log-Pearson III fit, the 10,000 year flood is estimated to be 32,000 m^3/s and for the GEV fit, the same value is 24,464 m^3/s. Furthermore, it is possible to assign return periods to existing calculated inflows such as the Standard Project Flood and various PMFs that can be found in References [19,22,27]. The results are documented in Table 3.

Table 3. Return periods in years for various floods, as generated by the distribution fitting process.

Flood Event	Peak Inflow (m^3/s)	Return Period LP3 Fit (years)	Return Period GEV Fit (years)
1986 Flood	6145	50	97
1997 Flood	8860	75	150
2017 Flood	5392	20	33
Standard Project Flood	12,459	250	610
PMF 1965	20,388	1360	4500
PMF 1983	33,046	>10,000	33,300
PMF 2003 (HMR 36)	25,202	3500	11,100
PMF 2003 (HMR 59)	20,530	1500	4800

The Standard Project Flood is mentioned to have a return period of 450 years [19], which is close to the average of the two distribution fitting results. However, the return period of the probable maximum flood is supposed to exceed 10,000 years, yet only the 1983 PMF achieved this for both distribution fits. Notably, the most current PMF was calculated in 2003 based on HMR 59 [27,54], and its return period does not exceed 5000 years for both distributions. Furthermore, according to this analysis, the return period of the 2017 flood is only 20 years for the LP3 fit and 33 years for the GEV fit. It should be noted that these flood figures are overall peaks, whereas the input for the fit is the slightly lower daily averages given by Reference [23], so these estimates are on the conservative side. In any case, these return periods should be viewed more as guidelines than as exact results. Nevertheless, lowering of the flood control elevation could allow Oroville Dam to still withstand these floods.

4. Discussion

Based on the above analysis, and after consulting dam inspection manuals [40] and reviewing the on-site investigation report [56], the following points stand out:

- From a structural standpoint, the main spillway chute appears to have initially failed due to uplift of its concrete floor slabs, caused somewhere between Stations +33 00 and +33 50 (2000 and 2050 feet of its rectangular section length, respectively). This uplift appears to have been caused by water accumulating below the chute floor, which was unable to be routed through the underdrains. This is evidenced by photographs showing them operating under pressure, which should never occur under design specifications. The authors findings agree with those of the Independent Forensic Team memos released recently [47,56].
- The rest of the damage to the main spillway was caused by high velocity flows due to the large amount of water that had to be routed through it to avoid erosion downstream of the emergency spillway.
- The fact that the Lake Oroville's surface elevation was at the nominal minimum flood control level, which was above that during previous major flood events, resulted in more severe conditions, even though the February 2017 inflows were not at a record high. Thus, a lowering of the minimum flood control level to 255 m is recommended. FOR et al. [41] revealed that this actually would not be a new requirement, but an adaptation to outdated assumptions made in the 1970 flood control manual [22]. Based on the main spillway rating curve [29], it would be feasible to maintain the dam reservoir at this level during wet seasons. Further research into the Feather River's Environmental Flow Components could also reveal crucial details of how floods impact the local ecosystem and how lowering the minimum flood control elevation might affect the existing balance [4]. Future research may use the available inflow and outflow data for this incident, as well as the most intense previous events to create spillway operation scenarios of their own and develop a more robust strategy for flood control.
- The current PMF value for Lake Oroville has a return period of less than 10,000 years based on the above analysis. It is recommended to either calculate a new 10,000 year flood for Lake Oroville using a probabilistic method, or use the 1983 PMF value which is suitably large. However, it is important to assign a return period to any resulting flood, as the whole concept of "probable maximum flood" is problematic [32] and its scientific content is disputed [31].
- The California Department of Water Resources' quick response to the incident and initiation of a full scale repair and reconstruction of the Oroville Dam spillways are admirable. However, under current design, the dam is only capable of withstanding the Standard Project Flood with a return period of 500 years without sustaining significant damage. In order to withstand a flood with a return period of 10,000 years without causing significant erosions to the downstream areas, the emergency spillway needs to be redesigned and fully armored with concrete. This has been repeatedly requested by local interest groups [41,43].
- In the United States, many are using this incident as an example of severe issues the country has with maintaining the gigantic number of high-risk structures it has built over the past century [42,57]. Indeed, the Oroville Dam itself has been around for half a century. Until a major problem occurs at a critical facility like this one, it is easy to get complacent and avoid or postpone critical maintenance procedures like routine inspections and small repairs. Rven when larger

problems or design flaws are pointed out [41], it is difficult to convince the authorities to fund large-scale repair projects. However, one would argue that such repair projects actually conserve money in the long run. The new Oroville Dam spillway is estimated to cost around $1.1 billion [58], which is significantly more than what would have been required for a full concrete armoring of the emergency spillway back in 2006.

In addition, the PMP–PMF analysis has several flaws. From a theoretical standpoint, the PMP suggests that there exists a theoretical upper limit of precipitation, which is simply not true. Nature is not bounded by numerical constraints, and the study of a brief history of available data cannot generate a true possible maximum value of precipitation. According to Reference [45], the only merit of the PMP value is that it is a large one. However, in some instances, this precipitation has been either exceeded shortly after it was published, and in others it has been considered absurdly high upon reexamination. On the other hand, constructing input timeseries of annual daily maxima from the available daily precipitation data is not a foolproof method either. As the daily maximum precipitation is a single value for each year, the resulting time series of annual maxima can be sensitive. For this reason, Appendix D contains the annual daily maxima series used as input for the distribution fit to promote further research and allow for cross-examination.

The concept of the Probable Maximum Flood is also highly controversial, for much of the same reasons as the PMP. Indeed, the fact that over the years various PMF studies for Lake Oroville have found largely varying values of probable maximum inflow and outflow does indicate that a true mathematical upper flood limit does not exist. Therefore, even the PMF is again associated with a certain degree of risk, however small. Especially due to the extent of the Feather River Basin and the large number of smaller reservoirs within it above Oroville Dam, it is difficult to generate a reliable design flood without taking multiple factors into account. At the very least, it is possible to assign a return period to existing design floods by using customary flood frequency analysis method.

5. Conclusions

The Oroville Dam 2017 spillway incident presents an interesting case study, as it is a failure of a dam's key structure that occurred under standard operating conditions, yet at an unfortunate time. It raises very interesting questions from a dam operator's perspective: What does one do when a spillway, a structure built to deal for emergency situations, fails just when it is needed? And in the specific case of Oroville Dam, is the auxiliary spillway a feature, or a mark of a critical flaw in its design? While it would indeed save the main dam from overtopping in the extraordinarily high flood event, in doing so it would likely not be able to hold for long, while its failure would flood an enormous area with more than 180,000 permanent residents. Furthermore, what has been thought of as "probable maximum flood" seems more probable then presumed, and it is definitely not a maximum.

An independent forensic team tasked with determining the causes of the spillway incident recently published summaries of their findings [47,56]. With the ability to conduct an on-site investigation, they were able to confirm some of the causes mentioned in this study as well as outline new ones. Namely, the redesign of chute's underdrain system apparently led to an inconsistent thickness in the concrete floor slabs, which resulted in cracks above the herringbone drains, allowing water to pass through the slabs and also potentially led to concrete spalling. Furthermore, the anchorage of the concrete to the foundation was in some places developed in weathered sections of rock, leading to pullout strength lower than the intended design.

After the incident, the California Department of Water Resources seems to have taken a different stand on the issue, being more open to suggestions about the construction of the new spillways [59]. Still, this response came at a rather late time and is being met with some criticism [42,43]. However, their stance on providing free access data to the public and attempting to communicate and cooperate with local residents and interest groups is definitely a step in the right direction. It must be stated that this study would not be possible without the large amount of digital information available directly from the Department of Water Resources and related websites.

If there is a lesson that must be learned from this incident, it is that even when a critical structure like Oroville Dam seems to operate up to standard, one small flaw can emerge at any time and result in a severe failure due to the sheer scale of the facilities and the conditions they are expected to consistently work under. While routine official inspections by the dam operators and independent authorities are a necessity, they are simply not enough as time goes by. Informal inspections of all related facilities must be conducted by dam operators on a weekly or bi-weekly basis, in accordance with existing guidelines [40], not with the intent of writing official reports, but simply to detect the telltale signs of imminent failure before the potentially worst outcome becomes a reality. If the dam operators had noticed the differences in the main spillway chute's floor slabs between mid and late January they might have been able to repair it in time and avoid the incident from occurring entirely, or at least mitigate its results.

Furthermore, this incident shows a possible lack of regulatory requirements based around the prevention of failures that could occur during normal operating conditions such as what happened at Oroville Dam. Even though no lives were lost as a result of the incident, some consequences on the local environment, economy, and communities might be felt in the years to come. In the end, while it takes a great amount of knowledge, research and responsibility to build a large dam, it takes much more to consistently operate one and protect it from damage.

Author Contributions: Conceptualization, A.K., A.T. and D.K.; methodology, A.K.; software, P.D.; validation, P.T., A.T., P.P., T.W. and D.K.; formal analysis, P.D. and T.I.; investigation, A.K.; resources, P.D., P.P., T.W. and D.K.; data curation, P.D. and T.I.; writing—original draft preparation, A.K.; visualization, A.K.; supervision, T.W. and D.K.; project administration, A.T. and P.T. All authors read and edited the paper before submission.

Funding: There was no funding for this work.

Acknowledgments: We are grateful to the three anonymous reviewers for their constructive comments which helped us to improve an earlier version of this manuscript.

Conflicts of Interest: The authors declare no conflict of interest.

Appendix A

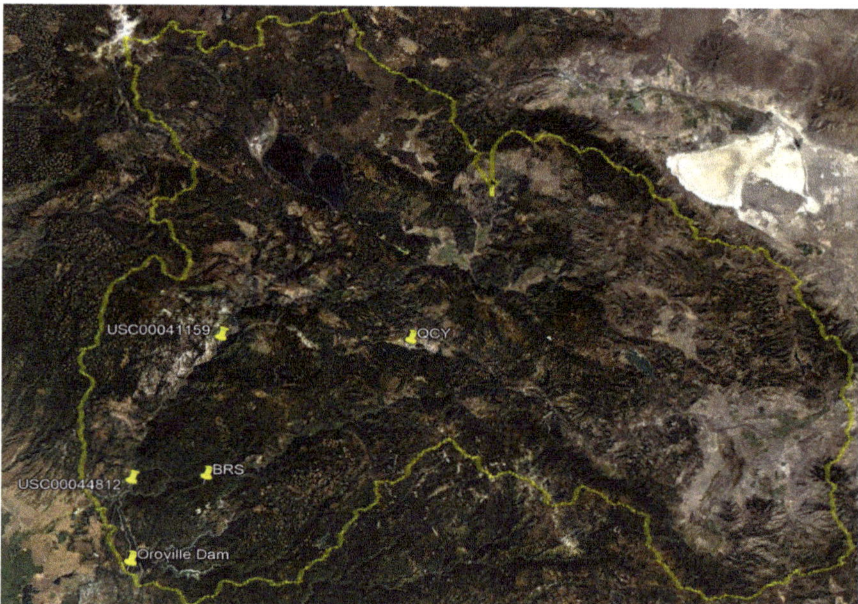

Figure A1. Map of Oroville Dam catchment and selected precipitation measurement stations [34–36].

Appendix B

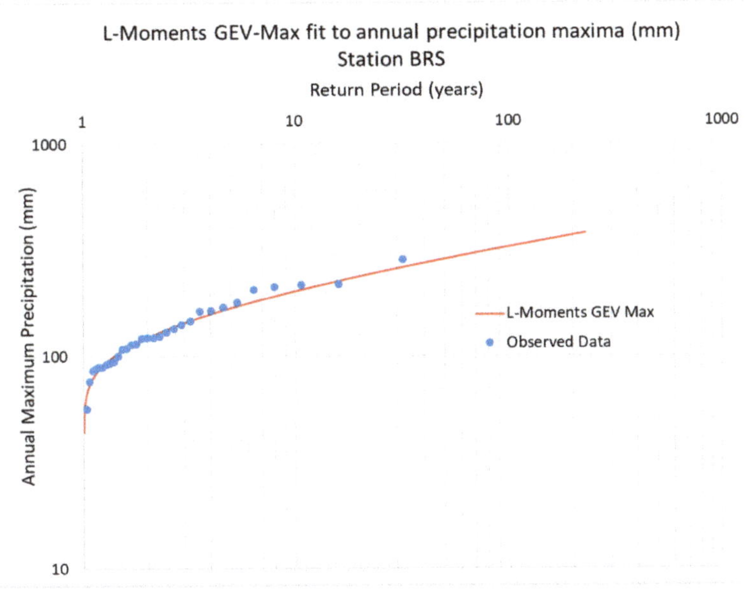

Figure A2. L-Moments GEV-Max distribution fit to annual daily maxima of precipitation measurements, Brush Creek station (BRS).

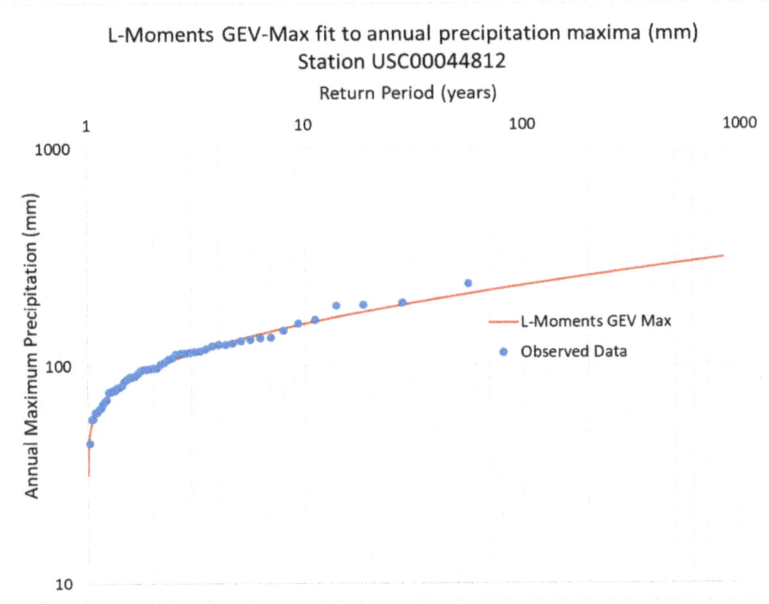

Figure A3. L-Moments GEV-Max distribution fit to annual daily maxima of precipitation measurements, station USC00044812.

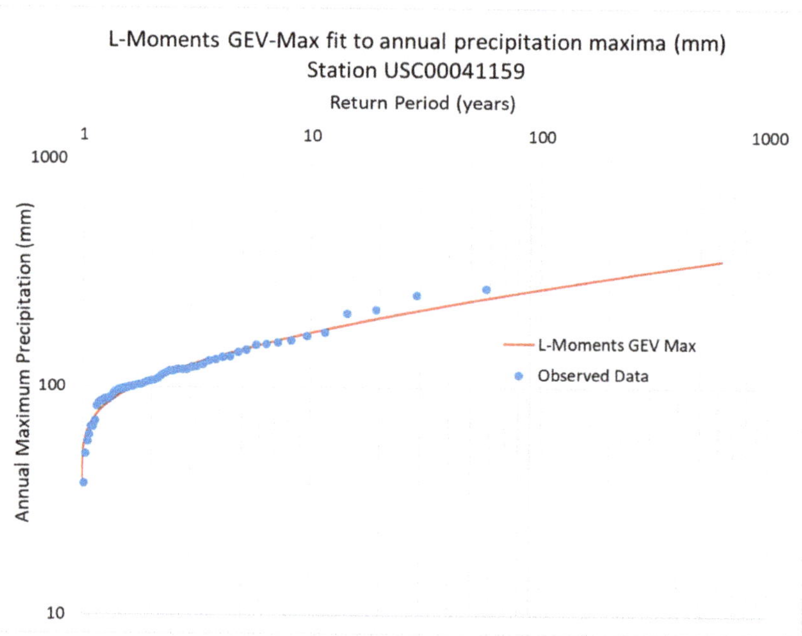

Figure A4. L-Moments GEV-Max distribution fit to annual daily maxima of precipitation measurements, station USC00041159.

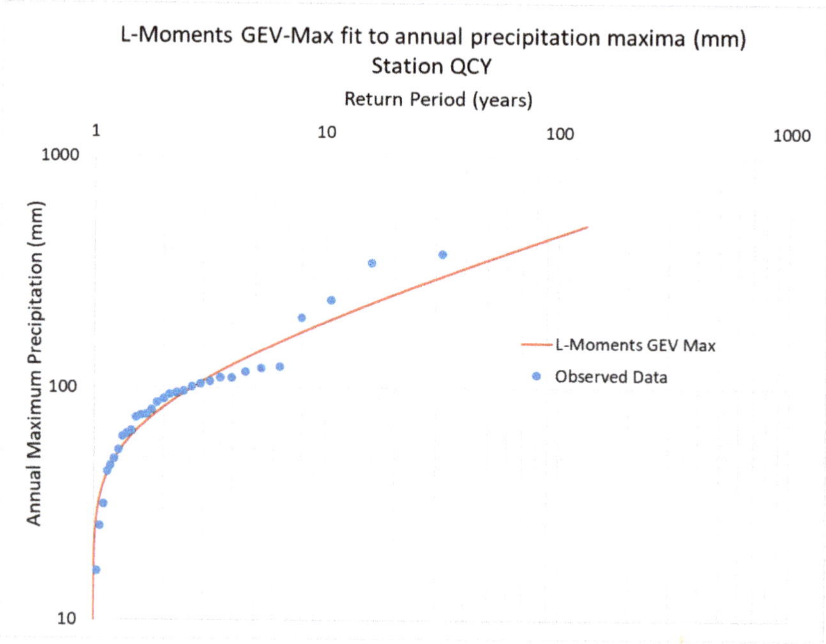

Figure A5. L-Moments GEV-Max distribution fit to annual daily maxima of precipitation measurements, Quincy station (QCY).

Appendix C

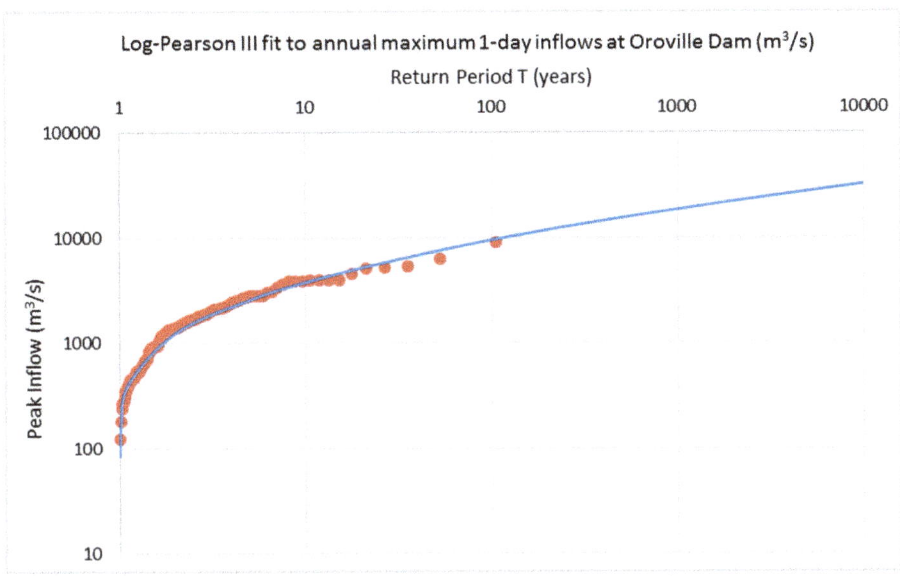

Figure A6. Log-Pearson III distribution fit to annual unregulated maximum 1-day inflows at Oroville Dam (m^3/s).

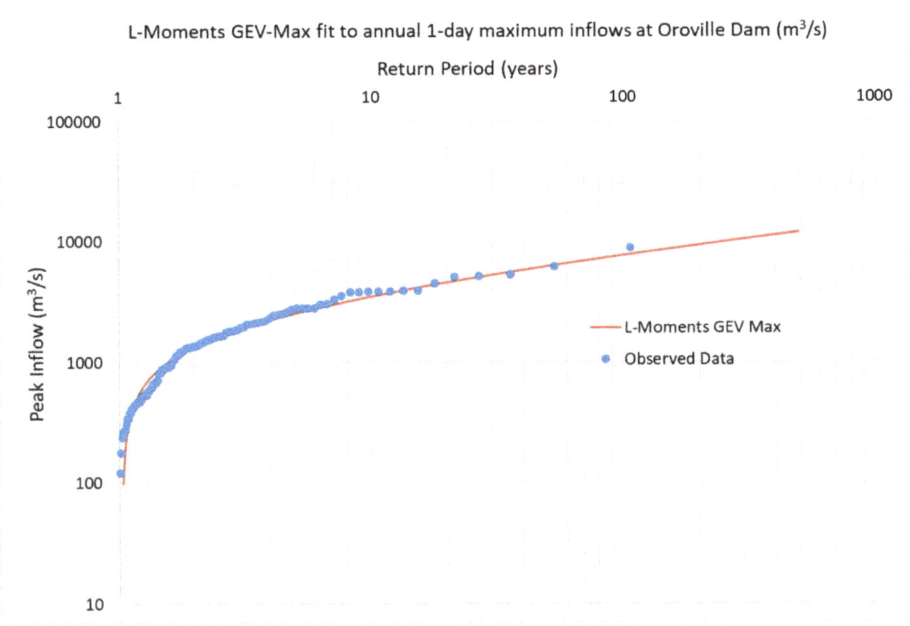

Figure A7. L-Moments GEV-Max distribution fit to annual unregulated maximum 1-day inflows at Oroville Dam (m^3/s).

Appendix D

Table A1. Annual daily maximum precipitation (mm), Brush Creek (BRS) station.

Year	Annual Daily Maximum (mm)	Year	Annual Daily Maximum (mm)
1986	217.4	2002	211.8
1987	121.9	2003	122.4
1988		2004	113.8
1989	89.4	2005	162.8
1990	93.2	2006	124.5
1991	122.4	2007	109.5
1992	89.4	2008	100.8
1993	170.2	2009	86.4
1994	76.5	2010	206.5
1995	141.5	2011	87.9
1996	179.1	2012	163.8
1997	285.2	2013	56.9
1998	115.1	2014	147.1
1999	94.5	2015	92.2
2000	130.3	2016	218.4
2001	108.0	2017	135.6

Table A2. Annual daily maximum precipitation (mm), station USC00044812.

Year	Annual Daily Maximum (mm)	Year	Annual Daily Maximum (mm)
1913	127	1941	116.8
1914	44.5	1942	114.3
1915	129.5	1943	162.1
1916	125.5	1944	88.9
1917	96.5	1945	92.7
1918	68.6	1946	61.5
1919	96.5	1947	113.8
1920	107.2	1948	64.3
1921	76.2	1949	70.4
1922	116.1	1950	144.8
1923	61.5	1951	95.3
1924	101.6	1952	104.1
1925	97.3	1953	115.8
1926	132.3	1954	109.2
1927	80.3	1955	189.2
1928	76.2	1956	125.5
1929	123.2	1957	119.9
1930	57.7	1958	89.7
1931	97.8	1959	77.5
1932	57.2	1960	66.8
1933	77.7	1961	79.5
1934	63.5	1962	239
1935	114.3	1963	133.9
1936	98	1964	156.2
1937	191.5	1965	88.9
1938	85.1	1966	87.4
1939	81.8	1967	134.9
1940	194.6		

Table A3. Annual daily maximum precipitation (mm), station USC00041159.

Year	Annual Daily Maximum (mm)	Year	Annual Daily Maximum (mm)
1959	98.8	1988	132.8
1960	168.4	1989	130.6
1961	96.8	1990	91.7
1962	271.8	1991	106.2
1963	125.5	1992	98.3
1964	254.5	1993	103.6
1965	117.9	1994	119.9
1966	109.2	1995	210.8
1967	142.5	1996	145.5
1968	97.3	1997	61.5
1969	121.9	1998	174
1970	115.1	1999	100.3
1971	100.3	2001	105.4
1972	50.8	2002	119.4
1973	160.5	2003	101.6
1974	102.4	2004	102.4
1975	82	2005	153.4
1976	66.8	2006	94.5
1977	70.4	2007	57.4
1978	136.1	2008	106.9
1979	135.9	2009	86.6
1980	157.5	2010	85.3
1981	154.9	2011	95.5
1982	218.4	2012	119.9
1983	117.9	2013	37.8
1984	88.9	2014	112.5
1985	86.6	2015	88.1
1986	123.2	2016	66.8
1987	88.4		

Table A4. Annual daily maxima of precipitation (mm), Quincy (QCY) station.

Year	Annual Daily Maximum (mm)	Year	Annual Daily Maximum (mm)
1989	121.9	2004	93.7
1990	80.3	2005	77.0
1991	95.5	2006	101.3
1992	38.6	2007	49.3
1993	201.7	2008	104.1
1994	43.7	2009	90.2
1995		2010	54.1
1996	239.8	2011	61.5
1997	86.4	2012	106.7
1998	96.5	2013	31.5
1999	46.2	2014	117.1
2000	65.5	2015	76.5
2001	63.5	2016	110.7
2002	110.5	2017	123.7
2003	74.7		

References

1. Koutsoyiannis, D. Scale of water resources development and sustainability: Small is beautiful, large is great. *Hydrol. Sci. J.* **2011**, *56*, 553–575. [CrossRef]
2. Efstratiadis, A.; Tegos, A.; Varveris, A.; Koutsoyiannis, D. Assessment of environmental flows under limited data availability: Case study of the Acheloos River, Greece. *Hydrol. Sci. J.* **2014**, *59*, 731–750. [CrossRef]

3. Tyralis, H.; Tegos, A.; Delichatsiou, A.; Mamassis, N.; Koutsoyiannis, D. A Perpetually Interrupted Interbasin Water Transfer as a Modern Greek Drama: Assessing the Acheloos to Pinios Interbasin Water Transfer in the Context of Integrated Water Resources Management. *Open Water J.* **2017**, *4*, 18.
4. Tegos, A.; Schlüter, W.; Gibbons, N.; Katselis, Y.; Efstratiadis, A. Assessment of Environmental Flows from Complexity to Parsimony—Lessons from Lesotho. *Water* **2018**, *10*, 1293. [CrossRef]
5. Lewin, J.; Ballard, G.; Bowles, D.S. Spillway gate reliability in the context of overall dam failure risk. Available online: https://pdfs.semanticscholar.org/4e93/2f4989c4670f1a256280aae93b5583f685b2.pdf (accessed on 11 January 2019).
6. Lempérière, F. Dams and Floods. *Engineering* **2017**, *3*, 144–149. [CrossRef]
7. Bhattarai, S.; Zhou, Y.; Zhao, C.; Yadav, R. An Overview on Types, Construction Method, Failure and Key Technical Issues during Construction of High Dams. *Electron. J. Geotech. Eng.* **2016**, *21*, 18.
8. Thompson, K.D.; Stedinger, J.R.; Heath, D.C. Evaluation and presentation of dam failure and flood risks. *J. Water Resour. Plan. Manag.* **1997**, *123*, 216–227. [CrossRef]
9. Xu, Y.; Zhang, L.; Jia, J. Lessons from catastrophic dam failures in August 1975 in Zhumadian, China. In Proceedings of the Geosustainability and Geohazard Mitigation, New Orleans, LA, USA, 9–12 March 2008; pp. 162–169.
10. Si, Y. The World's most catastrophic dam failures. *Qing* **1998**, *1998a*, 25–38.
11. Foster, M.; Fell, R.; Spannagle, M. The statistics of embankment dam failures and accidents. *Can. Geotech. J.* **2000**, *37*, 1000–1024. [CrossRef]
12. Vaníček, I.; Vaníček, M.; Jirásko, D.; Pecival, T. Experiences from the small historical dams failures during heavy floods. *IOP Conf. Ser. Earth Environ. Sci.* **2015**, *26*, 12–42. [CrossRef]
13. Papalexiou, S.; Koutsoyiannis, D.; Makropoulos, C. How extreme is extreme? An assessment of daily rainfall distribution tails. *Hydrol. Earth Syst. Sci.* **2013**, *17*, 851–862. [CrossRef]
14. Rahman, M.A. Damage to Karnafuli Dam Spillway. *J. Hydraul. Div.* **1972**, *98*, 2155–2170.
15. Moramarco, T.; Barbetta, S.; Pandolfo, C.; Tarpanelli, A.; Berni, A.; Morbidelli, R. Spillway Collapse of the Montedoglio Dam on the Tiber River, Central Italy: Data Collection and Event Analysis. *J. Hydrol. Eng.* **2013**, *19*, 1264–1270. [CrossRef]
16. International Commission on Large Dams (ICOLD). *Lessons from Dam Incidents*; ICOLD: Paris, France, 1974.
17. Committee on Dam Safety of the United States Committee on Large Dams (USCOLD). *Lessons from Dam Incidents, USA II*; American Society of Civil Engineers: New York, NY, USA, 1988.
18. Charles, J.A.; Tedd, P.; Warren, A. *Evidence Report—Lessons from Historical Dam Incidents*; Environment Agency: Exeter, UK, 2011.
19. California Department of Water Resources. *California State Water Project, Volume III: Storage Facilities*; The Resources Agency: Sacramento, CA, USA, 1974.
20. California Department of Water Resources. *Performance of the Oroville Dam and Related Facilities During the August 1, 1975 Earthquake*; California Department of Water Resources: Sacramento, CA, USA, 1977.
21. Jennings, C.; Strand, R.; Rogers, T. *Geological Map of California*. Available online: https://mrdata.usgs.gov/geology/state/state.php?state=CA (accessed on 11 January 2019).
22. US Army Corps of Engineers. *Oroville Dam and Reservoir, Report on Reservoir Regulation for Flood Control*; US Army Corps of Engineers: Sacramento, CA, USA, 1970.
23. Lamontagne, J.; Stedinger, J.R.; Berenbock, C.; Veilleux, A.; Ferris, J.; Knifong, D. *Development of Regional Skews for Selected Flood Durations for the Central Valley Region, California, Based on Data through Water Year 2008*; U.S. Geological Survey Scientific Investigations Report 2012–5130; U.S. Geological Survey: Reston, VA, USA, 2012.
24. Hickey, J.; Collins, R.; High, J.; Richardsen, K.; White, L.; Pugner, P. Synthetic Rain Flood Hydrology for the Sacramento and San Joaquin River Basins. *J. Hydrol. Eng.* **2002**, *7*, 195–208. [CrossRef]
25. Cudworth, A.J. *Flood Hydrology Manual*; U.S. Department of the Interior: Washington, DC, USA, 1989.
26. Oroville-Dam-lge.jpg (1200×786). Available online: https://dl0.creation.com/articles/p119/c11909/Oroville-Dam-lge.jpg (accessed on 9 November 2018).
27. California Department of Water Resources. *Oroville Facilities Relicensing, FERC Project No. 2100*; California Department of Water Resources: Sacramento, CA, USA, 2004.

28. Koczot, K.M.; Jeton, A.E.; McGurk, B.J.; Dettinger, M.D. *Precipitation-Runoff Processes in the Feather River Basin, Northeastern California, with Prospects for Streamflow Predictability, Water Years 1971–97*; U.S. Geological Survey: Sacramento, CA, USA, 2005.
29. US Bureau of Reclamation. *Hydraulic Model Studies of the Flood Control Outlet and Spillway for Oroville Dam*; US Bureau of Reclamation: Denver, CO, USA, 1965.
30. US Army Corps of Engineers. *Hydrometeorological Report No. 59*; US Army Corps of Engineers: Silver Spring, MD, USA, 1999.
31. Koutsoyiannis, D.; Papalexiou, S. Extreme Rainfall: Global Perspective. In *Handbook of Applied Hydrology*; McGraw-Hill: New York, NY, USA, 2017; pp. 74.1–74.16.
32. Koutsoyiannis, D. A probabilistic view of Hershfield's method for estimating probable maximum precipitation. *Water Resour. Res.* **1999**, *35*, 1313–1322. [CrossRef]
33. Papalexiou, S.; Koutsoyiannis, D. A probabilistic approach to the concept of Probable Maximum Precipitation. *Adv. Geosci.* **2006**, *7*, 51–54. [CrossRef]
34. *California Department of Water Resources Oroville Dam Hydrologic Data*; California Department of Water Resources: Sacramento, CA, USA, 2017.
35. Menne, M.J.; Durre, I.; Korzeniewski, B.; McNeill, S.; Thomas, K.; Yin, X.; Anthony, S.; Ray, R.; Vose, R.S.; Gleason, B.E.; et al. Global Historical Climatology Network—Daily (GHCN-Daily), Version 3. Available online: https://data.world/us-noaa-gov/e59ba4b2-a901-43d5-b759-3f7d7168d044 (accessed on 11 January 2019).
36. Google Earth. UC Davis Center for Watershed Sciences Map of Oroville Dam Catchment. Available online: https://en.wikipedia.org/wiki/Oroville_Dam (accessed on 11 January 2019).
37. White, A.B.; Moore, B.J.; Gottas, D.J.; Neiman, P.J. Winter Storm Conditions Leading to Excessive Runoff above California's Oroville Dam during January and February 2017. *Bull. Am. Meteorol. Soc.* **2018**. [CrossRef]
38. Pixel—California Department of Water Resources. Available online: https://pixel-ca-dwr.photoshelter.com/galleries/C0000OxvlgXg3yfg/G00003YCcmDTx48Y/I0000b5wb2ydgHxE/FL-Oroville-4842-jpg (accessed on 9 November 2018).
39. Oroville Dam Spillway Failure. Available online: https://www.metabunk.org/oroville-dam-spillway-failure.t8381/ (accessed on 24 December 2018).
40. Morrison-Knudsen Engineers, Inc. *Inspection and Performance Evaluation of Dams: A Guide for Managers, Engineers, and Operators*; Electric Power Research Institute: San Fransisco, CA, USA, 1986.
41. Friends of the River, Sierra Club, South Yuba River Citizens League. *Comments on Draft Environmental Impact Statement, Project 2100-134*; Friends of the River: Sacramento, CA, USA, 2006.
42. Stork, R.; Shutes, C.; Reedy, G.; Schneider, K.; Steindorf, D.; Wesselman, E. *The Oroville Dam 2017 Spillway Incident: Lessons from the Feather River Basin*; Friends of the River: Sacramento, CA, USA, 2017.
43. Schnoover, S. Oroville Dam: Group that Issued 2005 Warning still Wants Full Emergency Spillway. Available online: https://www.mercurynews.com/2017/09/20/oroville-dam-group-that-issued-2005-warning-still-wants-full-emergency-spillway/ (accessed on 9 November 2018).
44. Koskinas, A.E. The Oroville Dam 2017 Spillway Incident—Possible Causes and Solutions. Master Thesis, National Technical University of Athens, Athens, Greece, 2017.
45. Benson, M.A. *Thoughts on the Design of Design Floods*; Water Resour. Publ.: Fort Collins, CO, USA, 1973; pp. 27–33.
46. Federal Energy Regulatory Commission. *2014 Potential Failure Mode Analysis*; Federal Energy Regulatory Commission: Washington, DC, USA, 2014.
47. France, J.W.; Alvi, I.A.; Dickson, P.A.; Falvey, H.T.; Rigbey, S.J.; Trojanowski, J. *Independent Forensic Team Report, Oroville Dam Spillway Incident*; Independent Forensic Team: Leiden, The Netherlands, 2018.
48. Federal Energy Regulatory Commission. *2009 Potential Failure Mode Analysis*; Federal Energy Regulatory Commission: Washington, DC, USA, 2009.
49. Division of Safety of Dams, California Department of Water Resources. *August 3, 2015 Inspection Report*; California Department of Water Resources: Sacramento, CA, USA, 2015.
50. Division of Safety of Dams, California Department of Water Resources. *August 22, 2016 Inspection Report*; California Department of Water Resources: Sacramento, CA, USA, 2016.
51. Oroville Week of 1-30-2017—MNG-Chico. Available online: https://mng-chico.smugmug.com/Oroville-Week-of-1-30-2017/i-sZJBSXt (accessed on 9 November 2018).
52. Chaudhry, M.H. *Open-Channel Flow*, 2nd ed.; Springer: New York, NY, USA, 2008; ISBN 978-0-387-30174-7.

53. Falvey, H.T. Cavitation in Chutes and Spillways. Available online: https://www.researchgate.net/publication/259095887_Cavitation_in_Chutes_and_Spillways (accessed on 25 December 2018).
54. Interagency Advisory Committee on Water Data. *Bulletin 17B: Guidelines for Determining Flood Flow Frequency*; Interagency Advisory Committee on Water Data: Reston, VA, USA, 1982.
55. Seckin, N.; Haktanir, T.; Yurtal, R. Flood frequency analysis of Turkey using L-moments method. *Hydrol. Process.* **2011**, *25*, 3499–3505. [CrossRef]
56. Oroville Dam Spillway Incident Independent Forensic Team. *IFT Interim Final Memo*. Available online: https://damsafety.org/sites/default/files/files/IFT%20interim%20memo%20final%2009-05-17.pdf (accessed on 10 January 2014).
57. BBC US Dam Risk Prompts California Evacuation. Available online: https://www.bbc.com/news/world-us-canada-38952847 (accessed on 9 November 2018).
58. Sabalow, R.; Kasler, D. Oroville Dam repairs now exceed $1 billion and 'may be adjusted further' as work continues. *The Sacramento Bee*, 9 June 2018.
59. Sabalow, R.; Kasler, D. Is Oroville Dam ready for the rainy season? Main spillway fixed, but work remains. *The Sacramento Bee*, 11 February 2018.

MDPI
St. Alban-Anlage 66
4052 Basel
Switzerland
Tel. +41 61 683 77 34
Fax +41 61 302 89 18
www.mdpi.com

Geosciences Editorial Office
E-mail: geosciences@mdpi.com
www.mdpi.com/journal/geosciences

www.ingramcontent.com/pod-product-compliance
Lightning Source LLC
LaVergne TN
LVHW071255170726
843515LV00006BA/1408

9783039438334